LABOR MARKETS
AND SOCIAL POLICY
IN CENTRAL
AND EASTERN EUROPE

LABOR MARKETS AND SOCIAL POLICY IN CENTRAL AND EASTERN EUROPE

The Transition and Beyond

Edited by Nicholas Barr

Published for the World Bank
and the
London School of Economics and Political Science

OXFORD UNIVERSITY PRESS

Oxford University Press

OXFORD NEW YORK TORONTO
DELHI BOMBAY CALCUTTA MADRAS KARACHI
KUALA LUMPUR SINGAPORE HONG KONG TOKYO
NAIROBI DAR ES SALAAM CAPE TOWN
MELBOURNE AUCKLAND

and associated companies in

BERLIN IBADAN

© 1994 The International Bank for Reconstruction
and Development / THE WORLD BANK
1818 H Street, N.W.
Washington, D.C. 20433, U.S.A.

Published by Oxford University Press, Inc.
200 Madison Avenue, New York, N.Y. 10016

Manufactured in the United States of America
First printing September 1994

The findings, interpretations, and conclusions expressed in this study are entirely those of the authors and should not be attributed in any manner to the World Bank, to its affiliated organizations, or to members of its Board of Executive Directors or the countries they represent.

The cover photograph is of Riga, Latvia. By Curt Carnemark for the World Bank.

Library of Congress Cataloging-in-Publication Data
Labor markets and social policy in central and eastern europe :
the transition and beyond / edited by Nicholas Barr.
    p. cm.
    "Published for the World Bank."
    Includes bibliographical references and index.
    ISBN 0-19-520998-2
    1. Labor market—Europe, Eastern.   2. Labor market—
Central Europe.   3. Europe, Eastern—Social policy.   4. Central
Europe—Social policy.   5. Europe, Eastern—Economic
conditions—1991—.   6. Central Europe—Economic conditions.
I. Barr, N. A.
HD5764.7.A6L28   1994
331.12'0947—dc20

94-21174
                                                                                   CIP

History does not move at right angles. Nations are bound by their history, geography, and myths. To change the direction of events even in a revolutionary period requires tremendous will and perseverance.

—*Raymond Seitz, U.S. ambassador to the United Kingdom, speaking to the European Affairs Subcommittee of the U.S. Senate Foreign Relations Committee, April 4, 1993*

Contents

A Note to the Reader

AS A WISE PERSON ONCE OBSERVED in another context, the subject of this book is so large that perhaps a whole library could not hold all the books that could be written on it. This book is intended as one contribution to a continuing dialogue. The pace of change in the region proceeds unabated, and the time it took to produce the book did not allow it to reflect some of the most recent developments. The authors seek to provide policy tools for those facing the challenges of change and reform, and to encourage further research, analysis, and conversation. The field is ripe for such efforts. Many other books need to be written: books which have a longer perspective than is possible here, after only the first intense years of reform; books which report in more detail on early successes in the region; books which describe—country by country—the benefits of democracy, privatisation, and other fruits of reform; and books which express the voices of the people who are remaking this part of the world.

Foreword

THE PUBLICATION OF THIS BOOK marks the fifth anniversary of the fall of the *ancien régime*. It presents the achievements and errors of the post-communist governments, and offers suggestions about the way ahead.

The main purposes of the reforms, according to the authors, are higher living standards and increased protection of individual rights and freedoms. Labour markets and social policy are central to these aims, not just to protect people from unnecessary suffering but also because they are essential to the productivity of a modern industrialised economy. The authors point particularly to the Western European experience. Our countries are part of the same cultural and historical tradition, and the growth of the European Union is likely to assist (if not demand) a measure of convergence.

The material aspects of the reforms—higher living standards, freedom to travel, free access to the world's news media—are important. But they are not the *most* important. Higher living standards and democracy are only instruments; the true aim of the reforms is to *empower* individual citizens. The first way to do so is to give them knowledge and the freedom to use it. This involves a free press and a good education system. It also involves consciousness-raising—not the least of the tasks of the playwright.

Empowerment also has a more technical dimension. It comes very directly from better health. It comes from transferable job skills and well-functioning labour markets which allow people some power over their work. It comes from a measure of income security. Though some insecurity is inescapable in a market economy, extreme poverty and insecurity sap a person's identity and destroy his or her freedom. These topics are the subject of this book. They are important to the material aspects of the reforms. More important, they are central to giving the citizens of our reforming countries genuine freedom and some control over their own lives.

In my New Year's Address at the start of the reforms, I said that I dream "of a republic independent, free, and democratic, of a republic economically prosperous and yet socially just, in short, of a humane republic which serves the individual and which therefore holds the hope that the individual will serve it in turn." I still have that dream. This book will not bring it about. But, like the book's authors, I hope that it will help.

Václav Havel
Prague, March 28, 1994

Preface

THIS BOOK GREW OUT OF AN IDEA by Ralph W. Harbison, who argued that the large amount of experience gathered by World Bank staff during the early years of reform in Central and Eastern Europe deserved a more systematic and more readable treatment than is possible in reports which are often confidential, country-specific, and written under intense time pressure. He also dreamed up the idea of getting all the authors together to write simultaneously, negotiated with the managers of the World Bank Research Support Budget to provide the finance to make that possible, and is the only other person to have read every page of every draft.

The book is very much a team effort. Its overall shape is the result of long discussions during 1992 between those of the authors then based at the World Bank. The framework was then extended, with multilateral agreement, into a detailed outline for each chapter. The central part of the writing took place during the first two weeks of January 1993, when all the authors got together in a sea-front hotel in Brighton, England, using the first week to write the entire book in draft and the second to read and comment on each other's chapters.

More important than the time spent at the keyboard in our respective rooms were the late night discussions between gaggles of authors—World Bank staff, former ministers in and advisers to early reform governments, Western scholars of centrally planned systems, academic economists, political professionals, and civil servants. The teamwork and camaraderie grew out of this time of intense talking and writing, and also because, despite our diverse backgrounds, we all come from either or both of two stables: all the authors are current or former World Bank staff or consultants (Feachem, Fretwell, Harbison, Laporte, Schweitzer, Sipos, Thompson, Tomeš), or current or former staff or graduate students at the London School of Economics and Political Science (Crawford, Gomulka), or both (Barr, Estrin, Jackman, Preker, Rutkowski). (For more on the authors' backgrounds, see the section on contributors near the end of the book.)

Because of this communal writing time, each chapter bears the stamp not only of its author(s), but in all sorts of ways also of all the other authors. The intention is that the book should not be a collection of disparate chapters, but a single, unified volume in which all the parts fit together and reinforce each other. Successive drafts were greatly strengthened by comments during two extensive review meetings at the World Bank during 1993, comments from many other readers, and a hearing of the Social, Health and Family Affairs Committee, Sub-Committee on Labour and Employment, of the Council of Europe's Parliamentary Assembly in Paris in November 1993. The book is to be launched at an extended hearing of that Committee in Prague in October 1994.

The book is aimed at policymakers. The intention is not to write an instruction manual—the problems are far too complex for *anyone* to attempt such a task—but to offer an analytical tool kit which policymakers can apply to their own country. There is a strong emphasis on practicality because of the growing realisation during the writing that the success of the reforms depends at least as much on administrative and political skills as on policy design. A mouse whose mother lived so far away that he was always too exhausted to talk when he visited her consulted the owl about the problem. After some thought the owl observed that the problem was the length of the journey and advised the mouse to fly. When the mouse came to try out the policy he realised something was missing and went back to the owl to ask *how* he should fly. "Ah," came the lofty response, "I only design policy, I do not implement it." Given our limitations of time and experience, we have not always been able to avoid this mistake. But we have tried.

Finally, the reason we thought it important that this book should be written quickly and published early in the transition process is that the development of human resources is fundamental to the aims of the reforms. The achievement of those aims, however, requires technical analysis more than pious sentiments. It was once said of an elderly woman whose quality of life was restored by a coronary artery bypass operation that "the doctors, by treating her body as a mechanism, have restored her humanity." This book treats the political economy of the reforming countries as a mechanism. That, however, should not divert attention from the book's real purpose, which is to assist reforms intended to improve people's daily lives.

Nicholas Barr

Acknowledgments

WE HAVE MANY DEBTS. The first is to Gregory K. Ingram, who used his discretion as administrator of the World Bank Research Advisory Staff to award the research grant which made it possible for all the authors to get together in Brighton to write simultaneously. Additional financial support from the Suntory Toyota Centre at the London School of Economics and Political Science (LSE) is also gratefully acknowledged, as is the help of the LSE Research Grants and Contracts Office, which made most of the logistical arrangements for the writing period in Brighton.

Warm thanks are due to the many people who took time and trouble to comment on successive drafts. A number read and gave detailed comments on the book as a whole: Leszek Balcerowicz, Kemal Derviş, Robert Liebenthal, Alastair McAuley, Branko Milanović, Andrew Rogerson, and four anonymous referees.

Participants who offered comments on one or more chapters during two review meetings at the World Bank included Nancy Birdsall, Mary Canning, Robert Castadot, Claudio de Moura Castro, Fredrick L. Golladay, Ann Harrison, Stephen Heyneman, Virginia Jackson, Estelle James, Emmanuel Jimenez, Timothy King, Kathie Krumm, Jane Loos, William McGreevey, Jane Peretz, George Psacharopoulos, Bülent Sayin, Eugene Smolensky, Verdon S. Staines, James Stevens, Zafiris Tzannatos, and Christine Wallich.

Many other people read and gave detailed comments on one or more chapters: Sue Berryman, Martin Bobak, Maurice X. Boissiere, Antonio Campos, Giovanni Andrea Cornia, John Eaglehart, Guy Ellena, Gáspár Fajth, Jane Falkingham, Zsuzsa Ferge, Louise Fox, Ellen Goldstein, James Q. Harrison, Jeni Klugman, Jon-Eivind Kolberg, Ivar Lødemel, Michael Mertaugh, John Micklewright, Michelle Riboud, George Schieber, István Tóth, Anthony Wheeler, and Tatiana Zimakova.

Thanks are due also to students doing the graduate course on the Political Economy of Transition at the London School of Economics for spending most of 1993 and early 1994 asking thoroughly awkward questions which helped to debug earlier versions.

Invaluable research assistance was given by Barbara Dabrowska (chapters 1 and 7) and Renée Friedman (chapter 3).

Zuzana Feachem and Alexander Shakow assisted with and befriended the book from its earliest days.

The editorial-production team at the World Bank was led with panache by the indefatigable Alfred Imhoff, and included most particularly Elizabeth Forsyth, Brian Svikhart, and Michael Treadway. They worked unstintingly and with unfailing cheerfulness to a very tight schedule on the book's readability and appearance, and on the manifold tasks which lie between typescript and the final bound volume. On the other side of the Atlantic, June Jarman, with her customary wondrous and unflappable efficiency, organised the complex logistics of multiple authors in multiple countries. Others at the LSE who helped with logistics and word processing included Deirdre French, Alma Gibbons, Richard Stevens, and Elena Suhir.

My final thanks are to my wife, Gill, who has been involved from the start. We travelled together throughout Central and Eastern Europe. While I was meeting with officials, she was going into schools and helping with English lessons, reporting from day one that the reforms looked very different from the school room and bus stop than from the minister's office. Then, in Brighton and subsequently, she was our reality check. She read the entire book in draft, insisted that we remember the pain of the people in the reforming countries, and suggested a title for chapter 1 which recognised this.

We were given many very good ideas. We were not able to implement all of them—for reasons of space, for lack of expertise among the authors, or for lack of data. None of those thanked should be implicated in errors and infelicities which remain, for which mine is the ultimate responsibility.

Nicholas Barr

One

OVERVIEW: HOPES, TEARS, AND TRANSFORMATION

NICHOLAS BARR • RALPH W. HARBISON

GREAT HOPES FOR THE FUTURE SWEPT AWAY THE OLD ORDER which had stifled growth and crushed freedom for decades in Central and Eastern Europe (see figure 1-1). After 1989, democracy and free enterprise spread rapidly throughout the region. By early 1994, more than a third of economic output was produced by the private sector, and in a few countries growth had resumed after the output declines of the early transition.

Transformation, however, though necessary and desirable, has not come without tears. In December 1989 the main unemployment office in Warsaw paid benefits to five people. A year later, twelve months into their "big bang" reform, more than a million Poles were unemployed, and by mid-1993, 3 million. In Lithuania, output almost halved between 1990 and 1992. Poverty and infant mortality rose throughout the region, and death rates in Russia rose by nearly one-third. Everywhere, crime increased sharply. The region's people need no reminder that the early stage of reform is uncertain and painful—in some cases so painful that they look back on the old certainties with some nostalgia.

The Essence of Reform

The reforms, launched on a wave of optimism, have two central purposes: to raise standards of living, mainly by moving to a Western-style market economy, and to increase individual freedom and protect individual rights. This book is about labor market and social policies during this transformation. The book makes three main arguments. First, enhancing these countries' human resources—making labor markets more effective, improving education and training, reducing unemployment and poverty, and promoting better health—is fundamental to the reforms. Second, the state has a diverse and important role in these four areas. Third, the reforms will fail unless adequate weight is given to the political and administrative dimensions of policy implementation.

The Centrality of Human Capital

Effective human resources policies contribute to both the political and the economic success of the reforms. Democracy requires an educated citizenry; education empowers individual choice and so, in a different way, does good

Figure 1-1. *Central and Eastern Europe*

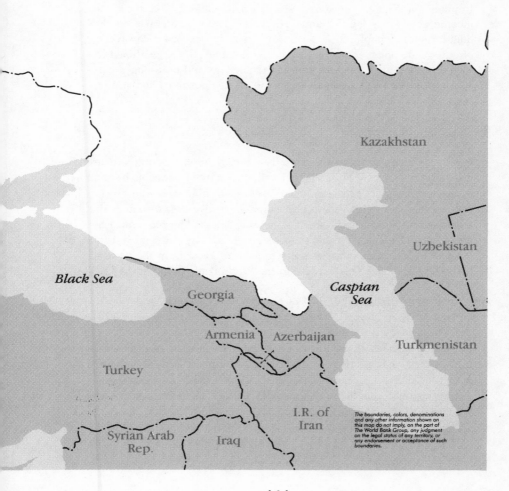

Russian
Federation

Kazakhstan

Uzbekistan

Black Sea

Georgia

Caspian
Sea

Armenia Azerbaijan

Turkmenistan

Turkey

I.R. of
Iran

The boundaries, colors, denominations
and any other information shown on
this map do not imply, on the part of
The World Bank Group, any judgment
on the legal status of any territory, or
any endorsement or acceptance of such
boundaries.

Syrian Arab
Rep.

Iraq

health; effective labor markets are essential to individual freedom. Policy in these areas is also directly linked to the political sustainability of the transition. The whole purpose of the reforms is to improve people's lives, so the reform effort will fail for political as well as economic reasons if it is unable to improve earnings opportunities, provide better education and more effective health services, and increase individual choice.

The common economic feature of these aspects of human resources is that they have a fairly direct relation to labor productivity and hence to economic growth. Although the causal links are complex and far from completely understood, the empirical evidence of such a relationship is strong. A study of East Asia argues that,

> from around 1960, the major distinction between [Hong Kong, the Republic of Korea, Singapore, and Taiwan (China)] and most low-income economies lay in human resource development. In secondary education for example, East Asian economies . . . exceeded the average of other developing economies by many multiples. They combined this high level of education with imported technology and returning expatriates to produce rapid productivity growth. [Leipziger and Thomas 1993, pp. 6–7]

Health contributes fairly directly to economic growth:

> It reduces production losses caused by worker illness; it permits the use of natural resources that had been totally or nearly inaccessible because of disease; it increases the enrollment of children in school and makes them better able to learn; and it frees for alternative uses resources that would otherwise have to be spent on treating illness. The economic gains are relatively greater for poor people, who are typically most hand-icapped by ill health and who stand to gain the most from the development of underutilized natural resources. [World Bank 1993d, pp. 17–18]

Effective labor markets, the relief of poverty, and economic growth are also linked in important ways:

> Rapid and sustainable progress on poverty has been achieved by pursuing a strategy that has two equally important elements. The first element is to promote the productive use of the poor's most abundant asset— labor. It calls for policies that harness market incentives, social and political institutions, infrastructure, and technology to that end. The second is to provide basic social services to the poor. Primary health care, family planning, nutrition, and primary education are especially important. [World Bank 1990b, p. 3]

Several arguments apply particularly to the countries of Central and Eastern Europe:

• Although the general level of schooling in these countries is good, the mix of skills, like the rest of the economy, is seriously distorted. In the absence of appropriate education, training, and, especially, retraining, shortages of the relevant skills will reduce the effectiveness of the reforms by hindering the supply response: economic growth clearly depends on the ability of firms to hire enough labor with the right skills. The point is worth emphasizing. It took the Federal Republic of Germany until 1955 to return to its prewar level of output, even though most of the necessary human capital—knowledge and skills in banking and credit, in managing competitive industrial firms, and the like—was in place. The greatest need was investment in *physical* capital. The situation in transitional economies is very different: it is not a matter only of physical investment (although those needs are daunting) but also of massive investment in adjusting the knowledge and skills of the population.

• The neglect of human resources in the present can lead to significant costs in the future. Failing to pay adequate attention to relieving poverty can be a false economy, particularly where the result is malnutrition, illness, and crime. Investing too little in public health (for example, in anti-smoking campaigns) and in health care leads to premature death. The economic impact is particularly severe when poor health strikes individuals in middle age, because they, as a group, shoulder the greatest family responsibilities. Failure to address their needs condemns whole families to poverty and imposes additional burdens on hard-pressed state budgets.

• Some forms of public investment significantly reduce public costs. Employment services may reduce the average duration of unemployment, and improved methods of finance and administration of health services can prevent medical costs from exploding. The scale of potential savings should not be underestimated: in 1990, partly as a result of the well-known problems of uncontrolled medical expenditure, *public* spending on medical care in the United States, which has essentially a private system, made up the same fraction of gross domestic product (GDP) as public spending on the *entire* national health service in the United Kingdom.

The Continuing Role of the State

The old system in Central and Eastern Europe, in which the state dominated all economic activity, is not sustainable. The early 1990s were associated with a major fall in output in all countries of Central and Eastern Europe, both those which introduced reform gradually and those which adopted shock therapy. The effects of this decline in output included a sharp fall in tax revenues and, at the same time, rising demands on the social safety net. The resulting fiscal crisis was one of the defining characteristics of the early transition. The second reason why the old system is unsustainable is because the move to private markets by definition implies a reduction in the role of the state, with less emphasis on its role as a public provider and more on its role as an enabler of private activity.

The fiscal crisis and the need for profound change in the role of the state give rise to two questions which recur throughout this book: What should be the *scale* of the state's activities—and hence the level of public spending—on human resources? And what is the appropriate *structure* of those activities; that is, which activities should remain primarily public and which should become largely or wholly private? It is important to keep these issues logically separate. The question of scale is largely a matter of budgetary balance. The question of the appropriate mix of public and private activities can be approached either in terms of technical arguments about economic efficiency or in terms of moral values, for example, about the extent of the state's redistributive activities.

The technical arguments rest on relatively recent developments in economic theory, which yield criteria for assessing where markets are likely to work well and where they are not. These market failure arguments have to be balanced against problems of government failure, which are in many ways the root cause of the failure of the old system in Central and Eastern Europe. We return to the topic later in this chapter, and the relevant microeconomic theory is set out at some length in chapter 2. A key conclusion is that, in many of the areas discussed in this book, the conditions necessary for private markets to be efficient are largely absent, and carefully designed state intervention can improve matters. Giving the state a major role in developing human capital can therefore be justified for *efficiency* reasons.

Most of the policy recommendations derive from these mainly technical considerations. To the limited extent that the recommendations in this book are based on values, much of the discussion is rooted in the experience of Western Europe. In part, this is because the inherited value systems in Central and Eastern Europe derive from the same culture. Also, and more concretely, many of the reforming countries hope eventually to join the European Union; if and when they do, they will have to adhere to its legislation, which embodies those values. As the arguments unfold, it will become clear that the technical arguments concerning market failure converge closely with the value systems of the Western European countries.

The Role of Politics and Administration

Just as important as designing good policy is paying attention to the political structures through which the electorate influences policy and to the administrative structures which implement it. The move to a market economy and changes in methods of government require enormous change in both political and administrative institutions. On the administrative side, industries now organized as public enterprises need to be privatized. Legal structures require substantial revision, and much new legislation is needed on the conduct of market relations. The administration of income transfers, health care, and education and training needs to be overhauled. These and other changes impose enormous strains on a very limited institutional capacity. Reform, at least in the early days, therefore should be as administratively simple as possible.

All these reforms, quite apart from their administrative demands, are also intensely political. Indeed the political problem is an integral and wholly unavoidable part of the reforms. As Balcerowicz (1993b) and others have argued, grossly distorted relative prices were in many ways the essence of the old system of central planning. For reform to succeed, therefore, the structure of wages and prices must change, which means that some relative wages and prices must fall. Successful reform thus cannot avoid creating losers, at least in the short run, and if the losers are politically powerful, they may be able to block or reverse reform.

Part of the problem is that democracy is new to the region, at least since the Second World War. Legislators and civil servants have little experience with democratic politics and yet have to cope with rising unemployment, falling output, increasing demands on declining government revenues, disappointed hopes of rapid increases in standards of living, and, in some countries, growing nationalistic pressures. These are problems which rational policy design alone cannot solve. Political stabilization is just as central to the success of the reforms as economic stabilization.

A Road Map of the Book

This book draws together the main threads of the first five years of experience since the wave of reforms began in 1989. Nearly all the authors have worked in one or, usually, several of the countries in Central and Eastern Europe. The book is ambitious in several ways. First, it covers a wide range of countries. This is deliberate. The nations of Central and Eastern Europe differ in their history, culture, and economic and social structure, but they have important characteristics in common: they share an inheritance of central planning and totalitarian government, their reforms have the same main objectives, and they have already made the same fundamental decisions: to move to a market system and to political liberalization. As a result, many of the problems are strategic and transcend national boundaries. These problems are the main focus of the book. Second, the book's thematic coverage is broad in that we look at all the major aspects of human resources policy. The linkages among them are so strong and the problems in many ways so similar that a unified treatment strengthens the argument. Third, the book was written early in the transition process. This, again, is deliberate. The book is not intended to be a definitive historical account; rather, it is forward looking and should therefore be timely. A few years from now, it will be interesting to examine empirically what has happened across the broad range of social policy.

In other respects, the book is more modest. There is relatively little discussion of central versus local issues, partly because these issues transcend the social sectors and partly because space is limited. Nor is there much discussion of ethnic and regional issues; data on these issues are scarce, and none of the authors has the expertise to approach such a difficult area.

The major purpose of the book is to help policymakers in the countries of the region. It attempts to draw out a policy strategy deriving both from the decisions already made about the overall direction of the reforms and from the nature of the transition itself. This is not to deny the importance of country-specific policy choices. As explained in chapter 6, the proposed strategy is general, but its application will vary from country to country in the light of their different policy objectives and differently evolving political and economic regimes. Thus the book is not an operational manual. It makes significant demands on its readers: policymakers in the region know much better than the authors which features of the transition apply to their countries and with what force; it is they who must judge the detailed applicability of the book's content to their own countries. The book makes demands also in that some parts are more relevant to some readers than others. Western economists, for example, might skip the exposition of economic theory; readers in Central and Eastern Europe might skip the discussion of the inheritance from the old system.

The needs of policymakers determine the shape of the book. The first two chapters set the scene. Chapter 2, unlike the rest of the book, is completely theoretical. It discusses the role of government in a market economy and sets out the economic theory relevant to decisions about the dividing line between the market and the state: it discusses why the state might intervene (and why it should not) and how it could do so. The analysis of the policy chapters depends crucially on that discussion, which persons not conversant with recent microeconomic theory should read, but which others may skip.

The next three chapters are concerned with the economic and political nature of the transition. Chapter 3 summarizes relevant aspects of the inheritance from the old regime. Chapter 4 discusses the major economic and political forces driving change, particularly those released by market forces. Chapter 5 describes the major constraints which policy design must take into account.

The second part of the book discusses policy design and implementation in different parts of the social sectors. Chapter 6 links the two parts of the book by drawing on the earlier chapters to set out a policy strategy which shapes all the subsequent recommendations. The discussion of labor markets in chapters 7 and 8 looks at wages and employment and at policies to ameliorate unemployment. The next two chapters analyze income transfers: chapter 9 concentrates on social insurance, and chapter 10 on family support and poverty relief. Chapters 11 and 12 discuss more direct forms of investment in human capital: (a) education and training and (b) health and health care. Chapters 7 to 12 have a common structure. Each chapter begins by discussing the inheritance and, in particular, the strengths of the old system (which should be preserved), its weaknesses (which the reforms are intended to improve), and perennial problems (which are likely to remain however successful the reforms). Not all aspects of the old system were bad, nor will the reforms solve all problems. The second part of each chapter discusses the forces driving change and, in particular, the political imperatives and the economic effects of the move to a market economy. Although these shaping forces share elements in common, they emerge in different ways in different chapters. The third part of each chapter discusses the pros and cons of the various policy options. The last section summarizes the main conclusions in the form of fairly specific recommendations intended to answer a hypothetical request from a minister as to the practical policy actions he or she should take.

Again, policy design is only part of the story. Policy must also be implemented, and even the best-designed and most effectively administered system will founder if it lacks political support. The administrative and political aspects of implementation specific to, say, education or health care are discussed in the chapters devoted to those topics. Chapter 13 considers broader aspects of implementation, such as the machinery of government and the quality of public administration. Although these issues far transcend the social sectors, they have a crucial bearing on the success of social sector reform.

The rest of this chapter offers a microcosm of the book. The next section discusses a series of central themes. Succeeding sections set out the book's central analytical framework, the resulting policy strategy, and some of the early achievements of the reforms. The final section suggests an agenda for policy action. Readers in a hurry can read this chapter alone or in combination with the concluding section (or the whole) of any of the chapters on policy or with chapter 6 plus any of the policy chapters.

The Central Issues

Three issues pervade the book and shape its recommendations. Alongside the implementation issues already mentioned are two pillars of the economic reforms. The first is stabilization, which consists of controlling public expenditure and containing inflation, mainly through macroeconomic policy. The second is restructuring, which consists of introducing private markets (with consequent changes in the profile of industrial production) and designing the legislation and regulation necessary to support them.

Stabilization and the Fiscal Crisis

A central fact of the transition has been the decline in output in all the reforming economies. As table 1-1 shows, output fell by differing amounts in different countries between 1990 and 1992, but the decline was substantial in every case. Among the countries where reform was well under way, the smallest declines, of 17 to 18 percent over the three-year period, were in Hungary and Poland, while the largest decline, in Lithuania, was nearly 50 percent.[1] The average fall in output was about 11 percent a year. For reasons discussed in chapter 4, these figures should be treated with caution. Nevertheless, output clearly fell far more than it has ever fallen in the industrial West since the Second World War. Alongside the decline in output were major problems of inflation, which was lowest in Czechoslovakia, where prices nearly doubled between 1990 and 1992, and highest in Lithuania, where something which at the start of the reforms cost 10 litas cost 1,750 litas by the end of 1992.

WHY IS STABILIZATION NECESSARY? One of the major purposes of stabilization is to prevent inflation from continuing on anything like this scale. An obvious question (and one often asked by citizens in Central and Eastern Europe and elsewhere) is why inflation matters so much. The answer is simple: inflation, beyond a relatively low level, tends to accelerate rapidly, and high rates of inflation have disastrous effects:

• Accounting for and monitoring economic activity through the measure of money become impossible. A firm's accounts bear no relationship to its output or its profitability, nor have public agencies any mechanism to control or even measure their expenditure. Borrowing and lending become impossible, unless people are allowed to use foreign currencies. Although inflation often encourages a flight into real assets, this is not much comfort in the former socialist countries because capital markets, including markets in housing, are poorly developed and pose enormous risks. Private saving becomes a high-risk activity.

• For these reasons, inflation tends to reduce saving and to create considerable uncertainty. The former harms domestic investment, and the latter reduces investment from abroad, thus preventing or at least hindering economic growth. Thus inflation can have major ill effects on output.

• The fiscal effects are also deleterious. In many countries, because of lags in the collection of taxes, inflation leads to a sharp fall in tax revenues, unbalancing the budget and generating further inflation.

• Because inflation destabilizes values, its redistributive effects are blatantly unjust. Public sector wages are seriously eroded and can become worthless. Partly in consequence, corruption becomes rampant. Profiteers prosper, while individuals who work, thereby contributing to output, face declining standards of living and see the value of their savings evaporate. Citizens resent govern-

Table 1-1. *Output Growth, Inflation, and Unemployment in Various Countries in Central and Eastern Europe, 1990–92*
(percent)

Country and year	Annual growth of GDP [a]	Inflation[b]	Unemployment [c]
Belarus			
1990	−3.0	4.5	0.0
1991	−1.0	83.5	0.0
1992	−10.0	969.6	0.5
Bulgaria			
1990	−12.0	30.0	1.0
1991	−23.0	338.0	8.0
1992	−7.7	91.0	15.0
Czechoslovakia			
1990	−3.0	10.0	1.0
1991	−15.9	57.7	6.6
1992	−8.0	10.8	5.1
Estonia			
1990	−3.6	17.2	0.0
1991	−11.8	210.6	0.5
1992	−31.6	1,069.3	1.6
Hungary			
1990	−3.3	28.1	1.6
1991	−9.9	33.4	8.0
1992	−4.4	23.1	12.2
Latvia			
1990	−3.5	10.5	0.0
1991	−8.3	262.4	1.0
1992	−32.9	958.1	2.5
Lithuania			
1990	−5.0	376.2	0.0
1991	−13.4	224.7	0.3
1992	−35.0	1,020.5	1.1
Poland			
1990	−11.9	585.8	6.3
1991	−7.6	70.3	12.0
1992	1.0	43.0	13.6
Romania			
1990	−7.4	4.2	0.0
1991	−15.1	161.0	3.0
1992	−15.4	210.7	8.1
Russian Federation			
1990	0.4	5.6	0.0
1991	−9.0	92.7	0.7
1992	−18.5	1,354.0	1.4
Ukraine			
1990	−3.0	4.2	0.0
1991	−11.9	91.2	0.0
1992	−14.0	1,445.3	0.3

Note: The data should be treated with considerable caution, as broad orders of magnitude rather than exact numbers, both because the construction of cross-country comparative data raises well-known problems and because, for the reasons discussed in chapter 4, measurement problems are greatly exaggerated during the early transition.

a. The unweighted average decline in output over the three-year period for the countries covered was 29.6 percent, or an average fall of about 11 percent a year.

b. Annual percentage change in the consumer price index. The unweighted average increase in prices over the three-year period for the countries covered was 3,347 percent. Excluding Lithuania, the average increase in prices over the three-year period was 1,959 percent, for an average annual rate of inflation of about 175 percent.

c. Annual average.

Source: Unpublished data from the International Monetary Fund and World Bank.

ments which allow such things to happen. These consequences have been seen over and over again, for instance in Latin America.

Inflation destroys the capacity of a market economy to allocate goods in any rational way or to create a politically tolerable distribution of income. It thus defeats the central objectives of the reforms. In the short term, stabilization policy must be able to prevent runaway inflation. This requires that real incomes be kept broadly in line with output, and if output falls, as shown in table 1-1, real incomes must fall by roughly the same amount.

CHOICES AND TRADEOFFS. Given the scale of the decline in output, stabilization makes hard choices inescapable. If resources (labor, machinery, raw materials, and land) are used for one purpose, they cannot be used for another. This is always true, but the problem is even more acute when output is falling. Policymakers must therefore decide how to divide output between different uses at a particular point in time (known as static efficiency) and how to use resources to make output grow faster (dynamic efficiency), for instance by spending less on consumption and more on investment. Choices have to be made in at least three dimensions:

• *Choices between alternative activities.* Within the social sectors this means that what is spent on medical services cannot be spent on family support, what is spent on pensions cannot be spent on education, and what is spent on hospital equipment cannot be spent on primary health care. Exactly the same choices arise between the social sectors and other areas: what is spent on human capital cannot be used for investment in roads or telecommunications.
• *Choices between public spending and private spending.* Higher government spending generally requires higher taxes, which implies less private spending.
• *Choices between present consumption and future consumption.* Higher consumption today means less investment today, hence slower growth in output and lower consumption in the future.

The central tradeoff concerns the competing claims of higher standards of living in the present (a source of strong short-term political pressure); higher standards of living in the future (a central objective of the reforms), which depend on past and current investment in human and physical capital; and fiscal restraint, which contains the size of the public sector so as to allow room for the developing private sector. In addition, most of Central and Eastern Europe must satisfy the claims of debt repayment.

If output is rising, it is possible to satisfy all four sets of claims simultaneously. This is, by definition, impossible if output is falling. Consumption can then be higher only if investment is lower; this puts present living standards into direct conflict with future living standards. Increased government spending (other than through borrowing abroad) can be financed through increased taxation, through domestic borrowing, or through an expanded money supply. Each of these methods has problems: excessive taxation reduces the incentive to work or to create new jobs; excessive domestic borrowing, by pushing up interest rates, reduces private investment and harms economic growth; monetary expansion creates inflationary pressures and may harm international competitiveness. All these solutions, therefore, conflict with the need to raise standards of living in the present, or in the future, or both.

Thus, if output is falling, real incomes must fall, and one of the most difficult decisions facing governments in Central and Eastern Europe is what to cut and by how much. Deciding how to resolve competing claims for resources by the ministries responsible for different parts of the social sectors, in principle a matter of economic efficiency, is also one of the major political tasks of government.

Restructuring: The Market and the State

The economies which emerged from the communist era were very different from those of the industrial West.

WHY IS RESTRUCTURING NECESSARY? In the communist era,

- The structure of ownership was heavily weighted toward the state.
- The structure of production was strongly biased toward heavy industry and agriculture.
- The structure of trade was distorted, particularly by excessive specialization and the consequent giganticism of production units (a small number of plants in Hungary, for example, produced buses for much of the communist world).

These structural imbalances were in part the cause of the declining, and eventually negative, rates of growth discussed in chapter 3. If economic growth is to resume, restructuring is required to correct these imbalances. Two changes are particularly important for the social sectors: greater power, but also more responsibility for the individual, and a new balance between the market and the state.

CHANGING ROLES OF THE INDIVIDUAL AND THE STATE. The reforms require a fundamental change in the responsibilities of the state, the enterprise, and the individual. Under the old system, the economic functions of the state and the enterprise overlapped, and the two together were largely responsible for satisfying the needs of workers. The state financed various benefits for individuals, both directly and through subsidies for basic commodities. Considerable state subsidies were also extended to enterprises, often in the form of soft budget constraints, whereby enterprises received nearly automatic financing of any deficits they incurred. Enterprises offered job security and paid wages which had little to do with market forces; they also administered many benefits, such as family allowances, through workers' pay packets and, in some countries, also through direct provision of in-kind benefits such as housing and preschool education.

In a market economy, enterprises are responsible for ensuring that production is efficient and for making profits which, in the long run, protect existing jobs and create new ones. Enterprise losses are not subsidized out of the state budget. Individuals, with the assistance of the state, have the main responsibility for finding jobs or pursuing other earnings opportunities. Reduced job security is an unavoidable price to pay for increased efficiency and higher standards of living. An effective labor market concentrates the costs of adjustment on a subset of the labor force: the unemployed. The state offsets this concentrated burden, at least in part, by redistributing income from the employed to the unemployed and other groups such as the elderly. In addition to its distributional role, the state helps the market run efficiently and complements or replaces market activity where private markets are inefficient or nonexistent.

Two key principles follow from these new responsibilities:

• Wages and employment are largely determined by the market in the interests of efficiency. Although the quest for efficiency is to some extent tempered by other objectives (for instance, those specified in the various conventions of the International Labour Organisation), the major purpose of the labor market is to assist the movement of workers into productive jobs.[2]
• The system of income transfers is the main method for pursuing distributional objectives, including protecting individuals and families from poverty.

These twin targets—using labor efficiently and treating people equitably—require twin instruments.

THE DIVIDING LINE BETWEEN THE STATE AND THE MARKET. The dividing line between the state and the market is in many ways the central issue in the process of economic reform. The analysis in chapter 2 establishes the enormous advantages which private markets have in large parts of the economy and thus the need for large-scale privatization of state-run enterprises. But there are exceptions to the rule that private is better. The former socialist countries have learned the hard way that one simple solution, central planning, does not work. It is a fundamental error, however, to assume that the answer is the other simple solution, private markets always and everywhere. A more balanced message is that state intervention is much less frequent in a market economy than under the old regime in Central and Eastern Europe and that such intervention as does occur is often different from the traditional intervention in communist systems, which consisted of public production and allocation.

The following points about the scope and nature of government intervention are stressed throughout the book:

• The distinction between the scale of and the structure of the state's activities is important: a budgetary crisis is not per se justification for privatization. That decision is more properly made on structural grounds. By the same token, commodities which the private sector produces efficiently should continue to be produced there, even if the state is running a budgetary surplus.

• Another crucial distinction is between the different types of government intervention. Intervention does not necessarily mean public organization of an activity. It can consist of regulation of private activity (for example, consumer protection in areas such as drug testing, where people have insufficient information to protect themselves). It can take the form of partial public funding in ways which establish an appropriate incentive structure for private activity (for instance, tax incentives which encourage contributions to private pensions). It can take the form of public funding of privately produced goods (for example, free prescription drugs) or privately produced services (for example, vouchers for education). Finally, it can take the form of income transfers to certain groups of people, such as the unemployed.

• Technical problems with private markets may justify government intervention, but do not automatically justify it. Government intervention is appropriate where two conditions hold: there is a market imperfection of some sort, and the situation with intervention is more efficient than the situation without it.

Table 1-2 indicates the range of government spending and its major components across several groups of countries: the industrial countries, the better-off countries of Latin America, some of the high-performing Asian economies, and Central and Eastern Europe. The choice of economies was largely determined by the availability of data, which have severe limitations. They are not always strictly comparable across economies, and they relate only to *central* government and therefore represent a lower bound for total public expenditure. At the low end of the range (Argentina, Japan, Mexico, and Thailand), total central government spending is between 15 and 20 percent of GDP; the top of the range, the Central and Eastern European economies excepted, is between 40 and 50 percent (France and Italy). The high levels of total central government spending in Central and Eastern Europe present a striking contrast. The low figure for Yugoslavia is not an accurate indicator of total public spending because most social spending took place at the subnational level. Certain components of social spending in some of the other economies in the region—for instance, health and education spending in Czechoslovakia— show low levels of central government spending for the same reason.

Social spending (social security, welfare, housing, health, and education) in the highly industrialized economies ranged from just below half of central government spending in Canada, the United Kingdom, and the United States to more than two-thirds in France and Germany. These figures, however, do

Table 1-2. Components of Central Government Spending in Various Economies, 1991

Economy	Percentage of total expenditure					Total expenditure as a percentage of GDP	GDP per capita (1990 dollars)
	Social security, welfare, housing	Health	Education	Defense	Other		
OECD members							
Canada	36.4	5.2	2.9	7.4	48.1	23.9	20,440
France	46.4	15.3	6.9	6.3	25.1	43.7	20,380
Germany, Fed. Rep. of[a]	48.9	18.1	0.6	8.3	24.1	32.5	23,650
Italy	—	—	—	—	—	49.6	18,250
Japan	—	—	—	—	—	15.6	26,930
United Kingdom	31.8	13.3	3.2	11.1	40.6	38.2	16,550
United States	28.7	13.8	1.7	21.6	34.2	25.3	22,240
Latin America							
Argentina	39.4	3.0	9.9	9.9	37.8	13.1	2,790
Brazil	25.5	6.7	3.1	3.5	61.2	35.1	2,940
Mexico	13.0	1.9	13.9	2.4	68.8	18.1	3,030
High-performing Asian economies							
Singapore	8.2	4.6	19.9	24.0	43.3	22.1	14,210
Thailand	5.9	7.4	20.2	17.1	49.4	15.5	1,570
Central and Eastern Europe							
Bulgaria	23.9	4.8	6.2	5.6	59.5	77.3	1,840
Czechoslovakia	27.0	0.4	1.9	7.1	63.6	55.6	2,470
Hungary	35.3	7.9	3.3	3.6	49.9	54.7	2,720
Poland	20.5	16.1	14.3	7.5	41.6	29.3	1,790
Romania	26.6	9.2	10.0	10.3	43.9	37.0	1,390
Russian Federation[b]	24.5	1.2	5.1	19.1	50.1	26.8	3,220
Yugoslavia	6.0	0.0	0.0	53.4	40.6	21.0	—

— Not available.

Note: The data should be interpreted with care. First, because of differences in coverage, the individual components of central government spending may not be strictly comparable across countries. Second, the data do not cover expenditure by state, provincial, and local governments; this may seriously understate or distort the statistical portrayal of how resources are allocated for the various purposes, especially in countries where subsidiary levels of government are responsible for many social services. For further discussion of the construction and limitations of the data, see the technical notes in World Bank 1993d.

a. Prior to unification.

b. For 1992. Data include extrabudgetary funds (Employment Fund, Social Insurance Fund, Pension Fund, Social Protection Fund) but exclude all local government expenditures, which constitute about 80 percent of spending in the health and education sectors.

Sources: World Bank 1993d, tables 1, 11; for the Russian Federation, unpublished data from the Ministry of Finance; for Poland, World Bank data, and Poland, Ministry of Finance 1992, p. 139.

not represent total government spending on social services: in Canada, the bulk of health care and education is financed, respectively, at the provincial or the municipal level; education in the United States is financed mostly at the state and local levels; education in the United Kingdom was largely financed at the local level when these data were collected, although reforms have since shifted a larger share of expenditure to the central government. Even the high figure for total central government spending in Germany is biased downward because most education spending takes place at the level of the *Länder*. In the Latin American economies shown, social spending is again a significant fraction of central government spending, but, with the exception of Brazil, the shares are lower than in most of the major industrial economies. In the two high-performing Asian economies, social spending, particularly in the category which includes social security, is a significantly smaller fraction of central spending than in the industrial economies shown. What is significant here is the high level of spending on education, alluded to above, which in Singapore made up nearly one-fifth of central government spending. What is noteworthy about social spending in Central and Eastern Europe is that it does not differ greatly as a proportion of GDP from social spending in the highly industrialized economies, whose per capita incomes are much higher.

Social spending by all levels of government in all the industrial countries of the Organization for Economic Cooperation and Development (OECD) is shown in table 1-3. In the countries less oriented toward the welfare state (Australia, Japan, Portugal, and the United States), public social spending in 1990 ranged from 16 to 20 percent of GDP; at the top end (the Netherlands and Scandinavia), it was 35 percent or more. Public social spending in Germany was exactly the OECD average of 27.5 percent of GDP. Historically, the state has been the dominant spender on income transfers and education in Europe for most of the twentieth century and certainly since the Second World War. Public spending on health in the OECD countries rose consistently as a proportion of total health spending from 1960 onward, reaching an average of almost 75 percent in 1990 (for the historical data see OECD 1988a, table 1, 1990b, tables 1, 2). The overall pattern was one of rapid growth in spending as a percentage of GDP during the 1960s and 1970s, followed by a much slower rate of growth during the 1980s.

These figures may represent a peak. A major debate about the welfare state is now under way in Western Europe. Prolonged high unemployment and the prospect of growing numbers of pensioners have led Western governments to look for savings. Reforms in the United Kingdom in the late 1980s made marginal reductions in the size of the state pension, phased in for persons retiring after 2000. The United States legislated a phased increase in the age of retirement. Searching scrutiny of expenditure and much public discussion led Germany to reduce some benefits in 1993 (unemployment benefits, for example, were reduced by 3 percent). As the editor of the newspaper *Die Zeit*

Table 1-3. *Public Expenditure on the Social Sectors as a Percentage of GDP in OECD Member Countries, 1990*

| | Income transfers | | | | |
Country	Non-aged population[a]	Aged population[b]	Health	Education[c]	Total
Australia	3.6	3.8	5.6	5.1[d]	18.1
Austria	4.0	15.0	5.6	5.5	0.0
Belgium	8.7	9.7	6.8	5.0	30.2
Canada	7.6	4.4	6.9	6.8	25.6
Denmark	14.4	8.1	5.2	7.2[e]	35.0
Finland	13.3	7.5	6.3	5.7	32.8
France	8.2	11.7	6.6	5.5	32.0
Germany, Fed. Rep. of[f]	7.6	9.9	6.0	4.0[d]	27.5
Greece	4.0[g]	12.8[g]	4.0[g]	3.4	24.3[g]
Ireland	8.7	5.9	5.2	6.2[d]	25.9
Italy	3.4	15.8	6.3	5.0[h]	30.5
Japan	1.8	5.0	4.8	4.4	16.0
Netherlands	13.1	9.9	5.7	6.9[d]	35.7
New Zealand	6.3	6.7	6.0	6.1	25.1
Norway	13.9	7.7	7.1	5.7	34.4
Portugal	4.5	6.7	4.1	4.2	19.5
Spain	6.1	7.9	5.3	2.2[e]	21.5
Sweden	13.9	11.7	7.7	6.5	39.8
United Kingdom	7.4	10.3	5.2	4.8[d]	27.7
United States	3.5	5.8	5.2	4.8[d]	19.4
Average for all members (unweighted)	7.7	8.8	5.8	5.3	27.5

Note: Data are provisional estimates by the OECD. Data for Switzerland are not available.

a. Includes unemployment compensation, employment promotion benefits, sickness benefits, disability pensions, disability services, family allowances, low-income benefits, indigenous persons benefits, housing benefits, other miscellaneous services and benefits, and administrative costs. Excludes all benefits to the aged and survivors and health expenditures.

b. Includes all old-age and survivors' benefits, that is, all transfers and services to the elderly and survivors. Data for Australia include pensions for veterans; data for the United States include occupational pensions for civil servants and exclude expenditures on health.

c. For 1988 unless otherwise stated.

d. For 1987.

e. For 1985.

f. Prior to unification.

g. For 1989.

h. For 1986.

Sources: OECD social data bank and OECD 1992c.

put it, "We can certainly make our welfare pudding with less cream and fewer eggs" ("Fewer Holidays in the Sun," *Independent (London)*, October 4, 1993, p. 19; on reform in New Zealand, see Boston 1993; "What Happens When You Scrap the Welfare State?" *Independent on Sunday (London)*, March 18, 1994, p. 17). Despite this determined effort to control expenditure, the *structure* of state involvement has not changed significantly, at least in the mix of public and private activity. Instead, the *scale* of activity has decreased marginally (for a survey of what Western governments have done, see U.K. Department of Social Security 1993).

The last conclusion is noteworthy. In both the United Kingdom and the United States during the 1980s, writers like Murray (1984) and political leaders like Ronald Reagan and Margaret Thatcher argued for a reduced role for the state. Nevertheless, a U.K. study concluded that

> the welfare state, and indeed welfare itself, is very robust. Over the thirteen years from 1974 to 1987, welfare policy successfully weathered an economic hurricane in the mid-1970s and an ideological blizzard in the 1980s. The resources going to public welfare were maintained; [and] welfare indicators continued to show a steady improvement. [Le Grand 1990, p. 350]

In the United States,

> the Reagan era ended with the welfare state substantially intact, though somewhat frayed around the edges. It is now tilted more toward its middle-class beneficiaries than it was a decade ago, but the broad contours remain essentially as they have evolved since the 1930s. [Peterson 1991, p. 133]

The descriptive picture in the OECD countries, then, is that (1) the scale of public social spending is under review, (2) its structure remains largely unchanged, and (3) even in the low-spending industrial countries, public social spending in 1990 was between one-fifth and one-sixth of GDP. The first of these observations is clearly the result of the macroeconomic situation and the demographic prospects. The second and third can be explained, at least in part, by market imperfections (see chapter 2), which make private financing and provision inefficient or impossible in a number of these areas. In the high-spending countries of Western Europe, moreover, social solidarity (also discussed in chapter 2) remains an objective. Why is public social spending 15 percent of GDP in some countries and more than 30 percent in others? First, objectives may differ: state social spending is lower where social solidarity is less of an objective (as is true in Australia and the United States). Second, the constraints may differ: spending is lower where income is lower and fiscal constraints are tighter (in Portugal). Third, social conditions may differ: social spending is lower in countries where enterprises play a larger social role (in

Japan)[3] or where the extended family can still be relied on to some extent (on the importance of social environment in policy design, see Kopits 1993).

Experience outside the West is somewhat different. Eight of the major East Asian developing economies have been among the twelve most rapidly growing developing economies in the world since 1960.[4] State social spending in Singapore and Thailand (the two economies from this group included in table 1-2) is both lower and more weighted toward education, particularly basic education, than is true in the industrial economies. This is not the place for a detailed discussion of the East Asian economies (see World Bank 1993a), except to note the broad context in which they operate. All of them have a more or less single, overriding objective, namely, economic growth, and a political consensus which by implication gives a fairly low weight to distributional objectives and to social solidarity. They tend to have fairly authoritarian regimes. They face fewer constraints than do the economies of Central and Eastern Europe: China apart, they have well-developed market systems, sophisticated banking systems, highly developed capital markets (by the standards of developing economies), and relatively stable prices. Their economies need no substantial restructuring which, together with their high level of spending on education, means that East Asian economies have a much more appropriate mix of skills than economies in Central and Eastern Europe. Finally, the extended family is still a significant part of their social structure.

The Analytical Framework

Many of the countries of Central and Eastern Europe have introduced market forces and liberalized their political institutions. Chapter 4 discusses those actions, the driving forces—the engine of reform—they unleash, and some of their outcomes to date. Outcomes are of two sorts. Some, such as greater efficiency and government which is more responsive to the needs of the electorate, are both deliberate and welcome. Others, such as the fiscal crisis, are unwelcome because they constrain the speed and success of the reforms. These and other constraints are discussed in chapter 5.

The Forces Driving Change

The move to a market system liberates two major economic driving forces: a widening distribution of income and a more explicit role for individual self-interest. The widening income distribution is a result of wage and price liberalization. Market-determined wages promote work effort and encourage workers to move into more productive jobs; workers respond by acquiring skills which are in high demand. Such results are a fundamental purpose of the reforms. There are also malign outcomes, however. Although some individuals are better off, others are worse off; to a disproportionate degree, the losers have been women, children, and the elderly. Because of the data problems discussed in chapter 4, the precise extent of change is controversial, but not its direction. The transition aggravates poverty because the fall in output lowers real wages, because the heavy subsidies on basic commodities are removed, and because unemployment rises. Policies to alleviate unemployment and poverty become correspondingly more important.

Alongside these economic driving forces are the forces released by political liberalization. Some of these forces emerged at the start of the transition; others came later. There was initially a reaction against the inheritance of central planning and totalitarian government; this reaction manifested itself in a swing toward private markets (a reaction against central *planning*), toward decentralization (a reaction against *central* planning), and toward democracy. In general, these tendencies are beneficial. They can, however, overshoot and produce less desirable outcomes. In the absence of regulation, some markets will be inefficient; decentralization can hinder the development of policies best formulated at a national level; and democracy can lead to legislative paralysis, for instance, where parliamentary representation is divided among many small parties. In the social sectors, such tendencies created pressures to privatize too fast in inappropriate areas (for example, moving too fast and too far to replace public funding of medical care with private medical insurance), led to poorly coordinated efforts to revise educational and training qualifications, and produced legislation which was either badly drafted or too complicated to be implemented effectively. Some countries subsequently experienced a political backlash against the reformers and sometimes against the reforms themselves, in part because of disappointed hopes that living standards would improve rapidly.

Constraints

Constraints arise in many guises and explain why the transformation is taking so long. Although the distinction is not watertight, the constraints are usefully divided into those which are primarily economic, those which mainly concern political effectiveness, and those which relate to institutional capacity.

ECONOMIC CONSTRAINTS. This class of constraint derives, first, from the massive misallocation of labor and capital inherited from the old system, necessitating a large-scale movement of workers and making high unemployment likely, if not inevitable. A second constraint is the fiscal crisis itself. Output fell fastest in the state sector, where most tax revenue is collected, and was only partly offset by increased activity in the private sector, where tax enforcement is not yet fully effective. Public expenditure, conversely, fell more slowly. Social spending was thus squeezed by a fiscal crisis precisely when demands on the social safety net increased. A third major constraint at the start of the transition was the deficient legislative infrastructure underpinning the operation of a market economy. Major gaps were evident in such crucial areas as the definition of property rights, the conduct of private enter-prises, regulation of the banking system and other financial institutions such as insurance companies and stock exchanges, regulation of working conditions, and consumer protection. This deficiency, although perhaps affecting other areas more strongly, also slowed reform of the social sectors.

POLITICAL CONSTRAINTS. There are two main types of political constraints: those connected with political attitudes and those arising from the machinery of government. Political attitudes toward unemployment are particularly important. Prior to 1938, countries like Czechoslovakia were substantially industrialized under a market system; reform has largely reconnected such countries with their own history.[5] The former Soviet republics are different: they have never been democratic, nor have they ever had a modern market economy. Unemployment is thus completely new, and the political capacity to carry through the reforms (particularly when coupled with the governance issues discussed below) cannot be taken for granted.

The machinery of government also constrained change in the early transi-tion. The powers and responsibilities of the legislature and the executive were somewhat fluid, and the judiciary was not always in a position to resolve any particular dispute. The electoral regime sometimes yielded a fragmented out-come, leading to government by coalitions which were sometimes unstable. The amount of new legislation required and the amount of detail in each law produced parliamentary overload. The problem was particularly acute for social sector legislation, much of which was politically highly charged. Alongside the constitutional arrangements (the rules of the game) was the separate constraint of an initial lack of experience in how democratic politics are conducted (that is, a lack of experienced professional players). One view, held by Balcerowicz (1993a), is that the early transition was a revolutionary period in which the first postcommunist governments had to spend their political capital initiating a comprehensive "big bang" liberalization and creat-ing a fait accompli. The opposing view is that, even allowing for the revolu-tionary situation, the new governments should have spent more time explain-ing their policies to the electorate.

INSTITUTIONAL CONSTRAINTS. During the early transition, institutional capac-ity imposed constraints which were compounded by the movement of many of the most able public servants to the private sector. The public sector resisted change, in part because change is painful, but also because the indi-viduals in power wanted to retain that power. Asking state institutions to organize the privatization of large parts of the state apparatus, for example, created a clear conflict of interest. A separate problem is that reformers tended to undervalue administrative skills. For instance, little effort was made to involve people with implementation skills in the design of policy for the social sectors. None of these problems is unique to the region, but they are posing enormous constraints during the transition because the scale and speed of the changes needed are so great.

Much of the existing human capital is inappropriate, and this remains a major constraint. As discussed in chapter 11, the problem is not generally one of too little human capital, but of the wrong type of human capital. "Oldthink" is pervasive. Insufficient weight is given to such key concepts as the tradeoffs discussed above and the importance of incentives. Training in modern technology is often inadequate, and partly as a consequence, attitudes toward modernization can be ambivalent. Missing concepts and the lack of key skills were significant impediments to implementation early in the reform.

The Policy Strategy

The policy strategy summarized below shapes the recommendations made throughout the book. It seeks to harness the beneficial results of the driving forces while recognizing the constraints. The key needs of the early transition were—and are—to address the fiscal crisis, to privatize large parts of the economy, and to focus on the practical implementation of these and other very major changes. These needs lead directly to a strategy with four parts: controlling the budget, liberating market forces, containing market forces, and implementing the reforms:

• *Maintaining macroeconomic balance is a necessary response to the excessive size of the state sector generally and to the fiscal crisis specifically.* It includes policies to reduce the size of the public sector, policies to balance government revenue and expenditure, and, although outside the scope of this book, policies to maintain a stable monetary system. Containing the explosion of social sector budgets and diversifying the finance of social services are key contributors to maintaining macroeconomic balance.

• *Building markets is the primary method by which to pursue higher standards of living.* It involves increasing consumer choice and diversifying supply. Policies to improve incentives include allowing wages to be determined largely by market forces and avoiding taxes which are so high as to discourage work effort or new employment. Employer contributions should be neutral with respect to employing men and women (thus, for instance, the cost of family benefits should fall on the taxpayer, not the employer). Building markets also includes actions to raise labor productivity and to encourage mobility between jobs and between skills.

• *Regulating market forces is the complement to building markets.* It is necessary for two entirely separate purposes: to address market imperfections and to redistribute income. Regulation, for instance in the form of drug testing regimes, is a necessary companion to policies which diversify supply. Policies to address unemployment and poverty include both income support and more direct action to help match workers with jobs. Action is needed to ensure access to basic goods and services, including nutrition, health care, and education. Finally, individuals should be able to insure themselves against loss of income (because of unemployment) and, separately, to redistribute income to themselves over their lifetimes. Although in principle these objectives could be accomplished through private insurance, private pensions, and the like, various problems including capital market imperfections require state intervention, at least to regulate private sector activity.

• *Implementing the reforms requires administrative skills and political experience as well as good policy design.* Like the legs of a tripod, all three sets of skills are necessary simultaneously.

Early Achievements

There are two reasons for arguing that, notwithstanding the problems described at the start of this chapter, the reforms are very much a move in the right direction. First, the past is no longer on offer, not least because of the increasingly appalling economic performance of central planning. Second, even in the short run, the reforms have produced some beneficial outcomes, and those benefits are real and sustained.

ECONOMIC ACHIEVEMENTS. The economic gains achieved early in the reform include increased availability of goods, wider choice, improved incentives, improved creditworthiness, and improved access to foreign know-how.

Price liberalization reduced excess demand and encouraged additional production of desired goods. Queues have largely disappeared, and so have black market prices. Many other ill effects of shortage have faded away: forced resort to imperfect substitutes (for example, buying cheese instead of milk if only cheese is available), forced saving, the humiliation and cost of time spent in queues, and bribes paid to people with access to goods in short supply. What is surprising are not the gains themselves, but the speed with which the introduction of market forces produced them.

Not only has the quantity of goods increased, but so have their range and quality. The combined effect of price liberalization and recession has been to switch from a shortage economy to one with low demand. This has given consumers more power to insist on, and sellers an incentive to provide, the goods which consumers want at reasonable prices. Easier access to foreign goods has improved matters further. Wider choice has also started to manifest itself in the labor market, in health care, and in education and training.

The introduction of market-determined wages and prices, together with liberalization of supply, has improved work incentives. Market-determined wages together with harder budget constraints have also given firms incentives to hire only as many workers as they need, and to hire only those with the type of skills they need, and have given individuals stronger incentives to acquire those skills. Economic reform has also liberated latent entrepreneurial talent, particularly in the emerging private sector.

Most of the countries of Central and Eastern Europe have improved their export performance and increased their international reserves. One result is that domestic currencies have become convertible, at least within the country of issue, and many of the countries have achieved a relatively stable exchange rate.

Finally, convertibility of domestic currencies and increased foreign investment have improved access to Western technology and skills. Significant benefits have been evident relatively quickly in some countries in telecommunications, banking, trade, and the mass media. Access to foreign books, including textbooks, has also improved.

POLITICAL ACHIEVEMENTS. Reform has also brought major political benefits. A free press, including the broadcast media, has developed rapidly in most of the countries. Just as opening the economies to foreign trade improved the incentives for domestic producers, so opening up the countries to foreign news (made possible, for instance, through access to television satellite dishes, facsimile machines, and the like) has created incentives for the conduct of government to be more open.

Free, multiparty elections have increasingly become the norm. These, together with a free press, have started to build a political culture, which is essential for effective democratic government. The protection of individual rights has also improved. This was less an outcome of constitutional guarantees (which, in a formal sense, existed under the old system) than of free elections and freedom of the press.

The Agenda: Priorities and Sequencing

Both the scale and the structure of state activity will change significantly. The role of the state will diminish. The data in table 1-2, together with the analysis in chapter 2, establish an overwhelming case for reducing the scale of governmental activity. This will occur in some areas more than others. Industry, financial markets, and the production of consumer goods will all become almost totally private sector activities, with necessary regulation being the only intervention.

The state has a significant and continuing role to play in the social sectors, although the form of that involvement will become considerably more diversified. There will be some public production, but on a reduced scale, a continuing funding role in some areas (see, for instance, World Bank 1990b, 1993d, on the state's role in poverty relief and health finance), and more state involvement in facilitating market activity through regulation and through financial and other incentives for certain types of activity. Public social spending will continue but will be substantially reshaped: some social programs will need to be slimmed down, in some cases drastically; others will need to be reformed fundamentally to improve targeting.

The appropriate boundary between the market and the state is a matter of continuing discussion. Economies in the region have different objectives and social structures (in part stemming from their cultures and histories) and face different constraints. As a result, differences between economies in the role of government will always remain. That role will be substantial if the Western European model is followed, and rather less so if the models of the high-performing Asian economies and the United States are followed. Key questions for policymakers include the extent to which even a much slimmed-down Western European model is affordable and the extent to which the circumstances of the high-performing Asian economies are applicable.

The priority given to different policy actions should be determined in the light of what can be afforded in the short run given the limits to both financial resources *and* implementation capacity. The recommendations in later chapters are divided into two sets: short-term actions which are vital to the success of the reforms in the social sectors and medium-term actions which require some work now, but whose effects will not be immediate.

Short-term Priorities

Short-term policy has to concentrate on surviving the early transition both fiscally and politically:

- In the face of high inflation and rising unemployment, short-run labor market policies should concentrate on containing inflation through tax-based incomes policies and on removing the major inherited legal constraints which limit the flexibility of the labor market.
- Given the fiscal crisis and sustained unemployment, the short-term priority for the system of income transfers is the relief of absolute poverty. This requires the development of effective systems of unemployment benefits and direct poverty relief. Resources should be concentrated on protecting at least the minimum level of the major benefits. Deciding on the exact level of that minimum will force policymakers to face the acute tension between the economic and the political sustainability of the reforms.
- The major priorities for education and training are to maintain standards in preschool and basic education and to restructure secondary and higher education to make them more responsive to the needs of a market economy.
- Policies for the health sector should ensure that the share of GDP devoted to it are maintained, and should avoid excessively radical restructuring during the early transition. Specifically, they should concentrate on providing immediate relief of critical shortages, adequate budgets for services to protect vulnerable populations, cost-effective interventions against preventable diseases, and hard budget caps.
- A shortage of administrative capacity is one of the most acute constraints to effective reform. Thus there is an imperative that policy design should be as simple as possible.

- Politicians should devote a significant amount of their time to persuading the public that the government's policies are desirable. They should also try to dampen unrealistic expectations.

Besides these positive steps, certain policy "black holes"—actions or inactions—should be avoided at all costs. These arise where any parts of the policy strategy are ignored:

- Postponing action to control the cost of the state cash benefits system, although politically tempting, destabilizes fiscal and monetary policy. The resulting inflation shows what happens if the first part of the strategy—controlling the budget—is ignored.
- Introducing private pensions without the necessary macroeconomic, financial, and regulatory infrastructure—an example of attempting to build markets without regulating them—overlooks the need for consumer protection and ignores the high risk of inflation, which could easily decapitalize domestic funds. For both reasons, badly designed schemes risk discrediting the reforms before they take root. The answer is not to abandon private pensions, but to introduce reforms in their proper sequence.
- Introducing health insurance without the proper regulatory structure is another example of failing to combine efforts to build markets with efforts to regulate them. The problem is that, where doctors are paid by an insurance fund, neither doctor nor patient has any incentive to economize on medical services, and the result is a cost explosion. A regulatory structure is needed to contain medical spending.
- Ignoring the political dimension of implementing the reforms is a problem which is at its most acute early in the transition. The most successful reformers, building on Western experience, took steps to prevent overoptimistic expectations, began early to build party structures, and devoted considerable effort to communicating with the electorate. Neglecting such activities can mean that reforms are not implemented for political reasons.

Medium-term Priorities

Policies with a medium-term dimension, even though their effects will not be felt immediately, should not be delayed:

- Investment in human capital is crucial to economic growth.
- As fiscal constraints start to relax, income transfers will need to take on functions other than poverty relief; in particular, the relationship between contributions and benefits should be strengthened.
- Action on pensions, education, and health care should include support for the emergence of private providers. The necessary first step is to design and put into place a regulatory structure which protects consumers, and information systems which assist consumer choice.

[27]

- An early start should be made on policy design and legislation in areas where implementation takes time, such as computerizing social insurance contributions, revising teacher training programs, and devising hospital reimbursement schemes which encourage the efficient use of medical resources.
- Upgrading administrative capacity, although its impact will not be felt in the short run, is also a priority.

More detailed recommendations can be found in the concluding sections of chapters 7 to 13.

The decline in output and the fiscal crisis highlight the inescapable tradeoff between current and future standards of living. The economic problem is to choose a policy mix which alleviates extreme poverty, allows stabilization, and maintains international competitiveness, while still devoting sufficient resources to investment. The political problem is to find a meeting ground between the pressure to raise standards of living now and the need to devote more resources to investment in the interests of economic growth. Government cannot escape making choices about how resources are divided between consumption and investment. Because that choice is very much a political one, the electorate must understand the issues involved. The tradeoffs are both economic *and* political.

Notes

1. These totals are the result of compounding, not adding, the figures. Taking Lithuania as an example, output at the end of 1990 was $(100\% - 5\%) = 0.95$ of its level at the start of the year. Over three years, therefore, output fell to $0.95 \times 0.866 \times 0.65 = 0.5348$; that is, output is 53.5 percent of its previous level and thus fell by 46.5 percent.

2. The International Labour Organisation conventions seek to establish employment conditions, broadly defined, which protect the interests of individual workers in ways which do not preclude the operation of market forces.

3. Recent reports, however, suggest that, in the face of global competition, Japan is moving away from lifelong employment security, which, in any case, applies only to some 30 percent of the work force.

4. The eight are China, the "Four Tigers"—Hong Kong, the Republic of Korea, Singapore, and Taiwan (China)—and the three newly industrializing economies of Indonesia, Malaysia, and Thailand.

5. Many of the non-Soviet countries had prewar experience with market systems, but their experience with democracy varied considerably. Poland was a dictatorship after 1926, and Hungary an oligarchy with a severely limited franchise. The countries with the longest experience of democracy between the two world wars were Czechoslovakia and Romania.

Two

THE ROLE OF GOVERNMENT IN A MARKET ECONOMY

NICHOLAS BARR

THE DIVIDING LINE BETWEEN THE MARKET AND THE STATE is in many ways the central part of the economic reforms. It is a fundamental error to imagine that capitalist economics means an absence of government intervention. The precise dividing line may differ (for example, the balance between public and private pensions is different in different countries), and forms of intervention can vary (in some countries health care is publicly produced, while in others it is privately produced but heavily regulated). Yet, as tables 1-2 and 1-3 show, the state plays a continuing and major role in all industrial countries. This outcome is no accident: as a practical matter it is the result of a process of historical trial and error; in addition, it can be both explained and justified by recent developments in economic theory.

That theory—the underpinning of the policy analysis in part II—is the subject of this chapter, which thus provides important background for readers whose familiarity with economic theory is limited. The first section discusses the economic objectives of the reforms. The central sections summarize the economic theory of private markets and the state. The final section discusses the policy implications of the theory, both generally and in the context of Central and Eastern Europe, including cases where the theory points firmly toward private markets and others where it suggests the need for state activity.

The state's role in the West, though substantial, is fundamentally different from the old system in Central and Eastern Europe: the state intervenes only in parts of the economy, and the intervention tends to be carefully targeted. The implications for Central and Eastern Europe are, first, that intervention should be of the appropriate *type* (as we shall see, government can intervene in ways other than central planning) and, second, that it should be in the appropriate *area*. Many countries in the West have (in one form or another) a national health service; none has a national food service. Why this is so and, more generally, how to decide whether the state should intervene and, if so, how are the main subjects of this chapter.

The Objectives of the Reforms

The overall objectives of the reforms are to achieve higher living standards and greater individual freedom. This section discusses the economic objectives in greater detail, particularly as they relate to the social sectors, and divides them into five groups: enhancing efficiency, protecting the living standards of individuals and families, reducing inequality, strengthening social integration, and increasing administrative effectiveness.

Enhancing Efficiency

In policy terms, efficiency has at least three aspects. The aim of macroeconomic efficiency relates to the division of total national resources between expenditure on human resources, on the one hand, and activities such as physical investment, on the other. Spending nothing on health care is obviously inefficient (since people would die unnecessarily from potentially curable conditions); spending the whole of national income on health care is equally inefficient (since people would then die of starvation). The goal of macroeconomic efficiency is to make sure that enough is spent on human resources, but not so much as to crowd out other important activities. A particular aim is to avoid distortions which lead to cost explosions, such as uncontrolled expenditure on pensions and soaring medical costs. The aim, in other words, is to choose efficiently between human resources and other activities in the face of the tradeoffs discussed in chapter 1.

Microeconomic efficiency concerns the division of total resources among the different cash benefits, different types of medical treatment, different types of education, training, and retraining, and different types of employment services. Chapter 12 argues, for instance, that too many resources tend to go into hospital care and not enough into primary care and public health activities. Another aspect of microeconomic efficiency is to help workers move to jobs in which they are most productive.

Incentives are a third aspect of efficiency. High rates of income tax, for example, may discourage work effort, and high payroll taxes may impede new employment. Thus where they are publicly funded, institutions should be financed in ways which minimize adverse effects on labor supply and employment.

Supporting Living Standards

Protecting living standards embraces poverty relief, insurance, and income smoothing. The aim of poverty relief is to ensure that no individual or family falls below some minimum level of income or consumption. The aim of insurance is to see that no one faces an unexpected or unacceptably large drop in his or her standard of living. This is a major objective of unemployment benefits and of most health-related benefits. The definition of a minimum and of what constitutes "unacceptably large" poses difficult questions for policymakers which are both economic and political. The aim of income smoothing is to ensure that individuals are able, in effect, to redistribute income to themselves at different stages in their life cycle. This is an important aim of pensions and also of family allowances. All three aims are important. A pervasive theme, however, is the fiscal crisis which accompanied the early transition. Later chapters therefore argue that the aim of poverty relief should be given priority over insurance and income smoothing until the fiscal situation improves. As a practical matter, this means concentrating resources on protecting the minimum level of unemployment benefits, pensions, and other major benefits.

Reducing Inequality

The reduction of inequality has two aspects. Vertical equity concerns the contentious issue of redistribution from rich to poor. The objective of poverty relief implies at least minimal redistribution. More generally, should the formula under which pensions are calculated favor lower earners? Should health care or education be financed out of taxation and given free to the user? Horizontal equity relates to the aim that individuals who are the same in all relevant respects should be treated equally. Once more, the definition of "relevant respects" can raise difficult policy questions. Differences in cash benefits should take account of family size. Differences in education and training should reflect only factors which are regarded as relevant (such as a person's previous education and work experience), but not irrelevant factors such as ethnic background.

Strengthening Social Integration

Social integration embraces broader and less tangible goals. Cash benefits, health care, and education and training should be delivered in ways which preserve the recipient's dignity. They should also foster social solidarity, an aim which in recent years has received little attention in the United Kingdom and the United States, but which remains an explicit aim of public policy in mainland Western Europe and is an important part of the inheritance of Central and Eastern Europe. As examples, benefits should as far as possible depend on criteria which are not related to socioeconomic status, such as retirement pensions and, in most countries, health care. Additionally, benefits should be high enough and health care good enough to allow recipients to participate fully in the life of the society in which they live.

Increasing Administrative Effectiveness

Administrative feasibility is concerned, first, with simplicity, in that the system should be as simple as is consistent with policy objectives and thus easy for the citizen to understand and for the manager to administer. A second aspect is effectiveness: administration should be timely and accurate, meaning that the benefit both (a) should be awarded in the right amount and (b) should go to all who are entitled to it. Third, insofar as possible there should be as little abuse as possible. Finally, administration should be as cheap as is consistent with the achievement of the other objectives.

The Argument for Private Markets

The conventional argument for private markets is that they automatically achieve the efficiency objective and, moreover, do so at very little cost, since outcomes are the result of individual actions based on individual information.[1] Thus there is no need for the expensive information gathering and complex paperwork which would be necessary to achieve efficiency under central planning. In addition, there is no need for government to prioritize activities, since the actions of individuals do so. The free market, according to this view, is a highly efficient, self-adjusting information system, and the state does not have as much information, nor the ability to acquire it as cheaply, nor the capacity to respond to it as quickly or effectively, as does the free market.

Efficiency has a very specific meaning in this context. Broadly, it means making the best use of limited resources given people's tastes and available technology. A key underlying concept is resource scarcity (also known as opportunity cost), which means that if resources (labor, capital, raw materials, land) are used for one purpose, they cannot be used for another. Since the quantity of all those resources is limited, output is also limited. It is not possible to produce enough to satisfy everyone's demands completely: policy should seek to satisfy people as much as possible; since resources are scarce, it should seek to use limited resources as effectively as possible. This is precisely what economic efficiency means. It has three dimensions:

• *Efficiency in production means that inputs should be combined to obtain the maximum output from given inputs.* This is what engineers mean when they talk about efficiency. It is about building a hospital to a specified standard, wasting as little concrete as possible, and having as few workers as possible standing around waiting for something to do. It is also about the choice of technique, taking the prices of inputs into account. Thus, for example, the construction of buildings tends to be labor-intensive in developing economies, where wages are low, and more capital-intensive in countries like the United States.

• *Efficiency in product mix means that the right combination of goods should be produced.* The fact that a hospital can be built cheaply is not, on its own, justification for building it. The resources involved could perhaps give the local population greater satisfaction if used instead to build a school; or the land could be used as a park, and the money saved by not building a hospital could be used to reduce taxes, allowing people to spend the money as they choose on other forms of consumption.

• *Efficiency in consumption means that consumers should allocate their income in a way which maximizes their satisfaction, given their incomes and the prices of the goods they buy.*

These conditions relate to the efficient use of resources at a point in time (known as static efficiency). Resources have also to be used efficiently over time. Dynamic efficiency is concerned with optimizing the growth of output or consumption. The pursuit of dynamic efficiency impales policymakers on the key tradeoff discussed in chapter 1, the division of resources between current consumption and investment.

An efficient outcome simultaneously conforms with all these requirements. It depends on external conditions (more resources are devoted to hotels on Bulgaria's Black Sea coast than in Murmansk); it depends on tastes (the French spend more on food than the English; the English spend more on gardens than the Germans; Hungarians consume more paprika than anyone else); it depends on the age of the population (more resources tend to be devoted to education in a country with lots of children than in one with few); it depends on income levels (private ownership of cars and personal computers is much more widespread in Western Europe than in Central and Eastern Europe).

The "invisible hand" of market forces, it is argued, achieves exactly this efficient result. As Adam Smith (1776) put it over 200 years ago,

> Every individual . . . generally . . . neither intends to promote the public interest, nor knows by how much he is promoting it. By preferring the support of domestic to that of foreign industry he intends only his own security; and by directing that industry in such a manner as its produce may be of the greatest value, he intends only his own gain, and he is in this, as in many other cases, led by an invisible hand to promote an end which was no part of his intention.

In today's terminology, if a product has numerous suppliers, competition will force them to produce in cost-effective ways (efficiency in production). Firms which are unable to achieve this go out of business. The exercise of consumer choice in the face of market prices enables people to choose the goods they want and ignore the goods they do not want. This maximizes their satisfaction (efficiency in consumption); it also means that the production of unwanted products stops (leading to efficiency in product mix). The role of market prices is critical in all this. A higher price signals to consumers that the product uses scarce resources, forcing them to economize on consumption. If people continue to buy even at high prices, producers understand that this is a commodity which people want and which they should continue to produce. In contrast, if people do not buy the product, producers receive a signal to reduce or cease production. Market forces thus give consumers and producers incentives to act efficiently. The role of incentives is a recurring theme throughout this book.

The Assumptions Necessary for Markets to Be Efficient

The problem for policy design is that markets do not always or automatically operate in this way. This section and the next discuss the conditions under which (a) markets generally and (b) insurance markets specifically have these beneficial results. Where those conditions fail, the state may have grounds to intervene for efficiency reasons; in addition, even where markets are efficient, the state may intervene in pursuit of other objectives such as poverty relief.

Types of State Intervention

Before discussing *why* the state might intervene, it is useful briefly to outline *how* it could do so. It is important to be clear that pure private markets, on the one hand, and central planning, on the other, are merely the two extremes of a continuum. Intervention is possible in many forms short of complete central planning. The state can intervene in four generic ways: regulation, finance, production, and income transfers.

REGULATION. In some instances, regulation may have more to do with social values than with economics (for example, regulation of the hours during which shops may be open). But much regulation is directly relevant to the efficient or equitable operation of markets, especially where consumer information is imperfect. Regulation of quality, mainly on the supply side, includes hygiene laws relating to the production and sale of food and drugs, laws forbidding unqualified people to practice medicine, and consumer protection legislation generally. Regulation of quantity more often affects individual demand, as with the requirement to attend school and compulsory social insurance contributions. Price regulation includes minimum wages. Total expenditure can also be regulated, as occurs with global budget caps for medical spending.

FINANCE. Finance involves subsidies (or taxes) applied to the prices of specific commodities or affecting the incomes of individuals. Price subsidies can be partial, such as book user charges in schools and tuition fees in public universities, or total, such as free drugs and free primary education. One of the main reasons for subsidizing some activities is to create incentives which encourage private sector activity (for example, tax advantages for certain types of expenditure, such as private education). Similarly, prices can be raised by a variety of taxes, particularly in areas which the state wishes to discourage, such as a tax on tobacco. Income subsidies raise different issues, which are discussed shortly.

PRODUCTION. Although regulation and finance modify the operation of markets, they leave the basic mechanism intact. Alternatively, the state can take over the supply side by producing goods and services, common examples being pre-university education, defense, and (in some countries) most health care. Finance and production are entirely different forms of intervention, both theoretically and in practice, and are commonly used both separately and in combination.

INCOME TRANSFERS. The previous three types of intervention all involve direct interference with the market mechanism. Income transfers do not interfere directly; rather, they enable recipients to buy goods of their choice at market prices (for example, pensions allow elderly people to buy food). Income transfers can be tied to specific types of expenditure, such as food stamps or housing benefits, or, like cash benefits, can be untied.

Necessary Assumptions

How valid is the assertion that the market mechanism automatically leads to an efficient outcome?[2] The answer is that it is often, but by no means always, valid. The efficiency of market outcomes generally rests on four sets of conditions, including (a) that consumers and firms are well informed, (b) that markets are genuinely competitive, (c) that there are no missing markets, and (d) that several other assumptions hold. These conditions are critical to analyzing the role of the state and therefore require detailed discussion.

PERFECT INFORMATION. Both consumers and firms must be well informed. Simple theory implicitly assumes that consumers know which goods are available and understand their quality. As a relatively recent body of literature on economic theory makes clear, the assumption can fail in two important ways. People may have imperfect knowledge of the quality or the prices of goods.[3] The literature thus has two strands. The first analyzes the effects of imperfect information about quality. Consumers might be badly informed, for example, about the quality of a school or about the appropriate type of medical treatment. Producers might be poorly informed about the quality of a worker applying for a job or about the riskiness of an applicant for a loan or for insurance. The second strand analyzes the effects of imperfect information about prices and wages.

The assumption that consumers and firms are well informed about the nature of the product and about prices is plausible for some goods, but less so for others. Markets are generally more efficient (a) the better is consumer information, (b) the more cheaply and effectively the information can be conveyed, (c) the easier the available information is for consumers to understand, (d) the lower are the costs faced by someone who chooses badly, and (e) the more diverse are consumer tastes. Commodities which conform well with these criteria are food, clothing, and such consumer durables as radios, refrigerators, personal computers, and automobiles. Health care conforms less well: consumer information is often poor; people generally require individual information, so that the process is rarely cheap (violating the need for cheap and effective information); much of the information is highly technical (violating the need for easily understood information); and the costs of a mistaken choice can be high. Education can raise similar problems.

Where the assumption of perfect information fails, several solutions are possible. The market itself may develop institutions to supply information, for instance, computer magazines and professional valuers who can help people assess the price of a house. In such cases, the product being sold is information. In other cases, the state may respond with regulations, for instance, hygiene laws for food. This is appropriate where the potential costs of a mistaken choice are high and where the information is sufficiently non-technical for the consumer to understand. Where information problems are serious, and where the necessary information is too technical to be readily understood by the average consumer, the market will generally be inefficient, and public production may be a better answer. We return to this issue in the discussion of policy.

Along with information about quality and price, individuals also need accurate expectations about the future in order to make rational choices over time. This is broadly true of food, since people know that they will need to eat tomorrow, next week, next month; it is not true of medical care, because people do not know whether or when they will suffer health problems. In many instances, the market can cope with this sort of uncertainty through the mechanism of insurance. Private insurance, however, can be inefficient or impossible for technical reasons, and such problems are especially relevant to medical insurance and private pensions.

PERFECT COMPETITION. Perfect competition must apply to the markets for all inputs and outputs and also to capital markets (that is, access to borrowing). The main, although not the only, condition which must hold is that there are many buyers and sellers; thus no buyer or seller controls more than a very small part of the market, and so no individual or firm can control the market price. This generally requires that no major impediments prevent new firms from entering the industry concerned.

COMPLETE MARKETS. A full set of markets would provide all goods and services for which individuals are prepared to pay a price equal to or greater than the costs of production. In some cases, however, this is not the case. Public goods, which are discussed shortly, are one example of commodities which the market will generally fail to supply at all. A second type of missing or incomplete market occurs when certain risks are uninsurable. Third, capital markets may in some circumstances fail to provide loans, an example in the

West being student loans (for which two problems exist: the lender cannot easily assess the risk of the borrower, and generally no collateral is possible for such lending). Fourth, there may be no futures market, which means that it may not be possible to make a contract now to buy or sell a commodity on given terms at some time in the future. Finally, a commodity may not be supplied because a complementary market is absent. This is a particular problem if large-scale activities need to be coordinated, as occurs in the case of urban renewal projects.

Missing markets can arise for two totally separate reasons: some commodities are not produced because the necessary market institutions have not yet been put in place (examples relevant to Central and Eastern Europe include the lack of the necessary legislation concerning private property and the lack of a stock market); other commodities are not produced, or are inefficiently produced, because the market is inherently unable to supply them (an example relevant to the social sectors is some types of medical risks).

OTHER ASSUMPTIONS. Three final assumptions are that there should be no public goods, external effects, or increasing returns to scale.

Public goods in their pure form exhibit three technical characteristics: (1) non-rivalness in consumption, (2) non-excludability, and (3) non-rejectability. Private goods are rival in consumption in the sense that one person's consumption is at the expense of another's: if I buy a cheese sandwich, one less sandwich will be available for everyone else. Excludability means that I can be prevented from consuming the cheese sandwich until I have paid for it. Rejectability implies that I can, if I wish, choose not to buy the sandwich. Not all goods display these characteristics, the classic example being national defense. If the air force is circling overhead, the arrival of someone from another country does not reduce the amount of defense available to everyone else (non-rivalness in consumption). Nor is it possible to exclude the new arrival by saying that the bombs will be allowed to fall on him until he has paid his taxes (non-excludability). Nor is the individual able to reject the defense on the grounds of pacifist beliefs (non-rejectability). Similar considerations apply wholly or in part to roads, television broadcast signals, and public parks. Public health, too, has important public goods attributes: if the water supply is purified, or clean air legislation enforced, nobody can be excluded from the benefits. The structure of laws and, more generally, the rule of law have important public goods characteristics as well.

An important distinction should be noted in discussing public goods. For a private good, the additional cost associated with an extra unit of output and the additional cost of an extra user are one and the same. If it costs $1 to produce an extra cheese sandwich, it also costs $1 to provide for an extra cheese-sandwich-consumer. But this identity does not hold for public goods.

The cost of cleaning up air pollution, for example, is positive and generally high, whereas the additional cost of having an extra person breathe the air is zero. This has important implications:

• Non-excludability means that, if a public good is provided at all, it cannot be charged for (this is known as the free-rider problem). In such cases, the market will generally fail entirely.

• Non-rivalness implies that the marginal cost of an extra user (though not of an extra unit of output) is zero. With a price of zero, no output will be produced, however, and the efficient price must be based not on cost, but on the value placed by each individual on an extra unit of consumption. Since this is impractical, the market is likely to produce an inefficient output.

Thus, public goods create one of two problems: either the market is inefficient, or it fails altogether. If the good is to be provided at all, it will generally have to be publicly organized. This will involve public funding, but not necessarily public production. For instance, the state may mandate and pay for a water purification program, but the work might be carried out by a private firm.

A second potential problem, external effects, arises when an act of person A imposes costs or confers benefits on person B, for which no compensation from A to B or payment from B to A takes place.[4] Suppose, for instance, that my factory pours out smoke over the neighborhood. This does not affect my *private costs,* which depend only on the cost of my machinery, raw materials, and wage bill. However, it does impose an *external cost* on my neighbors in the form of dirt, irritation, loss of amenity, health risks, and so on.[5] Thus the social costs of my productive activity exceed my private costs. The effect of externalities is to create a divergence between private and social costs and benefits. In these circumstances the market output will generally exceed the efficient output. The same effect also operates in reverse. If I am inoculated against a communicable disease, this not only reduces my chances of getting the disease but also benefits other people who will not catch the disease from me. Similarly, many types of education and training not only make the recipient more productive but also make other people more productive.[6] In cases like this, where an action by one person confers an external benefit on others, the market output is generally smaller than the efficient output.

What solutions exist? On occasion the market can resolve the problem. Coase (1960) shows that where the law assigns unambiguous and enforceable property rights, the externality problem may be solved by negotiation between the parties concerned. This is often impractical. Property rights may not be enforceable, as is the case with water pollution. The numbers of people involved may effectively rule out negotiation, as with traffic congestion.[7] In such cases, intervention may be justified either through regulation (mandatory water standards) or through an appropriate tax on the activity generating the external cost (a tax on petrol).

Increasing returns to scale arise when doubling all inputs more than doubles output. For reasons beyond the scope of this chapter (see Barr 1993b, chap. 4, for a fuller discussion), competitive pricing under increasing returns to scale leads to long-run losses. Two forms of intervention are possible in this situation: paying firms a lump-sum subsidy equal to the loss associated with competitive pricing or nationalizing the industry and paying an identical subsidy. The appropriate intervention is therefore a subsidy or public production or both.

The Assumptions Necessary for Insurance Markets to Be Efficient

Many people do not like risk and are prepared to pay to avoid or reduce it. When an individual takes out insurance, he is buying certainty (the certainty that if he is injured, for example, his medical costs will be covered).

The supply of insurance has an easy intuition. Suppose that 100 people decide to fly to Frankfurt to see a football match; each person has a suitcase whose contents (for arithmetic simplicity) are worth $1,000; and everyone knows from long experience that on average 2 percent of suitcases get lost in transit. Thus each person faces a potential loss, L, of $1,000, which occurs with a probability, p, of 2 percent. In those circumstances, it would be possible to collect 2 percent $\times$ $1,000 = $20 from each of the 100 people, that is, $2,000 in total; when the group arrives in Frankfurt, they find which two people had lost their suitcase and pay each of them $1,000 in compensation. This, broadly, is the way in which private insurance operates.

More formally, an actuarial premium, π, is defined as

$$(2\text{-}1) \qquad\qquad \pi = pL + T$$

where pL is the expected loss of the individual buying insurance, and T is the insurance company's administrative costs and competitive profit. π is the price at which insurance will be supplied in a competitive market.

The price of insurance thus depends on (a) the degree of risk and (b) the size of the potential loss. Car insurance premiums are high for a driver who is young or who lives in a high-crime area (both factors leading to a higher probability of loss); and they are high for someone who drives a Rolls Royce or Mercedes Benz (because the potential loss is large). A middle-aged person with a good driving record and driving a small Ford pays a much lower premium.

Although private insurance can operate efficiently in a great many areas, in others, technical problems can arise on the supply side. The efficiency of private insurance requires that the probability, p, in equation 2-1 meets five conditions. First, the probability of a given individual having his car stolen must be *independent* of that for anyone else. If my car is stolen, that should not affect the probability that yours will be stolen. What this means, roughly speaking, is that insurance depends for its financial viability on the existence in any year of a predictable number of winners and losers. If, in the extreme, individual probabilities were completely related, then if one person suffered a loss, so too would everyone else; actuarial insurance cannot cope with this situation. An important example concerns inflation which, if it affects any one member of an actuarial pension scheme, affects all members.

Second, p must be less than one. If $p = 1$, it is certain that the insured person's car will be stolen; hence, there is no possibility of spreading risks, and the insurance premium will equal or exceed the cost of a new car.[8] An example of this problem is the chronically or congenitally ill, for whom the probability of ill health equals one unless insurance is taken out *before* the condition is diagnosed. The problem can also arise for the elderly, for whom the probability of requiring medical care is very high. As another example, the private market cannot offer insurance against any medical problem which the individual already has at the time he or she applies for insurance. Preexisting conditions, in short, are uninsurable.

A third condition is that p must be known or estimable. If it is not, insurance companies cannot calculate an actuarial premium, and private insurance is impossible. For example, the private market generally cannot supply insurance against future inflation because the probability of different levels of price increases in the future cannot be estimated. The risk of some types of health problems can also be hard to estimate.

Fourth, there must be no adverse selection, which arises when a purchaser is able to conceal from the insurance company the fact that he or she is a poor risk. As the earlier example of car insurance made clear, efficiency requires high-risk individuals to pay a higher insurance premium than low-risk individuals. But if the insurance company cannot distinguish high- and low-risk customers, it has to charge everyone the same premium, based on the average risk. As a result, low-risk individuals face an inefficiently high premium and may choose not to insure even though, at an actuarial premium, it would be efficient for them to do so; where the problem is serious, the insurance market may fail altogether. This problem arises particularly in the case of medical insurance for the elderly. It can also arise if health care is an important part of employer benefits: firms with the best health care packages tend to attract workers with health problems, thus reducing the firm's competitiveness.

Finally, there must be no moral hazard. The problem can arise in two ways:
• Moral hazard can arise, first, when the customer is able costlessly to manipulate the probability, p in equation 2-1, that the insured event will occur. The chances of developing appendicitis are beyond an individual's control, and so medical insurance for this sort of complaint is generally possible. In contrast, the probability of becoming pregnant or of visiting one's family doctor can be influenced by individual actions and are therefore generally not well covered by private medical insurance. Where the problem is serious, the

supplier cannot calculate the actuarial premium, and private insurance is generally impossible.

• The second type of moral hazard, known as the third-party payment problem, arises when the customer can manipulate the size of the loss, L in equation 2-1. The problem is particularly relevant to health care. If an individual's insurance pays all medical costs, then treatment is "free" to the patient. Similarly, on the supply side, the doctor knows that the insurance company will pay his charges. Thus neither side of the market faces any incentive to economize: both doctor and patient can act as though the cost of health care were zero. This is inefficient. It causes overconsumption, creates upward pressure on insurance premiums, and leads to uncontrolled escalation of medical expenditure.[9]

The Invisible Hand Theorem

The validity of the invisible hand argument, discussed earlier, is thus hedged by stringent conditions. As a theoretical proposition, the market will allocate efficiently when *all* the assumptions in the previous two sections hold. In such cases intervention on efficiency grounds is neither necessary nor desirable. Where one or more of the assumptions fails, it is necessary in each case to consider (a) whether the market can solve the problem itself and, if not, (b) which type of intervention (regulation, finance, or public production) or mix of interventions might improve efficiency. It is also necessary to ask (c) whether intervention would be cost-effective.

As a practical matter, these conditions rarely apply fully; it is generally sufficient that they are broadly true. Competition, for instance, may operate with a relatively small number of suppliers; minor forms of consumer ignorance can often be overlooked; and state intervention is, in any case, warranted only if it can improve on an imperfect market outcome. Nevertheless, the market's efficiency advantages are tempered both by the possibility of market failure and, completely separately, by the fact that it can lead to outcomes which are regarded as inequitable. Thus Arthur Okun (1975, p. 119) puts things slightly differently from Adam Smith:

> The market needs a place, and the market needs to be kept in its place. It must be given enough scope to accomplish the many things it does well. It limits the power of bureaucracy. . . . So long as a reasonable degree of competition is ensured, it responds reliably to the signals transmitted by consumers and producers. It permits decentralised management and encourages experiment and innovation.

> Most important, the prizes in the marketplace provide the incentives for work effort and productive contribution.

> For such reason, I cheered the market; but I could not give it more than two cheers. The tyranny of the dollar yardstick restrained my enthusi-

asm. . . . The rights and powers that money should not buy must be protected with detailed regulations and sanctions, and with countervailing aids to those with low incomes.

Implications for Policy

The majority of goods in the West are produced and allocated by private markets (for a fuller discussion in the context of the industrialized countries, see Barr 1992, 1993b; Mueller 1989; OECD 1990b, 1992a; for a more philosophical approach, see Okun 1975). A number of examples serve to illustrate why.

Market Success

Food, by and large, conforms with all the assumptions necessary for markets to be efficient. People generally have sufficient information to buy a balanced diet; food prices are well known, not least because food is bought frequently; and most people know roughly how much they will need over a given period. Food production and, especially, distribution are competitive; and there are no major problems with externalities or public goods. A possible violation is consumer ignorance about the conditions under which food is produced and about its ingredients. The state therefore intervenes to regulate hygiene; it may also require companies to label ingredients and to put expiration dates on packaging. Such regulation can readily be understood by the public and enhances consumer information, leaving the private market to operate efficiently. It is not surprising that no one seriously advocates a national food service.

Clothing, too, conforms for the most part with the assumptions and, quite properly, is therefore provided by the market. It can, however, be argued that people are less well informed about the quality of clothing than about food, not least because they buy clothes less frequently. Yet fewer regulations govern the quality of clothing. One reason is that the costs of mistaken choice can be high with food (including food poisoning and, in the extreme, death) but are much lower with clothing. It can therefore be argued that, even though one of the assumptions may fail, intervention, though theoretically justified, is not cost-effective.

Consumer goods such as televisions, washing machines, kitchen appliances, and personal computers fit into the same pattern as food and clothing. The market supplies considerable amounts of information through consumer magazines, newspaper articles, and consumer programs on radio and television, and aggrieved individuals can seek legal redress. Minor consumer ignorance is ignored where the costs of mistaken choice are small. Where the potential costs of poor quality are larger (such as electrical appliances which might catch fire), the appropriate form of intervention is regulation.

Cars raise two sets of issues related to their production and their use. On the production side, the arguments are broadly similar to those for smaller consumer goods, a key feature being the extent of consumer information about quality. In particular, consumers cannot easily check that a car's brakes and steering are safe and its tires well designed. Given the high costs of a mistaken choice, regulation of such safety features in the West is stringent and continually evolving. So far as the use of cars is concerned, regulation mainly addresses the external costs imposed on others by an individual's unsafe driving (drink-drive laws) or by the operation of a car which is in unsafe mechanical condition (worn tires, faulty brakes) or is unacceptably noisy or polluting. As in the case of consumer goods generally, no efficiency case arises either for a subsidy or for public production.

Housing does not conform with all the assumptions. Building a house very cheaply (with materials which catch fire easily or with inadequate sewerage) endangers not only the owner but also the neighbors. This type of external cost justifies regulation on public health grounds. If I build a factory in my back garden which creates a lot of noise and dirt, I again impose a cost on others. This type of externality can justify planning controls. For these and other reasons, a very strong efficiency case can be made for substantial regulation of housing. Beyond that, however, there is little efficiency case for state intervention in the housing market. This suggests that housing should generally be allocated by the market, subject to well-designed regulation; poor people should be helped not by price subsidies, but by income transfers which can be tied (paid in the form of housing vouchers) or untied (paid in cash). Given the cost of housing, however, any change from subsidized rents to market rents should be phased in.

The analysis of efficiency applies equally to factor markets. To make clothes, for example, requires cloth and thread and sewing machines. The managers of clothes-making firms are generally well informed about the quality of inputs produced by different manufacturers of cloth and thread; sewing machines, cloth, and thread are produced competitively, and the problems of missing markets or of public goods do not arise. Car manufacturers, similarly, are well informed about the quality and price of different grades of steel, different machine tools, and so on, which, again, are generally produced competitively. Once more, these areas are best left to the market, possibly subject to some regulation.

Market Failure

Not all commodities, however, lend themselves to unrestricted market allocation. Where the necessary assumptions fail, markets may be inefficient or may not exist at all.

Health care and medical insurance fail many of the assumptions. Consumer information can be highly imperfect, since much medical treatment is complex and technical, and its effectiveness is uncertain. In addition, knowledge of prices is scant. Nobody knows how much health care he or she will need, and private medical insurance poses major problems. In addition, medical care connected with communicable diseases can generate externalities. The case for regulation is overwhelming. Information failures justify regulating the quality of medical care, including medical qualifications, and the quality of the testing, production, and sale of drugs; this is true in all industrial economies. The externalities associated with communicable diseases give an efficiency case for subsidizing prevention or treatment, such as offering free smallpox vaccinations or free treatment of sexually transmitted diseases.

Medical insurance also fails many of the conditions necessary for efficiency. There are two generic problems:

• Gaps in coverage arise where risks are uninsurable, such as chronic, preexisting medical conditions, medical insurance for the elderly, and any medical care connected with consumer choice, which raises issues of moral hazard. Thus, for some medical conditions, the insurance market is incomplete.

• Exploding costs are the result of the third-party payment problem. The problem is not restricted to private insurance; it arises with any funding system which operates on a fee-for-service basis (one in which medical suppliers, like other suppliers, are paid for each item they provide). In such circumstances, neither doctor nor patient has any incentive to economize. The incentive to supply excessive treatment exists even where patients pay some user charges. The Czech Republic introduced just such a system in 1992 without the necessary regulatory structure, leading to an entirely predictable cost explosion. Many Western countries had similar problems in the early 1970s. They responded via regulation: controls on total medical spending (Canada) or controls at the level of the individual hospital (Germany and the Netherlands). The effect of such regulation has been to control medical incomes without interfering with medical practice.

The serious problems of consumer information and private insurance together make a case (although not an overriding one) for public production of health care. The case for doing so would be if regulation of *quality* in the face of imperfect consumer information and of *quantity* in the face of third-party payment pressures were more effective with public than with private production. Since the case cannot be quantified either way, it is not surprising that medical care is publicly produced in some industrial countries and privately produced but heavily regulated in others. The finance of medical care, in contrast, is substantially public in all the highly industrialized countries except the United States because of the gaps inherent in private medical insurance.

Unemployment insurance raises similar problems. The heart of the difficulty is moral hazard. If someone is still unemployed after six months, either he has genuinely tried to find work, but no suitable jobs have been on offer, or he has not looked for work very seriously, being content to enjoy more leisure. The first of these is potentially an insurable risk, the second is not. The problem is that the insurer cannot easily tell which explanation applies. For that reason, private unemployment insurance, except in very restricted circumstances, does not exist in the West (for a fuller discussion, see Barr 1992, sec. V). A separate problem is that private insurance cannot cover preexisting conditions. Thus it is not possible to offer unemployment insurance to a person without a contributions record who is already unemployed, the most obvious example being an unemployed school-leaver. In both cases, the insurance market is incomplete.

Inflation raises two sorts of problems in the context of pensions. First, if any one pensioner experiences a given rate of inflation, then so do all pensioners. This violates the first insurance condition, that probabilities must be independent. Second, the future rate of inflation cannot be predicted. For pensions, it is, in principle, necessary to know inflation rates for sixty years into the future, because pension contributions by a twenty-year-old depend on the rate of inflation over the years until he or she is eighty, or even older. Thus insurance is based on an unknown probability, and a premium cannot be calculated. For both reasons, inflation is an uninsurable risk.[10] Again, there is a missing market.

Training, like health care, requires a careful blend of state and private activity. A worker who needs retraining may be badly informed about the quality of a prospective training course and about what skills the market is demanding. The first implies a need to regulate quality for reasons of consumer protection. The second suggests some form of guidance for suppliers of training, guidance which cannot come from the prospective trainees themselves. A possible source of advice is a tripartite training council, composed of representatives of government, employers, and workers. Subject to such guidance and the necessary regulation, there are no arguments against training being provided by the private sector and considerable advantages if the supply of training is competitive. A separate market imperfection is that training may create external benefits, giving an efficiency argument (quite separate from any equity arguments) for a partial subsidy.

Government Failure

Markets may fail. So too may governments, a fact of which people in Central and Eastern Europe are more painfully aware than most (the arguments are surveyed by Inman 1987 and Mueller 1989). Four explanations are offered for the extent of and growth in government activity: government's role in dealing with market failures, its role as redistributor of income and wealth, its response to the electorate in the form of coalitions of voters or pressure groups, and the role of bureaucrats. The government failure arguments point to the behavior of the electorate and of bureaucrats as important distorting influences.

The influence of the electorate operates in various ways. The coercion-via-the-ballot-box argument (Downs 1957) is that the (many) poor, on their own or in coalition with others, outvote the (fewer) rich to impose redistributive tax and benefit regimes. Writers like Buchanan and Tullock (1962) and Tullock (1970, 1971) argue that most transfers from the rich are captured by the middle class through their electoral power. Other arguments stress the role of interest groups, such as the poverty lobby. Interest groups are also argued to use their lobbying power to bring about redistribution through regulation, especially where the regulators are "captured" by those whom they are supposed to regulate (Pelzmann 1976; Posner 1975; Stigler 1971). According to this view, regulation (for example, of the medical profession) is a barrier to entry which allows doctors to make monopoly profits.

Distortions can arise also within government, in that public sector institutions may in part be run for the benefit of the bureaucrats themselves (Niskanen 1971). This type of government failure is very well understood in Central and Eastern Europe.

For one or more of these reasons, it is argued, the size of the public sector may be inefficiently large, or its composition may be distorted to meet the needs of the bureaucracy, powerful interest groups, or voters in marginal constituencies. At least in a Western context, however, these insights should not be pressed too far. Some explanations of redistribution (see, for instance, Friedman 1962, and Hochman and Rodgers 1969) offer an explanation of tax-financed redistribution which does not rely on electoral coercion. Interest groups may enhance efficiency (Becker 1983, 1985). Regulation may result in monopoly profits (as enjoyed by doctors in some countries), but it also offers protection to imperfectly informed consumers (for instance, the regulation of medical training).

The power of bureaucrats can be overstated and their motivation misunderstood (Dunleavy 1985). Pay increases or enlarged departments can be monitored; voters may be able to vote with their feet against high local taxation (Tiebout 1956); and bureaucratic self-interest can easily lead to *less* government (officials in Margaret Thatcher's Ministry of Finance won favor by cutting expenditure). In addition, bureaucratic control may be more effective where the state regulates the quality of a publicly produced good or service than where it attempts to regulate a private producer. In the case of social insurance, for instance, the state has information about tax and social insurance contributions and powers which would be regarded as draconian in the private sector; and countries where private, fee-for-service medical care is publicly funded find it more difficult to contain costs than those where medical care is publicly provided.

Nor are the various government failure arguments necessarily equally applicable everywhere. Tullock's (1971) claim that benefits go disproportionately to the middle class may be more true of the United States than elsewhere: in

Germany and Sweden the lowest income quintile in the mid-1980s received net transfers of about 10 percent of gross domestic product.

Where and How Should the State Intervene?

The most important contribution of the literature on public choice and government failure is the idea that analysis of government should consider the incentives facing politicians and bureaucrats. The outcome of the political marketplace is not, however, necessarily inferior to that of conventional markets. Just as markets can be efficient or inefficient, so can governments. Were it otherwise, we would advocate anarchism, a conclusion from which the government failure literature pointedly refrains. An important counterpoint to government failure is market failure.

Inman's (1987) survey concluded that,

> while democratic processes do not generally guarantee an efficient allocation of social resources, *we cannot go the next step and conclude that collectively decided allocations . . . are inferior to individually decided market allocations.* [Inman 1987, p. 727; my emphasis]

> Neither the institution of markets, or voluntary trading, nor the institution of government, or collectively decided and enforced trading, stands as *the* unarguably preferred means for allocating societal resources. Each institution has its strengths and its weaknesses. [Inman 1987, p. 753; his emphasis]

The New Right properly criticizes a naive predisposition toward state intervention at the slightest sign of problems in private markets, but to argue that public sector inefficiency *automatically* implies that private markets are welfare improving is to make the same mistake. Decisions about the proper borderline between market and state involve judgment, so that different interpretations are possible.

TOWARD A CONCLUSION. The starting point is to distinguish clearly between the *aims* of policy and the *methods* available to achieve them. Aims embrace efficiency, equity, and the various other objectives discussed at the start of the chapter. Methods include income transfers and direct interference in the market through regulation, subsidy, or public production. The proper place for ideology is in the choice of aims, particularly distributional objectives and their tradeoff with efficiency. Once these aims have been agreed upon, however, the choice of method should be regarded as a *technical* issue, not an ideological one. Whether a commodity like health care is produced publicly or privately should be decided on the basis of which method more nearly achieves previously agreed-upon aims. A rationale for choosing between methods can be given in the form of two propositions relating to efficiency and equity, respectively.

Proposition 1. If none of the assumptions about efficiency fails, the efficiency aim is generally best achieved by the market with no intervention. Where one or more of the assumptions fail, state intervention in the form of regulation, finance, or public production may increase economic efficiency.

Proposition 2. Subject to minor qualifications it is possible to argue the following:

- Only efficiency arguments can justify intervention other than cash redistribution; if no such justification exists, equity is generally best pursued by income transfers.
- If efficiency arguments point toward public production and allocation of a good or service, then distributional goals can be pursued by in-kind transfers, for example, redistribution in the form of free education or health care.

Care is needed not to overstate the presumption in favor of cash redistribution. In some cases (medical and nutritional supplements for pregnant women, infants, and young children), benefits in kind might achieve poverty relief more effectively. In the context of Central and Eastern Europe, however, the argument for cash redistribution is strong: cash redistribution is necessary in the face of an inherited paternalistic mind-set, and it empowers consumers in a budding market economy.

Under what circumstances, then, should the state intervene? It may want to do so for distributional reasons: this is largely a political matter. Insofar as efficiency is concerned, the existence of market imperfections is a necessary but not a sufficient condition for intervention; fiscal considerations and cost-effectiveness are also relevant. Intervention should occur in cases (a) where it increases efficiency compared with the position without intervention and (b) where it is cost-effective and affordable. The second part of the condition does not always hold. As we saw in the example of clothing, the costs of intervention may exceed the costs of the inefficiency it is intended to prevent. A more poignant example is the fact that in poor countries many beneficial health interventions are not carried out because they are not affordable.

POLICY IMPLICATIONS FOR CENTRAL AND EASTERN EUROPE. Vital commodities like food are privately produced in the West and, to a greater or lesser extent, sold at market prices, because they conform with the necessary assumptions. Efficiency, in such cases, should be pursued through private, market production and allocation, subject to regulation where appropriate. Equity should generally be pursued through income transfers, with poor people buying food at the same prices as everyone else and in the same shops. This is exactly what occurs in the West for the great bulk of commodities. The theory thus shows why the clear policy prescription in Central and Eastern Europe is to privatize these areas as quickly as is feasible.

The same theory also explains why governments in all industrial countries intervene. If the problem is one of relatively minor imperfections in consumer information, the relevant intervention, if any, is regulation. Where there are missing markets, particularly in the case of uninsurable risks, the state may have to step in, either to regulate or to subsidize private insurance or (more usually with unemployment insurance and medical insurance) to opt for a substantial element of public funding, as table 1-3 shows.

Although there is considerable agreement about the areas in which the state does *not* intervene, there is less convergence of the precise forms of intervention in areas where states do intervene. In all industrial countries most pre-university education is publicly funded and publicly produced, while health care, depending on the country, is produced in the public sector, the private sector, or both. The theory presented here is a useful tool, but it is not a mechanical formula for policy design.

The discussion of this chapter, in conclusion, suggests three strategic implications for Central and Eastern Europe:

- The private market has enormous advantages in very large parts of the economy. This is the case for substantial privatization of formerly state-run enterprises.
- The state has an essential and continuing role in creating an enabling environment for a modern market economy. This includes the legislation which underpins market activity, such as the establishment of clear property rights and the development of corporate law. It also includes legislation and regulation to protect people in areas where they do not have the information to protect themselves, such as consumer protection legislation, the regulation of financial markets, and employment law.
- Unrestricted private markets produce some commodities inefficiently or not at all. In such cases, public funding, public production, or both may be more efficient than other options. In important parts of the social sectors, that is precisely the case.

The main purpose of this chapter has been to suggest an analytical framework which policymakers can use to decide which areas should be returned quickly to the private market, where regulation is necessary, and in which areas the state should continue to have a substantial role. The objective is to protect the reforms by avoiding two undesirable extremes: moving too little by failing to move to a predominantly market system or moving too much by privatizing everything, taking no account of the significant technical problems with private markets for some commodities.

Notes

1. For a fuller discussion on the economic theory of private markets, see Barr 1993b, chaps. 4 and 5; Stiglitz 1988, chaps. 3 and 4. For a simple introduction, see Le Grand, Propper, and Robinson 1992, chaps. 1 and 10. For a more technical treatment, see Johansson 1991; Phlips 1988.

2. The claim is a strong one. It has three implications: that an equilibrium exists, that it is unique, and that it is stable, which means that the system converges toward a new equilibrium in response to any external shock.

3. The literature on quality has its roots in classic articles by Arrow 1963 and Akerlof 1970 and is surveyed by Stiglitz 1987. The literature on price is surveyed by Mortensen 1986.

4. More formally, an externality arises when A's utility or production function is interrelated with B's.

5. External costs are so called because the costs are external to the firm which produces them, that is, the firm does not have to pay for the costs of pollution it causes. The concept of external effects lies at the heart of modern analysis of environmental pollution and the development of policies to encourage producers and consumers to act in ways which reduce it. For a fuller discussion, see Baumol and Oates 1979.

6. For example, if I learn the same word-processing package as you do, we can swap diskettes; because I learn the package, you are more productive.

7. Where numbers are small, negotiation may be possible, but only at the expense of violating the assumption of competition.

8. From equation 2-1, if $p = 1$, then $\pi = L + T$, that is, the premium will exceed the insured loss.

9. To see the intuition of what is happening, consider the way people behave in a restaurant offering "all you can eat for $9.95" or free drinks, in comparison with the way they behave when faced with a conventional menu.

10. The discussion here is concerned only with explaining why *actuarial* insurance cannot deal with inflation in the context of pensions. Possibilities other than insurance are discussed in chapter 9.

Part I

THE POLITICAL ECONOMY OF TRANSFORMATION

Three

THE INHERITANCE

SAUL ESTRIN

THE COUNTRIES OF CENTRAL AND EASTERN EUROPE ARE DIVERSE in their traditions, culture, and history (see the glossary). However, they share a common element in their historical experience: at some time after 1917 they have all been ruled by communist regimes.[1] These regimes imposed similar governmental and administrative structures and relied on nonmarket mechanisms for the functioning of the economy. All these countries are now in the process of transition to democratic rule and free markets. This chapter outlines the main features of the communist system which are germane to this process of transformation and particularly to the development of human resources.

The first question is why these countries should reform at all. Reform is costly in both resources and emotional energy, so major systemic failure is needed to justify it. Such failure is exactly what we observe in Central and Eastern Europe. The communist system broke down, in the sense that it had increasing difficulty delivering those material and social goods and services to which such systems are in principle committed: fast growth, technological advance, rising per capita collective and individual consumption, and social solidarity. It is beyond the scope of this chapter to offer a full account of that breakdown and its causes (see Kornai 1990; Ofer 1987; Wiles 1962). This chapter seeks instead to summarize the key features of the communist economic system, with a particular focus on the development of human resources, to indicate how these characteristics led to economic and social collapse, and to suggest the ways in which the systemic inheritance and the consequential economic breakdown might influence and shape the process of transition.

The main economic feature on which this chapter focuses is the slowdown in growth after postwar reconstruction, culminating for many countries in negative rates of growth. There is a strong human resources dimension to this issue. Kornai (1990), among others, has argued that communist systems of central planning were very effective at mobilizing the transfer of labor from the countryside, where productivity on the margin was very low, to the urban industrial sector, where it was much higher (although the methods of shifting the workers do not bear close scrutiny). This *extensive* form of growth led to rapid industrialization and exhausted the hitherto substantial reserves of effectively "free" labor resources. In part because the old system failed to encourage habits of thought based on the notion of tradeoffs, however, the highly centralized planning bureaucracy was much less effective at the efficient allocation of scarce inputs, whether labor, capital, or raw materials such as energy. These deficiencies which hindered growth of an *intensive* sort were magnified by other widely discussed aspects of communist economic systems. Among the most important of these were failures in the incentive system for both workers and managers, associated with the failure either to use or to enforce financial constraints on enterprises (what Kornai 1980 has referred to as soft budget constraints), with the inability of central planners to ensure that technological advances are assimilated into production processes, and with deteriorating levels of labor productivity and discipline.

Several important caveats should be noted at the outset. The purpose of this chapter is to identify common aspects of the inheritance which have an important influence on the transition process. To do so within the space constraints requires placing greater emphasis on the degree of conformity in Central and Eastern Europe to the Soviet model than is merited by the facts. Some crucial simplifications are therefore made. First, not all countries of the communist bloc operated a central planning system by the end of the communist era. Most notably, Yugoslavia embarked on its own road in 1952, based on workers' self-management and increasing reliance on markets and trade with the West (see Estrin 1983). This chapter offers little of relevance to Yugoslavia's situation. Hungary abandoned planning in favor of "market socialism" with its New Economic Mechanism, introduced in 1968 (see Hare, Radice, and Swain 1981), but, as Kornai (1986) has convincingly argued, market socialist economies displayed many of the same characteristics as centrally planned economies. The reform process, particularly the emergence of a private sector, did, however, begin much earlier in Hungary than elsewhere. Similar provisos could be made to some extent for Poland, which in addition had extensive private agriculture.

Second, little emphasis is placed on the numerous attempts to reform the planning system itself. Major efforts were made to improve the Soviet planning system under Khrushchev in 1956, Kosygin in 1965, and Gorbachev in the mid-1980s (see Aslund 1991; Ellman 1989; Gregory and Stuart 1986), and these were imitated in much of Central and Eastern Europe (see Hare 1987). Although crucial to a balanced understanding of the communist era, these reform efforts had surprisingly little impact on the fundamental characteristics and performance of the communist economic systems featured in this chapter. They are invoked only when directly relevant to understanding the inheritance for labor markets and social policy.

Finally, as table 3-1 shows, the different countries entered and departed the communist period diverse in economic situation as well as geography. Key differences include international trade and level of development (see Ellman 1989). The U.S.S.R., being large and rich in resources, was able to pursue an

Table 3-1. *Basic Data on the Economies of Central and Eastern Europe, 1990*

Country	Population (millions)	Number of cars per 1,000 persons	GDP per capita[a] as a percentage of U.S. level	Exports to CMEA	
				As a percentage of total exports	As a percentage of GDP
Bulgaria	8.9	127	26	69	34
Czechoslovakia	15.7	186	35	60	25
Hungary	10.6	156	30	43	16
Poland	37.8	112	25	41	14
Romania	23.3	—	19	—	—
U.S.S.R.	290.9	46	31	55	8

— Not available.

Note: CMEA is the Council for Mutual Economic Assistance.

a. Purchasing power parity.

Sources: Bruno 1992; Russian Federation 1992; Murrell 1991.

autarchic policy, but the countries of Central and Eastern Europe were smaller and more open (see Brown and Neuberger 1968; Collins and Rodrik 1991). For Czechoslovakia or Bulgaria, most trade was integrated within the region of the Council for Mutual Economic Assistance, but Hungary and to a lesser extent Poland sought Western markets; by 1989, trade to the countries of the Organization for Economic Cooperation and Development (see the glossary) represented around 20 percent of gross domestic product in Hungary. Conversely, Czechoslovakia and the German Democratic Republic saw a Soviet-type planning system grafted onto a relatively developed industrial system. The traditional developmental strengths of Soviet planning were in principle better suited to less-industrial economies like Bulgaria or Romania, and this is reflected in the relatively narrow differences in income per head which were established by 1989, with a range from Romania to Czechoslovakia of less than 2:1 (calculated on the basis of purchasing power parity; see Murrell 1991; Summers and Heston 1991).

This chapter contains three sections. In the first, the main legacies of the communist way of thinking are discussed, notably the differences (relative to standard Western views) in how key elements in society and the economy were conceptualized and interpreted and the lack of awareness which much of the policy community had of many basic economic concepts. Although these countries share a common inheritance from the communists, they are also highly diverse in history and culture in ways which influence the process of transition. The general systemic inheritance is the subject of the second section, while the third introduces those elements of the inheritance relevant for the following chapters.

The Intellectual Heritage

The countries of Central and Eastern Europe were under communist governments for a minimum of forty years, and the Marxist-Leninist creed is an all-encompassing ideology which permeated every aspect of public and private life. At least until the 1950s, these societies were driven and motivated by ideology to a degree hardly imaginable from the comfort of countries which are not totalitarian. Ideology remained a potent force until the revolutions at the end of the 1980s. In some countries, the domination of a monolithic ruling ideology reinforced previous historical and cultural experience; in others, it was ruthlessly imposed and enforced. Marxist-Leninist ideology was therefore not a sham behind which lurked more traditional motives of national self-interest. It led the communist parties to manage their affairs in ways which were sometimes detrimental to their interests and which would otherwise be incomprehensible to Western observers. An example is the self-declared aim of economic self-sufficiency, which led the U.S.S.R. and later the Council for Mutual Economic Assistance to choose autarchy as the guiding principle of international trade. An important proviso, however, is that many of the shortcomings which we now identify with communism were present in the region prior to the Bolshevik revolution, most notably totalitarianism, extreme centralization, and excessive bureaucracy.

Communist Ideology

Peter Wiles (1962) has provided an elegant and detailed exposition of the ways in which communist ideology influences economic and social thinking. At the heart of the communist vision is the division of society into classes: land owners, property owners, the owners and managers of fixed (industrial) assets, and wage laborers. History is interpreted as a struggle for power between these classes. The Bolshevik revolution was undertaken in 1917 to place power for the first time in the hands of wage laborers—the working class—through their "representatives," the communist party. The imposition of communist rule in the wake of the victorious Red Army after 1945 was similarly interpreted by ideologues as a victory of the working class in the struggle with the owners of capital for control of the production process.

Marx's original analysis was a critique of nineteenth century capitalism, by which market relations were replacing slavery, feudalism, and other forms of economic management. Marx did not, however, base his criticisms on the inhuman conditions to which early industrialization led; rather, he sought a "scientific" analysis which could pinpoint the direction future reforms must take. He was concerned to show that beneath the apparently benign surface of voluntary exchange through the mechanism of money, which underlay the emerging market system, was exploitation as severe and degrading as slavery or feudalism. The labor theory of value purported to prove scientifically that the seemingly fair profits earned by the new capitalist class were in fact the fruits of their exploitation of wage laborers. The working classes were therefore as justified in fighting against their condition as had been previously oppressed groups in society.

Marxist economic science viewed the triumph of the working classes in two phases. In the second, communism, the technological capacities of the economic system would have progressed to such an extent that scarcity of goods would have been abolished. The allocation of scarce resources would not be a problem because the economy would be capable of producing sufficient to satisfy all needs. In the interim, there would be socialism. The fundamental problem of economic scarcity would still pertain, but the country would be ruled by the communist party in the interests of the working class, the "dictatorship of the proletariat." However, before 1917 the specifics of how to run an economy on socialist lines had never been properly formulated. Its subsequent practical elaboration can be argued to have had three principal aspects: central planning, job security, and subordination of the individual to the collective.

Central planning was not at first a part of the dogma, but it became so later. Once the communist party had seized power in the Russian Empire, the many possible interpretations of socialism were the cause of furious policy debates, including a powerful faction in favor of the market mechanism (see Ellman 1989). However, the crude anti-market ideology of much Marxist writing reinforced traditional state absolutism and the Stalinist vision of a speedy, quasi-militarized process of industrialization, giving final victory to bureaucratic central planners. This outcome was made more likely by the apparent failure of Lenin's market-oriented reforms, known as the New Economic Policy, in the 1920s. Private ownership of fixed assets and agricultural land was then quickly suppressed and market forces were relegated to a minor role. Only final consumer demand and labor supply were left to be allocated by the market. The allocation of other resources was to be based on physically denominated quotas set by the planners, and money was left to play a passive role.

Central planning answered the question of how to run a socialist economy, but not how the communist party should use its power in the interests of the working class. One solution was to create as large a working class as possible, an answer consistent with military and strategic pressures for rapid industrialization. According to Marx, inequalities would persist under socialism, although the authorities supported widespread social provision and greater equality between blue- and white-collar salaries. Widespread training, particularly for agricultural workers joining the industrial proletariat, was also a priority, and the system came to regard absolute job security and, in effect, the right to a job as virtually defining characteristics of socialism. This substantial degree of job security did not, however, significantly increase workers' control over the labor process.

The subordination of the individual to the collective was the third part of the dogma. Because communists viewed society as composed of classes rather than individuals, individualistic activity was inherently suspicious. A significant consequence for the subject of this book was the application of the principle of collective or group, rather than individual, welfare to the evaluation of consumption. Thus collective consumption was acceptable and even desirable; except in obvious cases like food or clothing, private consumption was individualistic and potentially on the slippery path back to capitalism. Workers could consume health, education, and even vacations together in groups typically organized around their factories, but the system was not oriented to supply individuals with consumer goods for their private consumption.

This vision of socialism was devised in the U.S.S.R., where it was applied ruthlessly beginning in the late 1920s from Leningrad to Vladivostok. It was also imposed by force of arms on the diverse countries of Central and Eastern Europe brought into the Soviet bloc after the Second World War.

The Legacies of Communist Ideology

The consequences of communist ideology for policies governing the development of human resources during the transformation can be considered from two angles. First, there are the conceptions about key elements in society and the economy, for example, the rule of law, the role of the state and the individual, or the economic function of wages and benefits, which are different from those common in the West. Second, there is the lack of awareness among policymakers and administrators of basic economic concepts such as opportunity cost.

Different conceptions arose in two important ways: the role of the state in relation to its citizens and the structure of rewards. Under communism, the state, the citizen, and the law related to one another in ways very different from their relations in the West. The government apparatus under communism was not elected to act in the interests of society, nor even in the narrower interests of its own voters. Public power was explicitly a dictatorship by the communist party in the interest of the proletariat, as interpreted and enforced in the party through the mechanism of democratic centralism.

THE ROLE OF THE STATE. The aim of the party was to maximize the growth of production. Most social institutions and pressure groups were structured to assist in that goal (see Bergson 1964; Ellman 1989; Gregory and Stuart 1986). Classic Soviet-type planning held sway in most of the communist world until the fall of the communist regimes.[2] Maximization of the growth of industrial output was based on virtually total and highly concentrated state ownership of property and was attained by centralized economic decisionmaking under the planners. Although economic entities called enterprises, or bank branches, did exist, they fulfilled few of the functions such organizations undertake in market economies. They had little independent authority and acted, in many ways, as administrative arms of the center. Enterprises followed plans which covered in considerable detail most aspects of the firm's technical, industrial, and financial behavior, as well as the allocation of investment and the distribution of the fruits of research and development from central research facilities (see Berliner 1957; Granick 1954). Branches of the central bank verified that transactions in the plan had actually taken place and at official prices. Management carried out the plans, which were evaluated by success indicators based on plan targets, many of which were denominated in physical rather than value units. Workers and even the trade unions were also dedicated to achieving targets and, thereby, the wider national plan. Reforms affected the structure but not the objectives or the excessive centralization of the system.

This is not to say that the central authorities were in practice able correctly to evaluate the productive capacities of firms or the inputs required to attain plan targets. In fact, from the outset, widespread bargaining occurred between planners and enterprise managers about the contents of plans: managers sought and planners tried to resist the inclusion of slack in enterprise plans (see Ericson 1991). The planners responded to the managers' systematic tendency to understate their productive capacity and exaggerate their input needs by using the so-called "ratchet principle," whereby output targets were always raised to match at least the level of previous best achievements. Managers searched for slack in the planning system, which also contributed to their tendency to stockpile considerable hidden reserves of inputs, including reserves of labor, and to underuse capacity.

Responsibilities in society were therefore focused on satisfying the needs of the plan and of the party.[3] Secondary responsibilities for the state also existed to maintain the welfare of workers through the provision of basic services— health care and education—and to provide employment or the necessary retraining required for placement. The totalitarian conception of the state and society in Soviet-type systems, however, left no space for decentralized decisionmaking or for satisfying the needs of special groups. It also offered surprisingly little role for the state itself, as distinct from the party. Citizens deeply mistrusted their government, and of course the party was profoundly suspicious of the population. Hence, on the one side were the gulags, the show trials, and the vast internal security forces; on the other was the cynicism of the populations toward the political process, particularly in countries such as Czechoslovakia, Hungary, and Poland, where reform efforts had been savagely suppressed. Party rule also undermined the concept of the rule of law, because communist parties governed directly by decree and used the legal system only intermittently and as a mechanism for reinforcing their control. The absence of a clear and broadly accepted concept of the rule of law is a major impediment to the operation of market systems, which rely on clearly defined property rights and enforceable contracts. All these characteristics—the lack of decentralization, the distrust of government, and the undermining of the rule of law—emerge repeatedly in later chapters.

THE STRUCTURE OF REWARDS. The role of wages and cash benefits in a socialist economy is the second major area of difference with the West. Two well-known principles can underlie the distribution of income. The first is the principle of desert, in which people earn wages determined by their contribution to society. If the labor market is competitive, free markets will distribute income based on desert. Alternatively, income can be distributed according to need. Judging needs to be broadly similar for all people in society implies strict egalitarianism in the distribution of income. Distribution according to need is also consistent with allocating higher incomes to groups perceived to have greater demands, for example, the disabled or large families. In Marx's communism, society would distribute according to need, but for most communist parties in office, distribution has been based on desert. Thus, in contrast to China under Mao, no attempt was made to disguise self-interest as the motivating force in socialist labor markets, although many efforts were made to manipulate material incentives, particularly in the area of wage determination.

Thus communist countries did allow income differentials to be established between occupations, sectors, and regions, although these differentials were typically narrower than those in the West (see Wiles 1974 and the discussion in chapter 7). In interpreting deserts, however, individual contributions were usually evaluated relative to the demands of the plan. To quote Lenin (see Ellman 1989, p. 209), "Distribution is a method, an instrument, a means, for increasing production." Wages were set centrally, and the Soviets in particular sought a scientific basis for wage differentials which led after Stalin to a system of national job evaluation. The idea was to grade all jobs and workers in such a way that actual pay was determined by a combination of occupation, grade, work norms, output levels, hours worked, bonuses, and regional coefficients set to encourage migration to unattractive areas or to offset labor flows to richer areas. Wages were also used to assist specific groups, for example women. The result of this procedure was a very large and inflexible table of earnings by sector, which grew more and more complex and incoherent over time. Pay also came to rely on bonuses. According to Gregory and Stuart (1986) in 1972, the basic wage, which was determined by the job evaluation system, accounted for 59 percent of total pay, bonuses for 20 percent, premiums for exceeding norms for 11 percent, and differentials for regional and special working conditions for 10 percent of average wages in Soviet industry. The "scientific" basis for determining wages therefore caused problems because it failed to ensure standardized wage rates for given jobs by industry. Notwithstanding these "objective" formulae, differentials were in practice kept very narrow for ideological reasons (see Wiles 1974).

Since the wage system was based on deserts, the principle of distribution according to need was satisfied through transfer payments and the provision of widespread free public services. The distinction between wages, cash benefits, and benefits other than cash was blurred since some cash and many noncash benefits were allocated through the place of work and based on the fact of employment. Wages also differed in nature from those in a market economy because of the critical role which state subsidies played in consumer prices (in Czechoslovakia subsidies amounted to around 15 percent of the wage bill) and in housing (so rents in 1991 amounted to less than 10 percent of living costs in Czechoslovakia). Social welfare at enterprises, such as canteens, holiday resorts, transport, medical services, and day care, also amounted to a significant proportion of the wage bill (around 5 percent in Czechoslovakia).

The early communist rulers were aware that education and training could have major economic benefits. Ellman (1989) notes that Strumilin advocated expenditure on education as a high-yielding investment in 1924, and calculated in 1931 a sixfold return to educational expenditures. The socialist economies all rapidly developed extensive education systems and on-the-job training programs. However, these were focused on the educational requirement of extensive growth and were not well adapted to changing needs in recent decades. Although wage differentials might have given manual industrial workers an interest in building their skills, the signals were too weak and distorted to encourage many other forms of human capital accumulation.

Labor relations, in particular low worker morale and low efficiency, became increasingly serious problems for several reasons. First, and most important, wages for most people were very low. The average level was often inadequate to support necessary consumption. The reason in part was that high levels of investment were needed to fuel the planners' overriding objective—growth— and these had to come at the expense of wages. Moreover, the wage system generated numerous petty distortions and injustices which undermined the potential incentive effects. The widespread shortage of goods and queuing also called into question the actual welfare gains obtained from higher pay. The emergence of a large-scale black market, while alleviating shortages, may also have demoralized the general public.

Workers in socialist countries also typically worked in enormous and unattractive factories and plants, because directing fewer firms simplified the planning process and increased central control. For example, Soviet giganticism meant that Czechoslovakia, Hungary, and even Poland had very few firms in the industrial sector in 1989, often only a few thousand. Huge enterprises in the West have often been riddled with disputes over labor relations, but such disputes were all suppressed in communist regimes. Management principles were explicitly based on Taylorism (that is, a "scientific" approach to management, which treated labor as little different from machines), with complete faith in the scientific principles of labor organization at the workplace and the total absence of countervailing influences. In this conception, trade unions were part of the state and party apparatus and represented merely another tool to ensure fulfillment of the plan. According to Ellman (1989, p. 189), "Workers have little control over their work and are very vulnerable to arbitrary measures by the bosses." In a totalitarian society devoted to maximizing production, taking it easy at work represented one of the few ways available to register dissatisfaction with the system. Poor labor morale and effort were also a consequence of people responding to declining real wages by holding more than one job. These problems were particularly prevalent in the U.S.S.R.; they were somewhat less acute in the rest of Central and Eastern Europe, particularly after the late 1950s.

LACK OF AWARENESS OF BASIC ECONOMIC CONCEPTS. Wiles (1962) argues that the Soviet-type planning system entailed a deliberate decision to sacrifice efficiency in the allocation of current resources in favor of rapid economic growth. Bergson (1964), in contrast, believes that static allocative efficiency was sacrificed without achieving a significant offsetting gain in growth. Whatever the motivation, central planning systems clearly were not concerned with making rational economic decisions on the margin and, in the past, had neither the information nor the expertise to do so. Rational economic decisions must be based on prices which reflect domestic or international scarcities. Prices played nothing but an accounting role in a Soviet-type planning system and were, in any case, not used as the basis for decisions. Thus investment decisions were evaluated on the basis of internal accounting procedures and prices, without reference to their implications for shadow cost, let alone for scarce energy, raw materials, and imported inputs. Production decisions were made without reference to profitability at world prices.

The reason for such an approach is simple but fundamental: under Marxist principles, scarcity was not an ideologically acceptable description of the relationship between inputs and outputs. *Communist ideology thus ruled out any notion of opportunity cost in the allocation of resources.* It follows that there was no concept of efficiency in the Western sense of the term nor an awareness of the tradeoffs discussed in chapter 1. A further consequence is that Soviet economics did not analyze individual behavior and contained no real behavioral assumptions. No efforts were made to predict the reactions of agents to changes in incentives, and there was little attempt to test the validity of economic propositions (and hence no econometrics; see the glossary), and no welfare economics.

Hare and Hughes (1992) suggest one indication of the consequences. They investigate the efficiency and competitiveness of industry in Central and Eastern Europe by applying Western prices to sectoral inputs and outputs in order to see which activities might benefit, and which might lose, from the price realignment associated with reform. They found that around two-thirds of output had a comparative disadvantage at world prices. Some sectors, accounting for more than one-quarter of output in some countries (measured at domestic prices), even produced negative value added according to their calculations; at world prices, in other words, the material cost of supplying the products, before considering either labor or capital costs, exceeded the revenues that could be generated from selling them. The methods have been disputed because they were applied at a branch rather than enterprise level and because the approach assumes that firms do not alter their pattern of input use in response to the new prices. But the study does provide preliminary orders of magnitude for the extent to which resources were misallocated under central planning. It also helps to highlight that a crucial inheritance of enterprises and the state bureaucracy is an orientation to production rather than profitability, particularly profitability at world prices. This has clear implications for the allocation of human resources.

The General Systemic Inheritance

This section explains how the planning system allocated resources and what the consequences were for economic performance, particularly with regard to labor. The discussion summarizes a large literature (see Bergson 1964; Ellman 1989; Gregory and Stuart 1986; Wiles 1962).

General Background

Communist ideology viewed the private ownership of property as the basis of worker exploitation. Communists, therefore, always nationalized most productive assets when they seized power and often made private ownership effectively illegal. In consequence, such economies have almost no capital market institutions and very few small- and medium-size firms of the sort typically privately owned in the West, although some had begun to reemerge in the market socialist countries by 1989. The effects on entrepreneurship, corporate governance, and labor incentives have been widely discussed (see Fischer and Gelb 1991; Kornai 1990).

In the classic Soviet-type planning system, nationalized enterprises were grouped into branches and placed in a hierarchical structure under ministries and ultimately the central planners. The economy was then run like a giant firm, with orders, typically denominated in physical units, descending from the planning organs down through the ministries and their subdivisions to firms and plants. In this way, the economic structure became locked into a rigid pattern with large firms themselves controlled by branch ministries which were defined by their industrial profile. Neither firms nor ministries had an incentive to challenge the structure or to diversify their activities.

In a market economy, the bulk of economic relations are horizontal, between different enterprises or between firms and consumers. Under planning, the flow of information and orders was intended to be vertical, from planners down to managers. Horizontal relations, which would imply some degree of decentralization of decisionmaking, were meant to be kept to an absolute minimum. The system was designed to do without, and often to suppress, individual initiative and effort. In practice, however, plan implementation required a considerable degree of horizontal interaction to counteract the informational deficiencies of excessive centralization. This took the form of informal exchanges between firms and the gradual emergence of a full-scale secondary market.[4] This is an example of the "paradox of intent," in which decisions arising from short-term opportunism proved to be entirely inconsistent with communist ideology.

The objective of the system was growth, and this was to be achieved in the early years of planning by the transfer of labor from agriculture to industry, in particular to heavy industry. Rapid industrialization was based on very high rates of investment. Shares of investment to net material product were typically around 30 percent in the U.S.S.R. and Eastern Europe in the 1950s and were still normally in excess of 20 percent as late as 1985. The growth was devoted not to satisfying consumer needs, but rather to expanding continuously the production of material inputs. A revealing statistic presented by Ellman (1989) is that, as in most Western countries, the bulk of production in the Russian Empire in 1913 was devoted to the supply of consumer goods, 65 percent as against 35 percent for producer goods. By 1946 that proportion was reversed, and *in 1985 only 25 percent of Soviet industrial output was of consumer goods*. The remaining 75 percent was of producer goods. Because it stressed extensive growth, the system also generated faster growth in relatively less-industrial economies, such as Bulgaria and Romania, than in countries which were already industrial, such as Czechoslovakia or the German Democratic Republic.

The fact that Soviet-type systems were inefficient in the allocation of resources does not imply that the plans were slack or that managers had an easy life. The planners' main tool in the heyday of planning was the material balance, a table listing the total sources and uses of each commodity. The potential sources were from production, imports, and stocks. The uses were for production, exports, stocks, or consumption. As Ellman (1989) stresses, it is important to realize that the planners not only failed to obtain "optimal" plans, in the sense of plans that maximized social welfare, but also failed to achieve "feasible" plans, in the sense of sources (supply) in the material balance which equaled uses (demand) for each product. When demand exceeded supply, as it frequently did, the planners did not seek to achieve consistency by reducing output targets; that would have conflicted with the objective of growth. Rather, they cut supplies to uses which were regarded as inessential, based on the application of the "priority" principle. Since there was no subsequent adjustment of plan targets, plans were rarely attainable.

This tautness of planning assumes great significance when one reflects that the institutional and incentive structure, which represented a large proportion of the salaries of workers and managers, was manipulated to ensure attainment of the plan. Enterprise directors in particular were under enormous pressure to attain plan targets, with as much as 35 percent of incomes depending on bonuses related to the achievement of such targets. Although plans were rarely feasible and always hard to achieve, financial constraints were seldom if ever binding. This asymmetry of pressures was a major cause of a number of widely recognized characteristics of planned economies:

• Low labor productivity was widespread because hoarding labor was in management's interest. Few penalties were imposed as a consequence of cost overruns or failures to meet output profit targets, but the spare resources could be critical to last-minute attempts to meet plan targets. The same applied to all other factor inputs, most notably capital and raw materials such as energy. Hence Brada (1989) shows that overstaffing in Czechoslovakia reached up to 15 percent of the labor force by the end of the 1970s. Firms also held excessive stocks: the Soviet ratio of inventories to net material product was around 50 percent in 1960 and 80 percent in 1985, as against less than 40 percent in the United States.

• The reliance on physical gross indicators of performance implied a persistent problem related to quality and product mix, since targets were rarely specified tightly enough to prevent skimping in the production process. Because sellers were never directly responsible to buyers, market forces could do nothing to eradicate the problem of poor quality.

• Because plan targets were based on quantitative production norms, managers were suspicious of technical advances which would imply higher norms. The system therefore failed to offer adequate incentives to innovate. This problem is also related to the "ratchet" effect which attaining current targets had on future plans and which was an important element of the bargaining between planners and managers.

• The widespread uncertainty resulting from inconsistent plans also led to duplication of productive capacity within branches and even firms. Thus enterprises, unable to rely on spare parts for minor repairs, ordered extra equipment to cannibalize for parts and built their own foundries and other service units.

• With such pressures to attain plans, and the accumulation in enterprises of excessive stocks of most productive materials, it was inevitable that a secondary market between managers would emerge for the supply of inputs. As noted above, the excessive centralization and tautness of the system had the ironic result that enterprises increasingly came to rely on semi-legal (or illegal) exchanges through the secondary market, with communist party officials often playing the role of market trader. The increasing role of the secondary market in the functioning of the economy further undermined the credibility of the rule of law.

The Slowdown in Growth

Socialist rule was premised on rapid growth, which was the primary raison d'être of the economic mechanism. The collapse of the political system stemmed from the fact that, after early successes, the growth rates of all the planned economies started to slow down and ultimately declined to zero or worse. The most striking feature of this slowdown is that it occurred at levels of income and consumption per head far below those attained in the West and below those now achieved in some of the newly industrializing economies of East Asia. The slowdown occurred in countries, such as the U.S.S.R. or Czechoslovakia, where the planning process remained largely true to its origins until the fall of communism; but economic failure was equally pervasive in the market socialist countries, such as Hungary, which had abandoned the centralized allocation of resources. This section describes and attempts to explain the slowdown, with the knowledge that this is a major problem for post-communist reformers.

According to the official figures presented in table 3-2, the Soviets attained very high growth rates during the 1930s and during postwar reconstruction (for a full discussion and alternative evaluations of Soviet growth rates, see Bergson 1964 and Ofer 1987). The slowdown commenced in the 1950s and motivated the Khrushchev reforms. The problem was not restricted to the U.S.S.R., however; table 3-3 shows that even using official data the slowdown was also severe in the rest of Central and Eastern Europe. Growth rates had typically declined from more than 6 percent on average in the 1950s and 1960s throughout the region to around 4 percent in the late 1970s and to close to zero, or in some countries negative, by the end of the 1980s.

Table 3-2.   *Official Growth Rates in the U.S.S.R., 1928–91*
(percent)

Year(s)	Growth rate
1928–40	14.6
1950–60	10.1
1960–70	7.0
1970–80	5.3
1981–85	3.2
1986	3.0
1987	1.0
1988	0.0
1989	−5.0
1990	−10.0
1991	−15.0

Sources: Ellman and Kantorovich 1992; Gregory and Stuart 1986; and unpublished reports from Planecon.

Table 3-3. *Changes in Net Material Product in Central and Eastern Europe,*
1971–89
(percent)

Country	1971–75	1976–80	1981–85	1986	1987	1988	1989
Bulgaria	7.9	6.1	3.8	5.3	5.1	2.4	−0.4
Czechoslovakia	5.6	3.7	1.6	2.6	2.1	2.4	1.3
German Democratic Republic	5.4	4.1	4.6	4.3	3.3	2.8	1.0
Hungary	6.2	2.9	1.3	0.9	4.1	0.3	−2.0
Poland	9.7	1.2	−0.8	4.9	1.9	4.9	0.0
Romania	11.5	7.0	4.5	7.3	4.8	3.2	—

— Not available.
Note: These data derive from official statistics and therefore must be treated with caution.
Source: United Nations 1990.

The most obvious explanation of the slowdown in the U.S.S.R. in the 1960s is that the possibilities for growth of an extensive form were exhausted. This is consistent with the slower growth always attained in the more-industrial countries of the Council for Mutual Economic Assistance and with the relatively more sustained momentum in Bulgaria and Romania. The system grew via rapid capital accumulation and the increasing use of labor in industry. According to Fischer (1991), the capital stock was being expanded at between 7 and 9 percent between 1960 and 1985, which probably implied that the system was rapidly entering the region of diminishing returns to accumulation. At the same time, the economy began to face labor shortages in the 1960s: most of the easily available labor located near the urban industrial centers had already transferred from the agricultural sector (although the share of agriculture in employment remained high), and participation rates were very high by Western standards (87 percent of the relevant age group in 1980 as against 71 percent in the United States). As a result, the expansion of the urban industrial labor force was restricted to around the underlying rate of growth of the population, which was less than 1 percent after 1970.

Empirical discussion has focused on the case of the U.S.S.R. Martin Weitzman (1970) has argued that the reason for the decline in industrial productivity lay in very low rates of substitutability between capital and labor. As noted already, there were great differences in the rate of growth of capital and labor inputs in the U.S.S.R. These would ultimately limit the growth of Soviet output if substitutability between capital and labor was low. Weitzman estimated the elasticity to be significantly below unity. In contrast, Desai (1985) uses industry and branch data to argue that the cause of the problem was not low substitution elasticities, but more general factors.[5] Similar findings have been made by Brada (1989) for other countries of Central and Eastern Europe. These authors explain the decline in growth as a result of systemic failures, most notably the inherent inefficiency of resource use, shortages of key raw materials such as steel and coal, bottlenecks in energy and transport, and a marked decline in labor morale and discipline after the late 1970s.

Other explanations of declining industrial productivity relate to the effects of sustained underinvestment in human capital, especially education and health:

- Ofer (1987) argues that the exclusive emphasis on attaining current plan targets led the system to neglect long-term problems, for example, infrastructure and the maintenance of physical and human capital.
- Particularly in the Soviet case, there is some reason to blame the deterioration of human capital which is associated with both alcoholism and a declining quality of health care. In the 1970s, the Soviet share of strong alcohol in total alcohol consumption was probably the highest in the world. Consumption of pure alcohol per head (persons more than fifteen years old) had risen from 4.4 to almost 12 units between 1955 and 1979 (Ellman 1989), with a network of sobering up stations being established to counter widespread alcoholism. In the 1980s up to 20 percent of personal income was spent on alcohol. Alcoholism has been associated with numerous social ills—disease, divorce, car accidents—and was the subject of a major campaign by Gorbachev in the late 1980s.
- In contrast with earlier periods, long-term human capital formation was neglected after growth began to slow down, regarding not only the extent of investment in education and health, but also the quality of provision, such as the course content and teaching methods in education and the neglect of promotion and prevention, compared with treatment, in health.

These issues are addressed in more detail in later chapters. Other relevant factors underly the slowdown in growth as well:

- The price of energy increased after 1974, and the situation was exacerbated by the inability of the planning system to respond to it. This was particularly unfortunate for the energy-importing countries of Central Europe, which faced serious deterioration in their terms of trade.
- There was an increasing need for industrial growth to be based on flexibility rather than the mass production techniques of the old heavy industries. The communist system was weak at motivating change and taking risks and was therefore particularly unsuccessful at introducing information technology.
- Perhaps as a consequence of growing internal political weakness, many communist regimes in the late 1970s lost control over the macroeconomy. Most countries in Central and Eastern Europe, including Yugoslavia, financed higher domestic consumption and investment by borrowing from the West. For Poland and Yugoslavia, the harsher international economic climate in the 1980s led to domestic wage pressures, budgetary deficits, and open inflation. In contrast, Czechoslovakia and Romania attempted to

repay their debts and thereby paid a high cost in growth and real wages. The situation in 1989 is summarized in table 3-4.

• Finally, the effect of the Western arms buildup in the 1980s, and the attempt by members of the Warsaw Pact to match it despite their deteriorating domestic economies, diverted scarce resources from much-needed investment in other areas. Although it is hard at this point to evaluate the impact of defense expenditures, they cannot, in the Soviet case at least, be ignored.

The Systemic Inheritance in the Social Sectors

The communist economies were not all bad. Table 3-2 shows that they were capable of generating high rates of growth when labor as well as capital was plentiful and before the cumulative dysfunctional elements of the economic system were fully revealed. This book is about the development of human resources, and some of the most important early positive features of the system were in this area, although even here there were always serious problems. It is useful to distinguish between the traditional system during the heyday of planning, say before 1970, and the sharp deterioration which set in after that date. This section is intended to serve as a brief introduction to later, more detailed material on labor markets, income transfers, education, and health. As elsewhere in this book, the discussion of the specific inheritances is divided into three parts: the strengths of the old system, its weaknesses, and those perennial problems which will always remain. As soon becomes clear, the good and bad features are often opposite sides of the same coin.

Table 3-4. *Level of Hard Currency Debt in Central and Eastern Europe, by Country, 1989*

Country	Absolute level of gross debt (billions of U.S. dollars)	Gross debt per capita (U.S. dollars)	Ratio of net exports to gross interest (percent)[a]
Bulgaria	10.0	1,149	16.5
Czechoslovakia[b]	7.2	480	6.0
German Democratic Republic	21.7	1,312	13.0
Hungary[c]	19.7	1,858	18.8
Poland	40.4	1,058	41.5
Romania	0.2	9	1.8
U.S.S.R.	49.5	170	3.7
Yugoslavia	17.6	743	17.6

a. Merchandise exports in countries with convertible currencies.

b. Newly released official statistics for 1989.

c. Includes an additional $2 billion in hard-currency debt to countries in the Council for Mutual Economic Assistance which the National Bank of Hungary omitted from its statistics until mid-November 1989.

Source: Various national statistical abstracts.

Strengths

The traditional Soviet-type system was particularly strong in the provision of individual security to the industrial labor force, as well as in the provision of relatively generous cash benefits, health care, and education.

INDIVIDUAL SECURITY. Socialist economies provided an extraordinary degree of job security by Western standards. Once workers had a job and had passed the probationary period, which was in most cases merely a technicality, their job was effectively guaranteed until they wished to leave. Legally it was very difficult to dismiss employees for redundancy or incompetence, and some groups (older workers, pregnant women, and the handicapped) were protected by law against dismissal. Even during downturns in output, workers continued to be paid in full, and enterprises were not permitted to fail, being supported by state subsidy. As noted earlier, however, this job security did not prevent low morale in the workplace.

Marxists were particularly critical of unemployment in capitalist countries and sought to minimize it under planning. Indeed, the "right to work" was often a constitutional provision. High levels of employment were achieved by giving enterprises and local governments quotas of new entrants to hire from the labor market, and the planners were required to find jobs for unemployed workers. This policy was enacted without reference to labor productivity or enterprise profitability. The Soviets claimed to have liquidated unemployment in the 1930s, and unemployment pay was abandoned as being unnecessary. Nonetheless, unemployment occurred outside the urban industrial sector, and frictional unemployment also persisted; in 1967 the average period spent between jobs in the U.S.S.R. was thirty-three days.

A second positive feature of the system was that per capita consumption of necessities and of basic services was higher than one would expect given the low levels of income per head and the high shares of investment. Gregory and Stuart (1986) report a calculation that put 1977 output per head in the U.S.S.R. at around 60 percent of that in the United States (valued at U.S. prices) with consumption only around 45 percent. The explanation is clearly the high Soviet ratio of investment to net material product. Despite these low per capita levels of consumption, widespread subsidies of goods classed as necessities, such as food, housing, education, and health care, meant that the shortfall was less marked in these areas. Thus per capita consumption of food and education was 54 and 75 percent of the U.S. levels, respectively. However, the Soviets were not keeping up in health care, if the United States represents an appropriate standard: health care consumption was only 37 percent of the U.S. level (in fact a lower proportion of the U.S. level than in 1955). Strikingly, per capita expenditures for consumer durables were only 13 percent of the U.S. level.

INCOME TRANSFERS. Cash benefits were well established and relatively comprehensive in most socialist countries. They included a generous system of sick pay and earnings-related pensions guaranteed by the state and nominally available at retirement ages often significantly below those in the West (sixty for men and fifty-five for women).[6] There were also child allowances and long maternity leave; the imperative of job security allowed employed mothers to stay at home for a relatively long period without losing their claim over their job and while receiving an allowance, for example, four years in Poland and three years in Hungary and Czechoslovakia.

DEVELOPMENT AND REDEPLOYMENT OF HUMAN CAPITAL. The education system in Soviet-type economies was an important element in the overall process of development; it was in large part through the education and vocational training systems that the planners tried to mould the skills of the labor force to the dictates of the plan. Children could start school at very young ages, in part to facilitate the speedy return of mothers to the labor force. Secondary education was compulsory, and typically a supplementary group of specialized educational institutions was devoted to technical education, language training, and so forth. Since education policy was a major element in manpower planning, liberal arts education was almost nonexistent, and education was oriented to engineering and science. The impact and biases of the Soviet-type education system can be evaluated for the Soviet case on the basis of table 3-5. The system provided a lot of education for its level of income, far more than the average for the developing world, for example, and somewhat more than for Italy. By the standards of industrial economies as a whole or the United States in particular, however, average levels of provision, for example of higher education, were modest. The impact of this relative backwardness on the process of industrialization was mitigated by the strong bias toward scientific and technical training.

Socialist economies were also rather progressive on certain gender issues and introduced those policies from a relatively early date. Females enjoyed

Table 3-5. *Indicators of Human Capital of the U.S.S.R. in International Perspective, 1980*

Country or group	Mean years of schooling	Number of scientists and technicians per 1,000 persons	University graduates as a percentage of age group	Science graduates as a percentage of all graduates
U.S.S.R.	7.6	128	5.8	48
United States	12.2	55	15.5	30
Italy	6.4	83	3.2	50
Industrial countries	9.1	139	9.1	35
Developing countries	3.5	9	1.1	31

Source: World Bank 1991.

wide access to education. Female participation rates were very high, and women's central role in the labor force was underlined by generous maternity leave with job security and by widespread provision of child care facilities at the workplace. A major burden for women, in contrast, was the heavy reliance on abortion as a means of contraception.

HEALTH CARE. The basic provision of health care was apparently adequate in many countries, particularly in the period before the full impact of the slowdown in growth was felt. Socialist economies offered universal access to free medical care and used food programs to prevent malnutrition. As an example, table 3-6 shows that despite the relatively low income per head, the U.S.S.R. in the 1950s already had levels of life expectancy at birth comparable more to industrial than to developing countries. The gap narrowed further until 1980. Such evidence of success is consistent with data suggesting that the socialist countries had relatively high proportions of doctors and medical staff per head of population.

Despite this success, an important dimension of the gradual deterioration of the economies of the socialist bloc was the worsening health situation. For example, the death rate in the U.S.S.R. has actually been rising since the 1970s, from 9.7 per 1,000 population in 1950 to 9.8 in 1980, with a very sharp increase, particularly in the Russian Federation, after 1989 (see UNICEF/ICDC 1993, fig. 7). This is partly because the population is aging, but it also reflects the progressive rise in age-specific death rates for males and females aged twenty to forty-four. These problems are related to the increasing consumption of alcohol, rising numbers of traffic and industrial accidents, excessive fat in the diet, and possibly also to the rising consumption of tobacco. The deteriorating situation has not been helped by a declining standard of health provision, including a shortage of drugs and worsening quality of equipment.

Weaknesses

Some of the problems in the social sectors, many of them intimately connected with the better features, have already been noted. Several additional weaknesses also merit attention.

Table 3-6. *Expectation of Life at Birth in Different Regions of the World, 1950–80*
(years)

Region	1950–55	1970–75	1975–80
Industrial regions	65.2	71.2	71.9
Developing regions	42.6	53.4	55.2
U.S.S.R.	61.4	69.4	69.6

Source: United Nations 1978.

ORGANIZATION AND ADMINISTRATION. Traditional socialist economies grew by investing heavily, and the share of expenditure devoted to collective consumption expanded considerably, mainly at the expense of private consumption. Investment expenditures were devoted primarily to industry, although employment in education, training, and health expanded steadily. Such activities were not priorities in the allocation of investment funds, however, because, following Marxist ideology, they were regarded as "non-productive," that is, they produced an output which was not a physical commodity, but a service. The consequence was a chronic tendency to underinvest in infrastructure and social overhead capital such as roads, schools, and hospitals. As these activities expanded, existing equipment and buildings were used more intensively.

This problem was already serious before the slowdown in growth but became more acute beginning in the mid-1970s because several socialist countries sought to resolve their economic problems by maintaining consumption at the expense of investment. Thus the share of gross investment in net material product between 1970 and 1985 declined from 29 to 24 percent in Bulgaria, from 25 to 18 percent in Poland, and from 24 to 10 percent in Hungary. The decline in investment was particularly marked in the non-productive sector and greatly exacerbated the poor condition and inadequate provision of infrastructure in the run-up to reform.

A second administrative problem for reformers has been that many benefits, both cash and non-cash, as well as much of the vocational education system, were administered through enterprises rather than through government authorities. In addition to the resulting problems of fragmentation and public sector inexperience with administering social services, such structures have made it hard for reformers to restructure or close firms until the social sectors have been reformed. Given the low status attached to social welfare activities, it is not surprising to find that administrative principles and practices are very old fashioned. Since reform of the social sectors is highly demanding in its need for investment and its administrative requirements, the absence of reform is easily explained.

POVERTY AND INEQUALITY. Despite socialist propaganda to the contrary, communist regimes were not especially egalitarian. The data on this topic are fragmentary and open to varied interpretation (see, for instance, Atkinson and Micklewright 1992; Bergson 1964, 1984; Wiles 1974). The balance of the evidence suggests that the U.S.S.R. had measures of inequality broadly comparable with those of the Scandinavian economies or the United Kingdom, but below those found in France or the United States. The inequality measures were also, perhaps, more equal than those calculated for countries at a similar stage of development. They did, however, vary considerably between countries (see Atkinson and Micklewright 1992), with Czechoslovakia apparently maintaining particularly low levels of measured inequality. The after-tax ratio of the top 5 percent to the bottom 5 percent of income per head in the 1960s was 3.0 in Sweden, 5.0 in the United Kingdom, but 5.7 in the U.S.S.R. However, it was 12.7 in the United States but only 4.3 in Czechoslovakia. McAuley (1979) calculates that although inequalities had been reduced significantly in the U.S.S.R. since the 1950s, even in the late 1960s some 40 percent of the population was living in poverty by the formal Soviet definition. Even after adopting a much lower norm (half of the "minimum material satisfaction budget"), McAuley still finds that 25 million Soviets (around 10 percent of the population) were living in poverty in 1967. He concludes that "there are significant lacunae in the network of support provided by the Soviet social welfare programmes" (McAuley 1979, p. 74).

Perennial Problems

It is worth stressing the magnitude and therefore the likely duration of the process of transition. Using figures for purchasing power parity, incomes per head in Central and Eastern Europe were perhaps between one-third and one-fifth of those attained in Western Europe. This situation has been exacerbated by the sharp declines in output associated with the transition, shown in table 1-1. Rollo and Stern (1992), using various assumptions, calculated how long it would take for these countries to return to their 1988 standard of living. Even on their optimistic scenario, the previous situation would not be restored in the countries predicted to grow fastest—Poland, Hungary, and the former Czechoslovakia—for more than ten years. Their pessimistic forecasts extend the period of recovery to twenty years or more. Even very rapid growth for a generation would not suffice to bring countries like the former U.S.S.R. to the living standards of the poorer members of the European Union (see the glossary).

Moreover, transition is likely to increase inequality and the proportion of the population in poverty in countries where the initial situation was already poor. These problems will undoubtedly be much exacerbated by the high levels of unemployment likely to be associated with the transition, a topic discussed in some detail in chapter 4.

Conclusions

This chapter has stressed some common elements of the communist economic system: ideology, excessive centralization, orientation toward the growth of production, and suppression of market forces. This concluding section briefly mentions some dimensions of the diversity of economies in transition.

The first dimension concerns cultural heritage. The Soviet bloc was a region with at least three cultural and historical backgrounds: Germanic (or

Austrian), for example in the Baltic states, Czechoslovakia, and Hungary; Balkan (Turkish), for example in Albania, Bulgaria, and Serbia; and Russian, for example in Belarus or Ukraine. These cultural inheritances exert influence today as traditional alliances and trading patterns reassert themselves. Second, communism lasted a much shorter period in Central and Eastern Europe than in the U.S.S.R., so that in countries like Bulgaria, the Czech Republic, Hungary, or Poland, it is possible to conceive of the restoration of previous property rights, institutional structures, and commercial norms which in the former Soviet republics have either atrophied or never existed.

Of comparable contemporary significance is the previous existence of the state itself. Bulgaria, Hungary, Poland, and Romania, for example, were independent countries throughout the communist era. The economy may have always been run on lines similar to those of the U.S.S.R., at least in Bulgaria and Romania, but separate institutions of central and local government, of public administration, and of the legal apparatus were in place. The same holds broadly for the successor states of countries which have broken up since 1989, for example the Czech Republic, Russia, and Serbia. These have typically inherited most of the former institutional framework, including the currency and the central bank. The collapse of communism, however, has created a number of new countries, all of which have to undertake the transition in the context of creating from scratch both nationhood and most political, social, and economic institutions. Some, such as the Baltic states, have historical precedents upon which to draw. Others, such as the Slovak Republic or Belarus, have little in their own recent history or experience to serve as a guide. Perhaps the *absence* of an inheritance in these entirely new states will prove as important for the reform process as the inheritances which this chapter has discussed.

Notes

1. This chapter focuses on the Soviet-type communist system typified by the dominance of the state bureaucracy. Other forms of communism, not discussed here, include the self-management variant of Tito's Yugoslavia and the Chinese model.

2. The main exceptions were China, Hungary, Poland, and Yugoslavia; and all countries experimented with reforms to the planning mechanism. Since the death of Mao, China has been moving toward radical economic but not political liberalization (see Ellman 1989; Ofer 1987).

3. In the market socialist economies, such as Hungary after 1968, the needs of the party were increasingly expressed directly through the government bureaucracy, especially that pertaining to banking and international trade, rather than through the instrument of the plan itself. Underlying institutional structures and behavior remained similar, however. Only in Yugoslavia during the late 1960s did regional fragmentation lead briefly to a relaxation of the tight grip of the party (see Estrin 1983).

4. The informal bargaining between managers and the government apparatus in the context of planned guidelines in effect came to replace the formal plan in Hungary and later Poland.

5. Desai's explanation centered on the determinants of declining total factor productivity, for example poor labor incentives, and resource misallocation associated
with the informational deficiencies of planning.

6. As discussed in chapter 9, these low nominal retirement ages were reached as
much on the exception as the rule because many special groups could and did retire
early with no loss of income.

Four

THE FORCES DRIVING CHANGE

NICHOLAS BARR

IMPORTANT HISTORICAL DIFFERENCES DISTINGUISH THE COUNTRIES of Central and Eastern Europe from one another, as chapter 3 makes clear. The same is true of their reforms. Programs vary with the size of the country and with its cultural, historical, and institutional inheritance; and the reform process may be path-dependent, in the sense that outcomes early in the process may influence the speed and direction of later policy choices. All the countries, however, share a similar inheritance of central planning and totalitarian government, and all are pursuing the twin objectives of higher living standards and increased individual freedom. According to Balcerowicz (1993a), three features of the transformation are common to these countries and also distinguish reform in Central and Eastern Europe from reform elsewhere. The scope of the change is massive, in that both the economic *and* the political systems are being almost totally overhauled; the situation is truly revolutionary. In contrast with that of most other revolutions, the method of change is for the most part peaceful. Moreover, he argues, the sequence of change is unusual in that political liberalization largely preceded economic reform.

At the heart of the revolution were two central decisions: to move from central planning toward a market system and to move from totalitarianism toward less authoritarian forms of government. This chapter is about those decisions, the driving forces they released, and the resulting outcomes. Understanding the forces driving change is important, since they largely explain why the outcomes are as they are. Greater reliance on market forces produces beneficial outcomes such as higher productivity and faster rates of growth, but it also leads to increased poverty, greater insecurity, and wider inequality. Increased democracy leads to greater individual freedom and greater responsiveness of government to the wishes of the electorate, but it also results in administrative overload, an increase in crime, and increased appeal to extreme nationalist political parties.

Outcomes are thus of two sorts. Some, such as greater efficiency, are wholly desirable; indeed, they are the major purposes of the reforms. These beneficial forces, however, may overshoot. The resulting less beneficial outcomes are noted as they arise and are taken up again in the concluding section of this chapter. The economic driving forces and their major outcomes are discussed first, followed by the political driving forces.

Economic Driving Forces

Markets have three ingredients: demand, supply, and prices, which link the two. The move to a market system liberalizes all three:

- On the demand side, individual choice becomes much more important, both for economic reasons (because of its central role in the efficient allocation of resources) and for political reasons (because of the backlash against the old system, which allowed little consumer choice).
- On the supply side, liberalization introduces competition by allowing new forms of ownership and by opening the economy to international competition.
- Price and wage liberalization connects the changes in demand and supply.

These changes release two sets of driving forces: a widening distribution of income and wealth and a more explicit role for economic self-interest.

Widening Distribution of Income

Economic liberalization has major distributional effects. It leads, first, to changes in the distribution of earnings. During the early transition, earnings differentials may be reduced because of anti-inflationary policies; subsequently they widen. Second, price liberalization changes people's real incomes, for example, by reducing the real incomes of individuals facing the largest price increases. Third, the broadening of property ownership widens the distribution of wealth and increases the incomes of wealth holders.

WAGES AND EMPLOYMENT. Increased diversity of earnings is a fundamental purpose of the reforms. Earnings which bear at least some relation to individual productivity encourage workers to improve the quality of their work, to move into jobs in which they are most productive, and to acquire new skills. The debate in the West has moved on to consider sources of labor motivation other than performance-related pay. In the context of Central and Eastern Europe, however, the issue is to rectify an inheritance in which wages bore more or less *no* relation to individual performance (someone talking about the old regime said that "work was somewhere we *went*, not something we *did* "). Thus performance-related pay improves incentives and, more generally, addresses the major problems of labor motivation inherited from the old system.

A key outcome of market liberalization, therefore, is a rising number of people with higher incomes. One way in which this occurs is that workers with skills valued by the market, such as speaking a Western language, can now command a market wage. This is precisely the result the reforms are meant to bring about. Second, the liberalization of property ownership, by extending the scope of private wealth, increases the incomes of wealth holders. Where wealth is acquired legitimately, this too is an appropriate outcome. However, higher incomes and higher wealth can arise also for malign reasons. Not all monopolies are bad: monopoly profits can encourage new firms to enter an industry. Sometimes, however, a monopoly of information or similar cause may enable a firm to prevent rivals from establishing themselves. This phenomenon arises in part because the reform of ownership takes longer than price and wage liberalization. Similarly, some individuals, often party officials from the old regime, are able to expropriate former state assets on advantageous terms, sometimes through dubiously legal methods.

Alongside higher incomes are lower incomes. In part, again, this is intended. People with few skills or low motivation face incentives to acquire skills and to work harder. That said, one of the major costs of the reforms is rising unemployment and increasing poverty. It should be remembered throughout that the importance of the work ethic in the inheritance, together with the security which the old system gave citizens, means that unemployment, poverty, and insecurity may be even more personally devastating in Central and Eastern Europe than in the West, where unemployment has a longer history.

UNEMPLOYMENT. Rising unemployment is occurring throughout Central and Eastern Europe. The old system deployed labor inefficiently, and the introduction of hard budget constraints and competition leads to displacement of workers as firms reduce their work force or close down. As a result, as much as 25 percent of the labor force needs to change jobs. Workers at greatest risk, at least in the medium term, are those in heavy industry and in areas with the greatest concentration of uncompetitive industries.

Displaced workers have four major destinations. One possibility is to find new employment. A relatively small number of workers will shift from their old job in a shrinking industry to a new job in a growing industry with no (or virtually no) intervening spell of unemployment. Such an outcome depends on the growth of new jobs and on the extent to which displaced workers are able through retraining to acquire the skills to fill them. A second option is to leave the labor market altogether: as discussed below, this is disproportionately the case with women and, depending on the generosity of the pension system, older workers, who may retire early. Emigration is a third possibility, although not one which has yet happened on a significant scale.

The fourth outcome is to be made unemployed. As a practical matter, it is difficult for large numbers of workers to change jobs without experiencing an intervening spell of unemployment. Some workers have skills which are no longer in demand. Others have skills which are in demand, but not in demand locally, and they are prevented from moving by inflexibility in the housing market. More generally, finding the right job takes time. A second, and separate, cause of unemployment is the decline in output. The result, as table 1-1 shows, is that unemployment has risen rapidly in countries like

Bulgaria and Poland and has become significant in almost all the countries in Central and Eastern Europe.

There is some evidence that, at least in the early transition, women bore a heavier burden of unemployment than men. In part this was because women's participation in the labor force was traditionally much higher in Central and Eastern Europe than in the West (in the U.S.S.R., because of high participation rates and the fact that women outnumbered men in the population, women were the *majority* in the labor force), but social attitudes also played a part. Although contrary to the ethic of the old system, the view persists of women as homemakers who should cede priority to men so far as employment is concerned. For either or both reasons, women were initially overrepresented among the unemployed in most of the countries in Central and Eastern Europe (Fong 1993; Fong and Paul 1992). In the Russian Federation in early 1992, women constituted 52 percent of the labor force, but 71 percent of the unemployed. Although outside the scope of this book, eastern Germany is instructive because it has the best data. From a low level in mid-1991, unemployment rose threefold for men and fivefold for women by mid-1992. Limited evidence in Hungary and Poland suggests that women, as well as becoming unemployed with disproportionate frequency, are also likely to remain unemployed longer than men. Similarly, in Russia, women constitute 71 percent of the unemployed, but only 61 percent of new job finders (Fong 1993). Male and female unemployment rates have converged somewhat as the transition proceeds, mainly because women are tending to drop out of the labor force (table 8-3).

POVERTY. Emerging poverty, a recurring theme throughout this book, is discussed in chapter 10, but it is useful to summarize some of the key issues here.

Poverty, although officially denied, existed under the old system (Bergson 1984, 1991; McAuley 1979) and became worse during the 1980s. Atkinson and Micklewright (1992, p. 178) report government estimates in the U.S.S.R. of 40 million poor people in 1989. Milanović (1991) found that nearly 25 percent of the Polish population was poor in 1987. Since poverty increased during the 1980s even though output was rising, it is not surprising that poverty continues to worsen throughout the region, in some countries dramatically (see UNICEF/ICDC 1993).

In the case of unemployment, the transition creates a largely new problem. With poverty, the reforms do not create a new problem so much as aggravate an old one. Because of political liberalization, the problem becomes more visible, and alongside this greater visibility, poverty increases for at least three reasons:

• Price and wage liberalization (an essential part of restructuring) widens the distribution of income. The relative prices which increase most are those which were previously the most heavily subsidized, particularly basic commodities such as housing, heating, food, and clothing, which form a larger fraction of the consumption of people with lower incomes than of those with higher. The withdrawal of such subsidies, though necessary on efficiency grounds, systematically harms the least well-off unless the authorities use some of the resulting savings for targeted income subsidies (an issue discussed at length in chapter 10).

- The decline in output reduces real wages, contributes to unemployment, and, because of its fiscal effects, reduces the ability of government to respond; and inflation erodes savings and also income transfers, which, for fiscal reasons, have often not kept up with price changes.
- Access to free health care and education can be reduced, particularly if supply shocks lead to shortages of critical drugs or if the fiscal crisis reduces the availability of textbooks.

Because of the transition, poverty is manifesting itself in new ways. Much of the burden falls on women. In Russia, women outnumber men 2:1 among the unemployed and pensioners; and 94 percent of single-parent households are headed by a woman. Experience elsewhere suggests that malnutrition is a growing problem, not necessarily because of an absolute shortage of food, but because of a deficient diet. Pregnant women, infants, and young children are particularly at risk: parents of young children tend themselves to be young and hence to have relatively low earnings; and mothers with young children have a higher than average risk of unemployment. Problems with health and nutrition for infants and young children can have long-term ill effects. Increases in maternal and infant mortality in Russia in 1992 (Fong 1993) are early warning signs of the impact of the transition. Health problems of other types are aggravated by poverty generally and by related problems such as poor housing, inadequate heating and clothing, and alcoholism. Drug shortages, which are a particular problem for the poor, have arisen in some countries. Educational problems are harder to quantify. Malnutrition affects the ability to learn. Shortages of school textbooks particularly affect the poor. Poverty can also aggravate homelessness, especially if rents are liberalized. For these and other reasons, poverty can also lead to the breakdown of families. Women, the young, and the old are the most vulnerable (see Burrows 1994).

Insecurity, though hard to quantify, is at least as important as poverty. It is one thing to know that income is low but relatively certain, as under the old system. It is quite another to face income which is falling and to be unsure how far it will fall.[1] Insecurity is problem enough for individuals who are able to find work but who do not know how much their wages will buy in the future or how long they will keep their job. It is considerably worse for those who cannot find work or for whom work is not an option, such as mothers of very young children, people with severe health problems, and the frail elderly. Such people must rely on state benefits, the real value of many of which has fallen, sometimes considerably.

DATA PROBLEMS. Poor data are a pervasive problem. They make it difficult to quantify the extent of poverty but also have much wider ramifications. First, there are well-known problems with determining the extent to which cross-country data measure the same thing. Second, measurement problems were greatly exaggerated during the early transition. Data on income are particularly suspect:

- Before the transition, figures on income took no account of the lack of availability of goods nor of the fact that access to goods was easier for some groups, such as the *nomenklatura*, than for others. The problem, then, is how to interpret changes in people's incomes.[2]
- The early transition led to sharp changes in relative prices, making it difficult to estimate inflation and hence difficult to assess what happened to people's real incomes.[3]
- A significant (but unknown) amount of income after the transition, particularly in the private sector, is not captured in official statistics.

For all these reasons, data on individual incomes, and also on gross domestic product, such as those underlying table 1-1, have to be treated with considerable caution.

A third set of difficulties, which are treated in a large literature, are specific to poverty (see Atkinson 1987, 1989, chap. 1, 1991a; Barr 1993b, chap. 6; World Bank 1990b, chap. 2, 1992c, chap. 1). These problems are particularly acute in Central and Eastern Europe, both because of the difficulties just discussed with the quality of data on income and because of significant gaps in the data on standards of living. The latter problem arises partly because the existence of poverty used to be officially denied and also because gathering such information is expensive. The precise situation varies from country to country. Data are relatively better in the Czech Republic, Hungary, and Poland, and Russia started to conduct surveys of living standards in 1992 (for a detailed discussion of data prior to the transition, see Atkinson and Micklewright 1992, chap. 3). A separate gap is the lack of data on the relative position of women. Quantitative evidence of the effects of the transition on poverty thus tends to be fragmentary, and policy often has to rely on more qualitative data.

The combined effect of these problems makes it difficult to assess the extent of poverty and even more difficult to measure changes in the number of people who are poor or in the depth of their poverty. Official statistics need to be treated with caution and interpreted with care (see Balcerowicz 1993b; Bratkowski 1993). The need to gather better information is urgent, for humanitarian reasons, because such information helps policymakers to estimate the costs of poverty relief and because poverty, alongside unemployment, has major ramifications for the political sustainability of the reforms.

An Explicit Role for Individual Self-Interest

The introduction of market forces involves the liberalization of demand and of supply. The first gives an increased role to individual choice. The second introduces a variety of property relationships and ownership options.

The increased role accorded individual choice coincides with the political objective of increased freedom and assists the efficient operation of a market system. In labor markets, for both reasons, individuals increasingly choose which skills they wish to acquire and which jobs they seek. Wage liberalization is crucial in this context, since market-determined wages inform individuals about which skills are in greatest demand. Similarly, in the context of education, training, and active labor market policies, individuals have increasing choice, for instance of schools, teaching methods, subjects studied, and training or retraining courses. Again, the exercise of such choice contributes to the aims of individual freedom and economic efficiency. As the reforms proceed, individuals will also have more choice over their doctor, hospital, and type of medical treatment. This is desirable both for its own sake and because, in a well-designed health-finance regime, patient choices can be used to give medical providers incentives to be efficient. In areas where consumer information is imperfect, choice can be guided by an appropriate regulatory structure, for instance, to ensure quality of training or medical services.

Supply liberalization introduces competition in various ways. Diversity of ownership contributes to the political aim of increased individual freedom and to economic efficiency. First, private ownership encourages competition among multiple providers. In well-defined circumstances (particularly the existence of well-informed consumers), competition contributes to efficiency by exerting downward pressure on prices and by improving the responsiveness of supply to consumer choice. Second, diversity of ownership allows different blends of public and private provision, depending on the commodity in question and on prevailing economic conditions. Third, alongside private ownership, competition in the economy as a whole is fostered by breaking up state monopolies and, through trade liberalization, by integrating the domestic economy into world markets.

Supply liberalization is acutely relevant to the social sectors. Diversity of ownership, by allowing employers to compete for workers, makes the labor market more competitive. This benefits workers, who no longer face a single major buyer (the state) in labor markets, and, it can be argued, thus improves their bargaining position. It also increases efficiency more generally, by allowing firms to signal their demand for labor through competing offers of pay and conditions, thus facilitating the movement of workers into their most productive use. Alongside such decentralization and competition, the state has a role both as a regulator of private labor markets and as an employer, although on a reduced scale, in areas like health care, education, and the administration of cash benefits.

In the context of education, training, and retraining, diversity of ownership allows private sector participation. In these areas, again, the combination of competition and a suitable regulatory regime contributes to efficiency. In health care, similarly, a range of suitably regulated ownership options can improve efficiency and responsiveness to consumer demand. Cash benefits are an area in which the state will remain a significant actor, as tables 1-2 and 1-3 show. As a complement to the state scheme, private pensions have a role in the medium term in empowering consumer choice, in fostering the development of capital markets, and (albeit more arguably) in furthering economic growth.

The argument in favor of markets is that they harness individual self-interest more effectively than any other method. The purpose of the reforms is to unleash those forces sufficiently to increase efficiency but, through regulation and other forms of state intervention, to contain their worst excesses. The benign aspect of relying more explicitly on individual self-interest is that individuals have an incentive to act in ways which contribute to the efficiency of the economy. These desirable effects are enhanced by a number of effects of the transition: relieving shortages enhances and empowers consumer choice, and opening the economy to international competition increases consumer choice and also facilitates the supply response.

Political Driving Forces

The core ingredients of the move to a market system are fairly clear-cut. The move toward democracy is less susceptible to simple analysis. It obviously involves the introduction of free elections with a fairly universal franchise. Beyond that, however, key questions remain. Should the electoral system be first-past-the-post, as in the United Kingdom and the United States, or some form of proportional representation (both systems are described in the glossary)? If the latter, should a threshold be set below which political parties are not allowed to occupy seats in the legislature?[4] At least as important as electoral arrangements is the constitution which underpins them. Is there a strict separation of powers, as in the United States? Which branch of government (legislature, executive, judiciary) has primary authority in any given area? If a conflict arises between different parts of government, does the constitution establish unambiguous decision rules?

This chapter makes no attempt to answer such questions (some of which are taken up in chapter 13; for an extensive discussion, see White, Gill, and Slider 1993). Instead, it takes the move to democracy (or at least increasing *glasnost*) as its starting point: such a move involves political pluralism, decentralization of political decisionmaking, and a new emphasis on the rule of law. These changes unleash two sets of political forces: people's hatred of the old system and their deep distrust of government, which was regarded not as of the people, by the people, and for the people, but as by the *nomenklatura*, which trampled on people's rights. More recently, some countries have experienced a separate anti-government backlash, as the electorate begins to feel the short-run effects of the reforms, particularly the decline in many people's standard of living.

Reaction against the Old System

The anti-government feeling generates pressures to throw out the good as well as the bad, and leads to lack of realism about the extent to which perennial problems can be removed. Increased political freedom unleashed a swing of the pendulum against the two key features of the inheritance: central planning and totalitarian government. The swing is toward (a) private markets, (b) democracy, and (c) decentralization. A fourth pressure, a strong desire in many of the countries to integrate with Western Europe (in part a reaction against inherited isolation), though noteworthy, is discussed only in passing.

PRIVATE MARKETS. The swing to private markets is one of the twin hearts of the reform. The roots of the pressure include distrust of government and the low institutional capacity of the public sector. Reformers face the problem of balance. There is a danger of not liberalizing the economy enough, thereby, at worst, blocking the reforms. They also face the opposite danger, of overshooting. The flaws of central planning being very obvious, the tendency in the early transition was to lurch toward another simple solution which does not work: private markets, always and everywhere. The backlash against state involvement had a range of ill effects, including deteriorating systems of health care, education, child care, and public transport. Unrestrained privatization, in the absence of suitable regulation, could result in the pillaging of state assets (which happened in the health sector in some countries) and other abuses of monopoly power. Nor are the citizens of these countries alone in this sort of behavior. Robert Strauss, U.S. ambassador to Moscow, in an on-the-record interview in early 1992, used the term "sleaze bags" to describe predatory Westerners taking advantage of such situations.

DEMOCRACY. The swing toward democracy is desirable for its own sake, because it increases individual freedom and gives government the incentive to be more responsive to the wishes of the electorate. Again, though, overshooting is possible. Democratic institutions have been put into place with remarkable speed. But reinventing a country is a huge task, and the fact that some problems remain is neither surprising nor cause for criticism. One reaction against totalitarianism was to democratize *everything*, partly out of a belief in democracy, partly out of a distrust of totalitarian government, and partly out of a desire to make the change irreversible.

A number of resulting problems emerged at the start of the reforms. One was legislative overload. The problem was an inevitable consequence of revolutionary change but, for reasons discussed in chapter 5, also had other, more avoidable causes. A second, more arguable, problem is that the division of power between executive and legislative forces tilts more toward the legislature than is common in the West. The balance between democracy, on the one hand, and the needs of effective government, on the other, is one of the central problems of political economy and has no easy answer. The inability of government to be effective is an important constraint on the design and implementation of the reform process generally, and of social sector reform specifically.

DECENTRALIZATION. The old system imposed one-party rule, and central management disregarded regional differences. Regional planning, far from respecting regional diversity, was a tool to impose uniformity. As part of the same process, minority differences were disregarded and, where visible, subjected to pressures to conform (one example was the treatment of ethnic minorities in Bulgaria and Romania). The process applied also to religious groups, such as the Muslims in Bulgaria and the Jews in Ukraine and Russia.

It is therefore not surprising that the relaxation of excessive centralization is one of the major forces driving the reform. The force manifests itself in different ways, including the move toward multiple political parties, toward power devolved to locally elected governments, and toward regionalism and nationalism. The benign aspects of decentralization are clear: political pluralism, increased responsiveness of government to the wishes of the local electorate, increased scope for the expression of indigenous culture, and reconnection with people's historical heritage.

Overshooting can arise in a number of ways. The optimal balance of power between central and local government is not determined easily and is an area of continuing controversy in the West. There is general agreement, however, that for some aspects of the economy national standards are desirable. Employment services and the structure of cash benefits, for example, should foster labor mobility, which requires at least coordination among regions, if not identical systems. Educational qualifications and occupational standards should be readily transferable. Entitlement to health care should not be constrained by local boundaries. Similarly, at a regional and national level, minimizing impediments to trade has major advantages.

Devolution of power may interfere with these objectives. More generally, a real tension exists between the freedom to pursue local objectives—including, for the first time, the right to establish a national identity—and the pursuit of common purposes which may be organized more effectively at a national level. Compromise is often possible. For instance, decentralization can operate within a central framework, enforced through regulation or through central government grants for approved local activity. The constraints resulting from centrifugal pressures are discussed in chapter 5.

Backlash against the Reforms

Although the matter cannot be proved, there is considerable anecdotal evidence indicating that what creates political backlash is not the outcome per se, but the outcome relative to people's expectations. The British population endured with stoicism considerable hardship during the Second World War, not least because Churchill promised "nothing but blood, toil, tears, and sweat." This suggests that political explosions are caused not by poor outcomes, but by disappointed hopes. At the start of the reforms, there was considerable optimism, not least in government policy statements, that the process of catching up with Western Europe would be rapid. In part that was true, in that some political objectives, particularly the introduction of democracy, were achieved fairly rapidly. So far as the economic objectives are concerned, the extent of the decline in output in the early transition surprised most observers even in the West. The slowness of recovery, the increased incidence of unemployment and poverty, and, more generally, the disappointment of the electorate's hopes of rapid economic gains led to a political backlash against the reforms in at least some countries of Central and Eastern Europe. This backlash helped various brands of extremists to gain momentum, posing a potentially serious danger to the new democracies. The role of politicians in preventing inflated expectations is therefore critical and is discussed in more detail in chapter 13.

Conclusions

The introduction of market forces and greater political freedom released a series of driving forces, including a widening distribution of income, a greater role for individual self-interest, and a political backlash against the old system. Some of the resulting outcomes contribute directly to the purposes of the reform: increased productivity brought about by the incentive effects of competitive markets, improved responsiveness of government to the desires of citizens arising from democracy, and the ability to decentralize to allow local electorates and cultural groups greater autonomy.

The driving forces also produce unhelpful outcomes, most of which reappear in the discussion of constraints in the next chapter:

- Rising unemployment and poverty emerge repeatedly, making stabilization and restructuring more difficult and posing a major political constraint.
- Market imperfections, in particular missing legislation, an incomplete regulatory structure, and exploitation of monopoly positions, hinder restructuring.

- The tendency for decentralizing pressures to overshoot forms part of the discussion of centrifugal forces.
- The problems which newly designed democratic systems often face (such as legislative bottlenecks) affect the machinery of government.

Notes

1. In terms of economic theory, a risk-averse individual may gain a higher level of welfare from a lower but certain income than from a higher but uncertain income. Thus, for some people uncertainty causes a loss of welfare *in addition* to the welfare loss associated with a lower level of income.

2. Suppose that before the transition, a pint of milk, when available, cost $1, and individual X had an income of $1,000. After the transition, milk is always available and costs $5, and individual X still has an income of $1,000. On the face of it, individual X is poorer, but if he or she could rarely get milk under the old system, matters are much less clear.

3. There are at least two sets of problems. First, with non–market clearing prices (such as existed under the old system), both Paasche and Laspeyres indexes may overestimate inflation and hence also overstate the fall in real wages and living standards. Second, in the face of sharply changing relative prices, the pattern of people's consumption changes, thus sharply changing the weights used to calculate a consumer price index. The use of a consumer price index to measure inflation in these circumstances is, to say the least, problematic.

4. In the Federal Republic of Germany, for instance, a party must win at least 5 percent of total votes cast before it can hold a seat in the legislature. The purpose of such a threshold is to limit the number of parties in the legislature, the idea being that this makes it easier to form stable coalitions.

Five

CONSTRAINTS ON CHANGE

NICHOLAS BARR • STANISLAW GOMULKA
IGOR TOMEŠ

SOME FEATURES OF THE EMERGING ECONOMIES constrain the speed of reform and, since the longer-term path of reform is influenced by earlier events, also its direction:

• The initial conditions are important. Some inherited institutions and attitudes are ill-suited to the needs of a market economy and are therefore exposed as constraints by the transition. Examples include missing institutions like financial markets, gaps in legislation, and inappropriate supply relationships resulting, for instance, from the extent of state ownership.

• A second set of constraints arises from the transition itself. The driving forces discussed in chapter 4 have both good and bad outcomes. Some of the bad outcomes, such as rising unemployment, are in part caused by the transition. Others, such as pressures to decentralize, relate to aspects which were suppressed under the old system and which are therefore released by the transition.

• External shocks are a further source of constraints, the major examples being the collapse of the old trading relationships and recession in the industrial countries during the early transition.

• The final set of constraints is inherent, meaning that they face all economies, including those of the industrial countries. An important example is market imperfections. Another is the need, because of resource constraints, to make difficult choices between different types of social spending and between social spending and other areas such as investment in physical capital.

Many of the constraints discussed in this chapter (see also Gomulka 1994) derive from some combination of these sources. The decline in output, for example, is the result of inherited distortions, of the policies needed to contain the inflationary pressures released by the transition, and of the collapse of trading arrangements. Some of the constraints (most clearly the fiscal crisis) have an obvious bearing on the social sectors. Others, such as those related to institutional capacity, though applying more widely, are also directly relevant to the discussion in later chapters. Yet others, such as constitutional arrangements, far transcend the social sectors. They are included here because the success of social sector reform is inextricably linked with the success of transformation generally. The need for fiscal restraint, for example, may lead government to seek a reduction in expenditure on pensions; if the electoral system leads to a fragmented Parliament, however, the government may find it impossible to enact the necessary legislation. Several countries, for such reasons, have had difficulty passing their budget laws. Though the distinction can become blurred, this chapter divides the constraints into three categories: economic, political, and institutional.

Economic Constraints

Two major types of economic policy underpin the transformation: stabilization and restructuring (see the glossary). Each faces a series of major constraints.

Constraints Connected with Stabilization

Stabilization is made difficult by key features of the inheritance, by the sharp decline in output, and—separate though connected—by the fiscal crisis which, to a greater or lesser extent, has arisen throughout the region.

THE INHERITANCE. The results of the inheritance were low output and, in many countries, foreign debts and a large accumulation of unspent savings (the so-called monetary overhang). The slowdown in growth occurred at relatively low standards of living compared with those in the highly industrialized countries. One of the responses, particularly in Bulgaria, Hungary, Poland, and the U.S.S.R., was to import more investment goods from the West using Western credit. Although in some instances bringing short-run gains in economic performance, these imports did nothing to help the long-run problem. Their major legacy, particularly in the countries just mentioned, was the large foreign debts shown in table 3-4. A second response to the growth slowdown was a relaxation of fiscal and monetary policy. Under mounting social and political pressure, the authorities increased consumer subsidies with no matching increase in taxes. Given the shortage of goods, one consequence was the buildup of forced savings. The resulting monetary overhang added to inflationary pressures when prices were liberalized.

THE OUTPUT COLLAPSE. Output fell sharply in many countries (see table 1-1). Over the three-year period from 1990 to 1992, according to official statistics, gross domestic product (GDP) fell almost 50 percent in the Baltics, about 40 percent in Bulgaria, 35 percent in Romania, about 27 percent in Czechoslovakia and the U.S.S.R., and about 17 percent in Hungary and Poland. The average fall in output (which is the unweighted average across the countries) was nearly 30 percent. These figures should be treated with caution. The difficulty is no longer the falsification of official statistics, but rather their lack of coverage, particularly of small-scale activity in the private sector, which has increased in importance yet is not fully captured in official statistics. Nevertheless, the output losses suffered in many countries in the early transition were comparable in magnitude to those suffered by the Western economies during the Great Depression of the 1930s.

There are two broad schools of thought about why the decline in output was so large in some countries during the early transition (see Desai and Estrin 1992; Fischer and Gelb 1991; Gomulka 1991, 1992; Gomulka and Lane 1993; Kornai, forthcoming; Laski and others 1993; Nuti and Portes 1993; on the Polish experience, see Sachs 1993). The first argues that the major causes were the inheritance, price liberalization, and external shocks. Given the previously distorted price structure, relative prices could be expected to change in the face of price liberalization. A particularly important distortion was the artificially low price of energy. Part of the decline in output was the result of reducing the scale of activities which were loss-making at market prices, especially those with negative value added.[1] Another influence was the size of the private sector. Inheriting a large private sector, which is more responsive than the public sector to external shocks, tends to reduce the extent of the fall in output. An additional factor was the collapse of the trade arrangements of the Council for Mutual Economic Assistance (see the glossary) and, in the former U.S.S.R., of trade among the republics. The output decline, according to this argument, was larger the greater the distortion of the inherited price structure, the smaller the inherited private sector, and the larger the dependence on trade with the Council for Mutual Economic Assistance.

The second school of thought argues that, in addition to these factors, the output collapse was to a significant extent caused by the reform policies themselves. It argues that fiscal and monetary policies were excessively restrictive, especially in Czechoslovakia and Poland, and that domestic markets were opened to foreign trade too fast. Along the same lines, it suggests that restructuring should be pursued more slowly, for instance, by continuing to extend some government assistance to state enterprises in the short term. The last issue is taken up in chapter 7, which compares policies to keep workers in their old job in the short run with those which allow them to become unemployed.

Discussion of the cause of the output decline, given its impact on people's lives, is clearly much more than academic. If the second line of argument is true, the transition could perhaps take place at a lower cost in unemployment and lost output. That view, however, may be optimistic, in that output losses much greater than those experienced in Czechoslovakia and Poland have been experienced in countries like Bulgaria and most of the U.S.S.R., where fiscal and monetary policies were not so restrictive. Thus the option of a relatively painless transition may not be available. Despite the importance of the question, there is as yet no generally agreed-upon answer.

THE FISCAL CRISIS. Interpreted here as a large budget deficit, the fiscal crisis arose to a significant extent out of the collapse in output. This was particularly true when that collapse was coupled with political constraints which made it difficult for government either to reject the claims of particular interests and reduce expenditure to match revenues or to increase their revenues to pay for desired expenditure. One result is downward pressure on the resources available for the social sectors. The crisis may not be inevitable, as the broad budgetary balance of the Czech Republic in 1992 suggests, but it is difficult to avoid (for a detailed discussion, see Miszei 1994).[2] In 1992, Poland, Hungary, and the Russian Federation had budget deficits of between 5 and 10 percent of GDP. In some countries, for instance the Ukraine, the deficit was considerably higher. Policies were put in place in late 1992 aimed at containing the budget deficit to 5 percent of GDP in Poland, to 6 percent in Hungary, and to 8 percent in Russia. The crisis occurred because tax revenues fell faster than GDP, while expenditure fell more slowly. Each aspect requires discussion.

During a recession in the West, the fall in tax revenues is generally greater than the decline in income because tax systems are progressive.[3] This is also part of the story in Central and Eastern Europe, not least because the profits of enterprises form a major part of the tax base (see Schaffer 1992 for the case of Poland), but it is useful to take things further. Two factors are central: what happens to the balance between wages and profits and what happens to the level of real wages. On the first issue, countries in Central and Eastern Europe taxed profits more heavily than wages. Thus any change from profits toward wages reduces revenues. So far as wages are concerned, the decline in output produces a fall in the real wage base for two reasons: the number of wage earners declines as unemployment rises, and the real wages of people in employment fall. Real wages fall because stabilization requires that real incomes must decline to keep domestic demand broadly in line with output. Historically, real income in Central and Eastern Europe had two sources: monetary income, such as wages, family allowance, and the like, and benefits in kind, such as cheap housing. The decline in real income could in principle occur through a reduction in benefits in kind, but in practice, some of these benefits, such as subsidized heating, cannot easily be reduced in the short run. Thus most of the decline in real income will have to occur through a reduction in real wages.[4]

Separate from the decline in the wage base, the amount of tax which can be extracted from each million forints (or koruni or roubles) of wage payments declines for at least three reasons. First, output falls most precipitously in the state sector, which is where the old system could most easily collect taxes. As a result, an increasing number of state firms start to delay or default on their tax and social insurance payments. Second, the growth that does take place tends to be in the private sector, where tax enforcement is not yet fully developed. Third, the combined effect of inflation and lags in the collection of contributions can result in a serious loss of real revenue (the Polish social insurance authorities estimated that in 1990 about 10–12 percent of their revenue from contributions was lost in this way).

Tax revenues fall, therefore, for several reasons: the real wage base declines, the yield from a given wage base declines, profits decline, the balance between profits and wages changes, and systems of personal income taxation are either still developing or not yet in place. The relative size of these factors, and their timing, varies from country to country. Their overall effect in Central and Eastern Europe was, however, that tax revenues, even if they did not fall immediately, eventually fell, and did so even more sharply than output over the early part of the transition.[5]

The other root of the fiscal crisis is that expenditure cannot easily fall as fast as revenues. First, rising unemployment and emerging poverty increase the demands on the system of income transfers; and in some countries, medical expenditure, particularly on pharmaceuticals, is rising rapidly. Second, some countries face the demands of servicing the international debt out of declining GDP, and the problem of debt service worsens as a country's exchange rate declines. Expenditure may also rise for reasons connected more with politics than with the decline in output: enterprise subsidies fell slowly in some countries; and in some countries, Poland for example, social expenditure, particularly on pensions, increased sharply. Again, there is significant variation between countries in the relative importance of these causes. Throughout Central and Eastern Europe, however, demands on the budget increased at precisely the time that revenues were falling.

Implications for the social sectors emerge repeatedly in subsequent chapters. Fiscal constraints affect the level of unemployment benefits and are particularly acute for pensions, given the prevalence of early retirement. In some countries, including most of the former U.S.S.R., these constraints are overwhelming, raising major issues of what level of poverty relief is affordable. Reduced health and education budgets lead to sharp declines in expenditure on medical supplies, schoolbooks, and the like. The resulting problem is twofold: devising a package to contain the government deficit and finding a way to obtain parliamentary and public support for the necessary measures. The problem is critical because of the tradeoffs between competing claims on output: the claims of individual consumption (a political imperative), those of investment in physical and human capital (a prerequisite for economic growth), and, in some countries, those of foreign debt repayment. When output is declining, it is not possible to meet all three sets of claims. Sustainability of the transition thus involves a balancing act of considerable delicacy. The link between economic constraints and political constraints is at its most acute for this aspect of the reforms.

Constraints Affecting Restructuring

Restructuring has to contend with three sets of economic constraints: inherited misallocation, poor labor motivation and inappropriate incentives, and market imperfections. The reforms are seeking to deal with the first two problems; the third, in Central and Eastern Europe as elsewhere, is inherent.

THE INHERITED MISALLOCATION OF RESOURCES. The structure of ownership was much too heavily weighted toward the public sector. The structure of production was unbalanced, being tilted toward heavy industry and agriculture at the expense of consumer goods, light industry, and services. The structure of trade was distorted by the arrangements of the Council for Mutual Economic Assistance, particularly the overspecialization and consequent giganticism of production units. The results included inefficiency so great that for some products total value added was negative, making output dramatically uncompetitive at world prices. Thus large numbers of workers need to move to more productive jobs. Such moves, however, are constrained both by the inherited mix of skills and by the misallocation of housing. As yet, no real housing market exists to address the problem. Since housing under the old system was often part of a person's employment package, housing constrains both the mobility of labor and the restructuring of enterprises.

POOR MOTIVATION AND INAPPROPRIATE INDIVIDUAL INCENTIVES. Both these problems were part of the inheritance but, at least in the public sector, continue in different guises during the transition. Poor labor motivation under the old system arose from low wages and bad labor relations. Neither the stick of threatened job loss nor the carrot of higher wages was present to increase productivity. On the first, shortages led to a sellers' market, giving service providers no incentive to respond to consumer demand and creating fertile ground for corruption. On the second, wages under the old system comprised a system of administratively determined tariffs (something like 60 percent of total pay) plus premiums, which in theory were related to individual performance but which in practice were often divided fairly equally across the work force.

The incentive structure militated against exposure of poorly designed policy. Decrees were issued from the top and implemented at lower levels. When reporting back up the chain of command, lower-level officials had little incentive to report anything but success, and middle-level officials had little incentive to investigate the claim, since they, in turn, would have to report failure to top-level officials. Thus there was an implicit conspiracy: decrees were issued, and everyone then pretended that they had been implemented (something similar went on in labor markets, giving rise to the view among workers that "we pretend to work, and they pretend to pay us").

During the transition, adverse incentives continue in the public sector. Not least because of the fiscal crisis, administrators, medical personnel, and teachers are poorly paid in comparison with the private sector and at least some state enterprises. The result is declining morale and a tendency for some of the most able individuals to leave the public sector. In addition, because wages are low, many workers have second and third jobs and little motivation to perform well in their primary job (medical personnel and teachers are important examples). For that reason, the Czech Republic introduced a 25 percent pay increase in 1992 for administrative personnel who gave up second employment and private business. Poland, too, introduced a substantial pay increase in 1992 for senior personnel in central administration.

MARKET IMPERFECTIONS. As discussed in chapter 4, economic liberalization releases the potential for enormous efficiency gains but also reveals market imperfections. The problem can be particularly severe with insurance, where major risks, such as unemployment, inflation, and important medical risks, are difficult or impossible to insure in private markets. In the presence of these technical problems, the state may intervene in various ways, so the extent of government intervention in the industrial countries should not be surprising.

Thus the workings of a market economy, largely for efficiency reasons, are underpinned by a substantial body of legislation and regulation. The initial deficiencies in Central and Eastern Europe were twofold: important types of regulation were absent, and much existing regulation was inadequately enforced. So far as the first problem is concerned, there were two sorts of lacuna. First, at the start of the transition, gaps were evident in the structure of laws establishing key institutions such as the ownership and transfer of private property, the conduct of private enterprises, the operation of a Western-style banking system, the operation of financial institutions such as private insurance companies and stock exchanges, employment conditions in the private sector, and the development of a system of unemployment benefits appropriate to a period of high unemployment. A second and separate gap was the lack of a regulatory structure to prevent market failure. Examples in the West include hygiene laws connected with the production and sale of food and the testing and sale of drugs, laws relating to the abuse of monopoly power, laws concerning the conduct of financial institutions such as private pension funds, and regulation of total medical spending.

The Western economies operate effectively because they have achieved a balance between the incentives arising from an appeal to individual self-interest and the restraining effects of regulation to prevent abuse, for example of monopoly power. The exact balance between these forces is the subject of continuing debate, but such regulation is indisputably needed to prevent the fraud and abuse which would otherwise occur. The absence of an effective regulatory structure hinders the reforms, not least by putting at risk the political support for the move to a market system.

Political Constraints

Political constraints on the reforms can usefully be divided into two sorts: (a) attitudes and pressures unleashed by political liberalization which may impede the reforms and (b) weaknesses in the machinery of government which affect the ability of ministers and officials to make and implement decisions.

Attitudes and Pressures

Political pressures derive from three major sources: inherited attitudes, a deep distrust of government, and centrifugal forces.

INHERITED ATTITUDES. A country's history shapes attitudes in important ways. Prior to the Second World War, countries like Czechoslovakia, Hungary, and Poland were to a greater or lesser extent industrial market economies, well acquainted with market systems generally and with unemployment in particular. None of this was true in most of the U.S.S.R., where unemployment is a completely new development and where the politics of unemployment are therefore of extreme importance. The political capacity to carry through reforms in the face of high unemployment (particularly when coupled with the governance issues discussed in the next section) cannot be taken for granted.

DISTRUST OF GOVERNMENT. Outside the U.S.S.R., government was seen as an alien body imposed by the Soviets, thus calling its legitimacy into question. Separately, throughout Central and Eastern Europe, the old regime had contaminated the law for political ends: individual rights were not protected sufficiently, and such protection as existed was not properly enforced.

In a democratic society, individual rights and obligations and basic guarantees of their enforcement are established by acts of Parliament; detailed implementation is the responsibility of the executive, usually on the basis of regulations. Under the old system, in contrast, many enactments were by decree, and significant conditions of individual eligibility were often not included in primary legislation but left to executive discretion.[6] This was true throughout the social sectors: labor conditions, incomes policies, cash benefits, health care, and education. Separately, violations of individual rights were generally less harshly punished than violations of the interests of the party. In all countries of Central and Eastern Europe, for instance, theft of state property was more heavily punished than theft of private property.

Even where sanctions existed, they were often ineffective. In Czechoslovakia, local government was empowered to inspect labor conditions, but no effective inspectorate was ever built up. Throughout Central and Eastern Europe, the effectiveness of safety inspections at the workplace was often hampered by the propensity to regard an adverse decision as an attempt to sabotage the plan. Thus dangerous plants were rarely, if ever, closed. The same was broadly true of hygiene inspections. The result was a very poor record on occupational safety.[7] A further problem of enforcement was the inadequate procedure for appealing decisions. In most countries, appeals were generally conducted by the authority which made the original decision. Executive decisions were rarely subject to judicial review. The result was an over-mighty bureaucracy, with serious limitations of individual rights.

Lies were pervasive:

> Life in the [old] system is . . . thoroughly permeated with lies: govern-ment by bureaucracy is called popular government; the working class is enslaved in the name of the working class; the complete degradation of the individual is presented as his ultimate liberation; depriving people of information is called making it available; the use of power to manipulate it is called the public control of power, and the arbitrary use of power is called observing the legal code; the repression of culture is called its development; the expansion of imperial influence is presented as support for the oppressed; the lack of free elections becomes the highest form of democracy; banning independent thought becomes the most scientific of world views; military occupation becomes fraternal assistance. Because the regime is captive to its own lies, it must falsify everything. It falsifies the past. It falsifies the present, and it falsifies the future. It falsifies statistics. It pretends not to possess an omnipotent and unprincipled police apparatus. It pretends to respect human rights. It pretends to persecute no one. It pretends to fear nothing. It pretends to pretend nothing. [Havel 1978; reprinted in Havel 1991, pp. 135–36]

Thus it is hardly surprising that there was widespread distrust of govern-ment. In the West, it is possible to trust the *process* by which governments come to power (though perhaps with reservations about some of the individ-uals thereby elected). The old regime in Central and Eastern Europe gave governments no such legitimacy. Distrust went both ways. It is not necessary to look as far as secret police to see that government distrusted its citizens. Examples include the amount of paperwork claimants for state benefits had to complete, the concept of self-declaration supported by sample audits being unknown.

Distrust of government constrains the transition. Without well-enforced individual rights, citizens feel threatened by economic liberalization, particu-larly in the face of rising unemployment and falling standards of living. The population tends to suspect widespread self-interest among officials, including reformers. This is one of the roots of the backlash against the reforms dis-cussed in chapter 4.

CENTRIFUGAL FORCES. When expressed as devolution of power and decen-tralization of administrative authority, the resulting liberation of individual, local, and regional initiative is desirable in itself and contributes to the process of reform. Excessive decentralization, however, can lead both to fragmenta-tion and to fiscal problems. The existence of multiple political parties (a consequence in some countries of centrifugal forces) is not itself a problem but becomes one if it interferes with the effective implementation of the reforms.

Regionalization, too, may overshoot if it prevents the development of institutions which are best coordinated nationally; different regimes for unemployment benefits or lack of widely accepted training qualifications, for example, can impede the mobility of labor. Decentralization of health care and education can also go beyond what is efficient.

Excessive decentralization also has fiscal implications: the smaller the locality, the greater the need for central government to make redistributive transfers between richer and poorer areas. Such transfers raise major issues, including the ability of central government to raise sufficient revenue and the need to decide how to divide transfers among localities. The latter task requires a method of assessing needs by locality, a task which makes heavy demands on institutional capacity and is likely to be highly political. If local authorities administer a service which is largely financed by central government, a separate problem is the need to design the transfer in a way which minimizes the incentive to spend excessively.

The message is not that devolution is necessarily a constraint. The real message is fourfold:

• The extent of devolution and decentralization is a complex matter.
• Answers will vary across countries, as they do in the highly industrialized countries.
• Central planning had all manner of highly undesirable consequences, and some decentralization is clearly desirable. It does not, however, *necessarily* follow that the optimal solution is to decentralize as much as possible.
• Decentralization requires an appropriate fiscal regime.

The Machinery of Government

Alongside constraints arising from political pressures, a separate set of problems constrains the ability of governments to implement the reforms: erosion of the rule of law, constitutional arrangements which can hinder effective government, and lack of experience in the conduct of politics in a democracy.

EROSION OF THE RULE OF LAW. Under the old system, the distrust of government, together with the tendency of government to override legal provisions for political reasons, produced a lack of rule of law in the Western sense that laws (by and large) are respected, enforced, and obeyed. In the U.S.S.R., many social benefits were under the control of the official trade union, and union officials frequently used these benefits as instruments of labor discipline and as forms of bribery and favoritism. The result was a disrespect for law so pronounced, according to the then–prime minister of Bulgaria early in his country's reform process, as to undermine the moral values of a civil society:[8]

By imposing a great number of impractical and often illogical restrictions and limitations that were inevitably and routinely breached or disregarded, [the communist authorities] also cultivated in the individual a feeling of being an offender. . . . This served to blur the line between the permitted and the forbidden. . . . Honest and moral people would think nothing of stealing from the state or cheating the authorities. [Dimitrov 1992]

The disrespect for legal processes is a severe constraint. Market systems rely heavily on the rule of law because decisionmaking is decentralized and economic relations founded on the principle of voluntary exchange. Individuals undertaking exchanges need to know that promises have a contractual basis, which provides some guarantee through the courts that the exchange will take place on the terms originally agreed to or that the party in breach of the contract will be penalized. A clear definition and enforcement of rights and obligations is a precondition for a well-ordered competitive environment, particularly one intended to attract foreign investment. Legal guarantees of this sort either did not exist or had atrophied in most countries of Central and Eastern Europe under the old regime. Even where a legal apparatus was put in place as part of the reform process, problems remain: people governed for years in a police state are unlikely to turn to the judiciary to resolve grievances, and the legal system is often understaffed and, as a result, may not be able to prevent widespread infringements. Both sets of problems are likely to be a constraint for the time being.

EVOLVING CONSTITUTIONAL ARRANGEMENTS. A key aspect of governance concerns constitutional provisions—the rules by which politics are played and political conflicts resolved—and, as part of those rules, the respective powers and responsibilities of the legislature, executive, and judiciary. In the West, these have evolved over many years and continue to evolve. In Central and Eastern Europe, they have had to be invented almost overnight. Constitutional changes in the early transition were a clear reaction against the inherited system. The importance of the legislature is in many ways a direct consequence of the heavy dominance of the executive under the old system. Even countries like Poland, Hungary, and the Czech Republic, notwithstanding several years of experience with democratic government, are finding problems with the system; and the problems are even greater in countries like Russia, which are larger and whose reforms began somewhat later.

The respective powers and responsibilities of the legislature and executive may not be clear and explicit; or where they are clear, they may work badly. In the early days of the reforms in Russia, for instance, presidential decrees and legislative acts could be in direct conflict with one another, with no constitutional rules for establishing which took precedence.[9] In such cases, the judiciary may not yet be in a position to resolve the dispute.

The electoral regime may yield indecisive results. The first democratic election in Poland was conducted under a system of proportional representation with no threshold, leading to a legislature with a fragmented party structure which hindered the formation of a stable coalition.[10]

Even when the powers of the executive are clear, the executive might not be able to enforce them, particularly where administrative structures lack the capacity to enforce the law. The problem is most acute in newly reformed countries. In Russia in 1992, for example, the central authorities experienced difficulties in persuading localities to remit tax revenues to the center. The problem exists on a smaller scale in other countries. In Hungary, the pressures for decentralization initially made it difficult for the central authorities to require localities to adopt a common code for categorizing pharmaceuticals.

An additional problem is the potential for legislative overload. In part the problem is unavoidable, given the ambitious reform program (the Czech and Slovak legislators had to pass more than 250 substantial pieces of legislation in the first year of the transition to launch the first phase of economic and social reforms). A second source of overload, in part a result of the distrust of executive authority, is the inclusion of considerable detail in primary legislation. For example, at least in the early days of the reforms, the level of benefits such as pensions and unemployment benefits was frequently included in primary legislation, so that changes in benefits (because of inflation, for example) required extensive parliamentary debate. The combined effect of large numbers of new laws and of the amount of detail in each can be a legislative bottleneck, slowing down many reforms both within and outside the social sectors.

A final issue—the power of the authorities in a democracy—is more subtle and needs to be interpreted with care. The old system was based on imposed discipline. Democracy, because it relies on the consent of the population, is based largely on self-discipline. At least during the early transition, however, democracy was often interpreted as the right to do anything—a reaction to totalitarianism which confused democracy with anarchy. People were no longer terrified of the police: although this outcome was in many ways desirable, crime rose and, in the absence of incentive-based wage structures, disciplinary problems increased in the workplace.

Problems of governance are an inevitable consequence of the transition. The issue is *not* that democracy is a constraint on government: that is precisely the *purpose* of democracy. The problems being discussed here concern constraints on the ability of government to implement reform policies even where there is general support for those reforms.

THE DEMOCRATIC DEFICIT. The issue of governance concerns the need to have clear rules of the game. The so-called democratic deficit concerns the lack of experienced professional players of the game of democratic politics. The early transition was a revolutionary period, and many of the usual rules of politics could not apply. As discussed in chapter 13, a key political aim is to restore "normal" politics as soon as possible. Even then, the political problems of transition are acute. Politicians facing a fiscal crisis have to reduce expenditure. They can do so only if they enjoy sufficient political support, and the development and maintenance of such support require an ability to manage the political dimensions of policy, particularly where it creates a significant number of losers. These are major problems even for experienced Western politicians, so that the difficulties facing politicians in Central and Eastern Europe should not be surprising.

Virtually no politician in the early reform process had any practical experience with democratic politics. Many early ministers were former academics who did not intend to stay in politics. Few understood how Western politicians operate, in particular (a) how they use the printed and broadcast media to inform the public, to shape the political agenda, and to learn what the electorate thinks and (b) how they use polls, encounter groups, and similar devices to gauge opinion in different constituencies. In the West, such skills are possessed by successful politicians, by leaders of pressure groups (such as Friends of the Earth), and by the best public relations experts. The lack of political experience manifested itself in a number of ways: the absence of top-down communication, scarcity of feedback, lack of political accountability, and inconsistent time scales of economic reform and political pressures.

Top-down communication has been absent since the start of the reforms, when relatively little weight was given to information and persuasion. Under the old system, the government deliberately kept the people poorly informed and used information as a device for political control and manipulation. Many, but not all, post-reform politicians underestimated the critical importance of informing and persuading the electorate in a democracy. One of the key roles of politicians is to inform and shape public opinion. Implementation of good policy requires the government not only to get policy right but *also* to educate and persuade the electorate of that fact. The second leg is critical if policies are not to fail for lack of political support. Lack of publicity and information during the design and implementation of policy can create a serious political constraint by reminding people of the common practices of the old system.

The issue of communication during the early transition is controversial. One view (Balcerowicz 1993a) is that communication costs precious time and wastes short-lived political capital. That may be true during the early, revolutionary phase of the reforms. However, once that phase is over, if not earlier, communications become vital. Although the matter cannot be proved, it might be argued that one of the ingredients of the political problems which eventually faced the first Polish reform government was that it did not devote enough time to communicating with the electorate. In contrast, the Czech elections of 1992 were won by a party which had made a considerable and sustained effort to communicate with the citizenry. In the medium term, if not in the short term, the absence of communication constrains the reforms because government cannot gain sufficient support for the necessary policies.

Political communication in a democracy moves in two directions: from the top down and from the bottom up. The second aspect is just as important as the first. Feedback mechanisms in the early transition tended to be partial and rudimentary. At least one senior (and otherwise politically astute) minister expressed surprise when an adviser wanted relevant newspaper cuttings on his desk every morning. The issue is not just the *quantity* of feedback (a few years into the reforms, politicians in some countries complained that they were deluged with criticism from opposition parties, economists, television programs, and trade unions) but also its *completeness*. Ministers need to know the views not only of the intelligentsia (typically the group with access to the media), but of *all* groups which might vote for the party in power. The shortage of such broad-based feedback in the early transition meant that ministers often lacked information about which policies were acceptable and which were not and about where the major sources of support and opposition were. The problem is declining in countries which have had at least one full electoral cycle since the start of the reforms.

Political accountability is weak. Many government officials do not understand the concept of a political mandate; that is, they do not understand that they are accountable to the electorate for past promises. A manifestation of the problem is the way in which policy can change very sharply with a change of minister.

The problem of inconsistent timing arises because the time scale for economic reform is medium term, but all the pressures of democratic politics are short term. This creates a vulnerability to populism which is greater the more painful are the policies. Government was granted some respite at the start of the process, given the high political capital of the first reform administrations. Over time, though, the problem became acute, particularly because of the need for fiscal stringency, and especially in countries which had never experienced unemployment. The heart of the political process is to cope with this dilemma.

In part, these problems are diminishing. Many countries have had at least one election since the start of the transition and are building up their political experience. Countries which have initiated reforms relatively recently, however, would benefit from advice from Western political parties, trade unions, and pressure groups on the practical operation of democratic politics.

Institutional Constraints

Institutional constraints arise through inadequate legislation and law-making, through inappropriate attitudes and priorities within administration, and through deficiencies in physical and human capital.

The Legal Framework

Countries in Central and Eastern Europe inherited important gaps in their legal and regulatory framework. In addition, much existing legislation was inappropriate to the needs of a market economy. Legislators had little time to reform substantial parts of the legal order. Legal expertise was (and remains) in short supply, affecting the lawyers who draft the bills, the legislators who adopt them, and the administrators who implement them, many of whom lack experience both with the substance of the new laws and with democratic procedures. The shortage of legal skills arises because totalitarian systems, for obvious reasons, reduced education in law: lawyers are a low priority where decisions do not require a democratic legislature and where their implementation does not require judges. Thus many laws were passed by "students" of the reform who were learning on the job.

LACK OF THE CAPACITY TO DESIGN A LEGISLATIVE PROGRAM. At the start of the transition, the pressures for speed were immense. A health minister asked early in the reform process, "Can't you get these lunatics [his ministerial colleagues] off my back; they keep on saying 'You've been health minister now for six weeks, where is our health care reform?'" The speed of the legislative process led to gaps between laws, inconsistency between laws, and sometimes inconsistency within a single piece of legislation. Problems arose also because of faulty sequencing of the development of laws. Because of time pressures, legislators sometimes adopted the tactics of the old system, delegating crucial parts of the more difficult and politically controversial decisions to executive discretion and failing to develop adequate sanctions against fraud and abuse. For all these reasons, legislation often had to be revised soon after adoption.

LACK OF A CAPACITY TO IMPLEMENT LEGISLATION. Effective implementation requires both acceptance of and support for the new legislation. Without acceptance, implementation is reduced to the imposition of law through the use of superior power, one of the inherited problems which the reforms sought to remove. This problem is connected with the democratic deficit (particularly the importance of maintaining two-way political communication) and with the human capital deficiencies discussed below.

A common problem is that insufficient time is allowed for administrative needs. The existing institutions are expected to prepare the administrative rules for implementing new legislation, to establish the necessary procedures, and to revise their own organizational structures to match the needs of the new law. To accomplish this, however, administrators need to understand the new legislation, accept it, and have the capacity to implement it. Without sufficient time, implementation will fail to meet the intention of the legislators. Similarly, citizens may not have sufficient time or information to adjust. Much of the new legislation involves new concepts, new institutions, and new rules and procedures. Problems arise where people are not given sufficient time to understand what is being proposed, an example being the panic caused by introduction of the new medical insurance scheme in the Czech Republic: people knew they had to register but did not know how or where.

A final problem is the lack of sufficient judicial capacity for enforcement. Where there is no judge, there is no law. Most of the new laws increase the work load of the judicial system, a natural consequence of the transition from a totalitarian system, which creates little work for judges. However, there is no immediate prospect of a substantial increase in staffing (there are not enough lawyers), nor in the supporting personnel or the physical infrastructure of judicial institutions.

The Administrative Structure

The structure of administration can constrain change in important ways: enterprises will need to be relieved of some of their administrative responsibilities, there is resistance to change within parts of the administration, implementation skills are undervalued, and the concept of decentralization is insufficiently understood.

THE ADMINISTRATIVE ROLE OF ENTERPRISES. Administration under the old system had a hierarchy within the central structure based on two criteria: (1) the importance of the institution for protecting and strengthening the regime and (2) the size of its budget. The result, as explained in chapter 3, was the high priority given to the army and police, to the productive sector over the unproductive sector (mainly human resources), and to investment goods over consumption goods. These priorities were particularly evident during times of fiscal constraint, when the institutions which were administering the social sectors were even more starved of resources than usual. In consequence, the administration of parts of the education, health, and social welfare systems was transferred to enterprises. A resulting fundamental constraint on reform is the need to relieve enterprises of most of these social responsibilities at precisely the time when the public sector faces an acute fiscal crisis.

RESISTANCE TO CHANGE. The old system was characterized by top-down central direction, often with political interference in administrative matters. One of the results was a passive administration which was reluctant to change. Change was resisted, first, because it is painful. The needs of the transition include a reshaped administration to do away with the redundant administration of central planning and new administrative structures designed to assist the operation of the private sector. Such restructuring makes heavy demands on the skills of administrators and often forces them to be retrained or to leave. When the new agenda is entrusted to the old establishment, old practices are applied to the new tasks. The paradox is that generally no one else can carry them out. A second type of defensiveness arises from efforts to retain institutional monopoly in the face of economic liberalization. Asking state institutions to organize privatization, thereby dismantling large parts of the state apparatus, creates an obvious conflict of interest. In the social sectors, the reforms provoke administrative defensiveness by demonopolizing health care, education, and, to some extent, cash benefits.

UNDERVALUATION OF IMPLEMENTATION SKILLS. In part because of the political nature of significant numbers of administrative appointments, officials under the old system tended to be more concerned with carrying out orders and less concerned with the outcomes of their activities. The resulting lack of interest in outcomes persists. At least in the social sectors, there is little assessment of whether a particular policy is working or of whether it is being effectively administered. The result is a lack of information on which to base future policy or future administrative policy and a lack of administrative accountability. Policy is based on the assumption that previous policies have been effectively carried out, which may well not be the case.

Since there is little interest in outcomes, implementation skills are undervalued, and policymakers rarely involve individuals with implementation skills in the formation of social policy. The problem arises repeatedly. An anecdotal but illustrative example is the flag of one newly independent country, whose initial design was so complicated that it proved impossible to manufacture by machine and had to be amended. Similarly, the initial speaker in one newly elected Parliament had to be replaced when it became clear that someone with a severe hearing problem was not best-suited to the job (Radio Free Europe/Radio Liberty, "Chaotic Opening Session of State Duma," *Daily Report* 7, January 12, 1994). Of greater importance, the time scale for increasing the levels of complex cash benefits is often unrealistic, and health care reforms can be too complex to implement effectively within a reasonable length of time.[11]

LACK OF CAPACITY TO DECENTRALIZE. The needs of central political control under the old system meant that power was rarely devolved to lower levels of administration and that the behavior of lower-level officials was monitored centrally. Although central administration was often small, it was powerful. Planning remained a top-down affair, with local government little more than an arm of the central administration.

Although the central administrative structures were largely dismantled during the early transition, and the central planning mechanism largely abolished, the distinction between (a) devolving power to a lower level of government with its own elected legislature and executive authorities and (b) assigning tasks to lower-level administrative branches of central government was not always clear. The mentality of centralized decisionmaking took time to change. The winners of early elections sometimes reverted to old practices by trying to use central power to enforce the reforms from the top down. Such behavior was based on the mistaken idea that electoral victory gave government a blank check.

Although concentration of power may ease the adoption of policies, the top-down mentality creates continuing problems: it prevents the widespread discussion which assists the development of good policy, it hinders speedy decisionmaking at lower levels, and it reinforces the grass-roots pressure to decentralize. The resulting social isolation of the advocates of the reform program may lead to opposition not only to them personally but also to the whole idea of the reforms.

The problem is more general than an ability to decentralize. What is lacking is the ability to influence opinion in situations where decisions are to be made and implemented *without* the use of top-down power. Again, this problem is linked clearly with the earlier discussion of the democratic deficit.

Human and Physical Capital

The physical infrastructure in Central and Eastern Europe is obviously deficient. So, too, is the human capital infrastructure: skills are outdated, old ways of thinking are pervasive, and there are unrealized communication problems because new concepts need time to take root.

DEFICIENT PHYSICAL INFRASTRUCTURE. Buildings are in bad repair, which is a particularly important constraint in the health sector, transportation is generally old, and communications are a bottleneck. Telecommunications have insufficient capacity and are grossly ineffective, and computing capacity is totally inadequate to meet the needs of a modern economy.

OUTDATED SKILLS. Inappropriate human capital is pervasive. The problem is not primarily one of *too little* human capital but of the *wrong type*. Selection criteria under the old system were mainly political. The quickest route to an administrative post was to have worked in the party organization. This was true of all white-collar professions. Similar selection criteria determined admission to high school and university, especially to the most prestigious schools where future administrators were educated. Personnel structures therefore tended to have two strands: the politically motivated and politically well-connected senior staff and the loyal professional junior staff. The former needed the latter to implement orders without exposing the superior's lack of professional expertise. The problem arose even in professional institutions like hospitals, schools, and universities.

In part because of these selection criteria, outdated skills are a widespread problem. No significant training in modern management was available under the old system because, for at least two reasons, there was no intention to modernize administration. First, neither senior staff, who did not want to have better-qualified subordinates, nor junior staff, who had no financial or other incentive to upgrade their skills, were interested in modernization. Training abroad was generally restricted to technicians working with foreign equipment. Second, administration, which in the 1950s was considered bourgeois and alien to the working class, was deliberately downgraded. Much damage was done by purges of administrative staff and by wage policies favoring manual workers. Thus teachers and public officials had little or no contact outside the communist world. With a few exceptions in medicine, sciences, and some technical areas, their knowledge rapidly became out of date.

OLDTHINK. The problems of old ways of thinking are pervasive and will not be overcome without effective education and publicity. The first major problem is that important concepts are missing. As discussed in chapter 3, a consequence of communist ideology is the absence of any notion that scarce resources have an opportunity cost (that is, the cost of resources used for one purpose is the other uses to which they could have been put). Without such a concept, there is no recognition of the sort of tradeoffs discussed in chapter 1 and hence no possibility of dealing with them. Another conceptual gap in the old system is that under state ownership, no clear distinction was made between the property of the state and that of the enterprise, and hence there was no clear concept of taxation.

A second problem is that old priorities remain. The old economic system was based on the priority of communal interests over individual rights and on the vision of being perfect at all costs, even if lies were necessary to preserve that vision. These values became internalized in individual behavior: subordinates hide unpleasant news from their superiors, and this continued at least during the early transition.

Third, old fears create defensiveness. Under the old system, economic and political difficulties were "solved" by reorganization, which was used as a way of hiding failure. The cost of such reorganization was paid by the personnel of institutions, who were manipulated like inventory. Thus citizens fear reform and tend to distrust anything new; for example, while the managers of social insurance institutions seek to computerize their activities, their staff hampers the installation and operation of the new equipment.

COMMUNICATION PROBLEMS. Difficulties arise both from lack of training and from missing concepts. The lack of a common educational background, aggravated by the inability to communicate in Western languages, creates difficulties in communicating with Western experts. Training received abroad and training from foreign experts are thus often less effective than had been hoped. A more subtle manifestation of the same problem arises because missing concepts lead to missing words. This creates major translation problems, both for oral interpretation and for translation of written documents. The term cash benefits, for example, initially proved a consistent problem for interpreters, since there was little understanding of the Western distinction between wages, which are mainly related to individual productivity for reasons of economic efficiency, and income transfers, which are mostly paid out of the state budget for distributional reasons. These, largely unrecognized, communication problems act as a barrier to the speedy and effective transfer of information and know-how.

The major institutional constraint is the lack of appropriate skills, resulting from the procedures for selecting administrators, from the persistence of oldthink, from isolation from the Western professional world, from communication problems, from the lack of the necessary equipment, and from the lack of individual incentives either for increasing skills or for high-quality performance. Manifestations include a lack of knowledge of modern administrative theory and practice, a lack of technical know-how, and, frequently, also a hostility toward modern administrative techniques and technology. Administrative methods in Central and Eastern Europe have changed little since the end of the Second World War. They are generally cumbersome, unnecessarily complicated and slow, and undermined by distrust, with exaggerated (but ineffective) monitoring of abuse. In the early transition, administrative personnel in most of Central and Eastern Europe generally relied on pencil and paper; typewriters were often the highest technology available.

The combined effect of legal problems, of resistance to change, of undervalued implementation skills, of outdated administrative methods, and of missing skills has been to create major constraints in institutional capacity. Since much social sector reform raises complex issues of the mix between public and private, the problem is particularly acute in that area. The administration of cash benefits needs to be largely removed from enterprises. Health sector reform involves both new ways of organizing the supply of medical care and more complex forms of finance. An important part of the agenda for the reform of education is decentralization. In all these areas, the constraints on institutional capacity limit the speed and effectiveness, and hence the political viability, of the reforms.

Notes

1. In the early transition in most countries, the fall in output was fairly uniform across broad sectors, although it varied considerably at a more disaggregated level (see the discussion in chapter 3).

2. The Czech situation is open to different interpretations. On one view, it shows that economic transformation can occur in conditions of relative economic stability and with relatively low levels of unemployment. On another interpretation, such an outcome occurs only because the real pain of transition has been postponed. It is pointed out, for instance, that in Finland, which faced an external shock similar to that experienced by the Czech Republic, unemployment rose relatively quickly from 4 percent to more than 17 percent.

3. To illustrate the point with a simple example, suppose income tax is paid at 25 percent on income above $2,000. Someone with an income of $6,000 will therefore

pay $1,000 in tax. If, during a recession, his income falls to $4,000, his tax bill will be $500. Thus, if income falls by one-third, tax revenues will halve.

4. If output falls, there has to be a decline in present consumption, in investment (and hence, generally, in future consumption), or in both (see the discussion in chapter 1).

5. In Poland, for example, real wages as measured by official statistics fell so much at the start of the reforms that profits were maintained and hence tax revenues did not fall initially. In the following year, in part because of the success of the anti-inflationary policy, profits fell (see Schaffer 1993). Thus the fiscal crisis became a major problem only in the second year of the reforms.

6. In the West, the *principles* of eligibility are generally included in primary legislation; *the level of benefits* is generally set under regulations. The respective roles of primary and secondary legislation are discussed in chapter 13.

7. Poor occupational safety led to high expenditure on sick pay and disability pensions and, more generally, to poor health (see chapters 9 and 12).

8. More anecdotally, when a minor scuffle broke out between two parliamentarians in the Russian Duma in early 1994, one of the participants was quoted as saying that, should he become president, his first act would be to imprison the other.

9. As an example, "Russian Economics Minister Andrei Nechaev told Interfax . . . that the government has ordered the Central Bank to reverse its decision canceling debts owed by enterprises. But in *Izvestiya*, Central Bank Chairman Viktor Gerashchenko defended his controversial instructions. . . . He conceded that the debt forgiveness had disadvantages 'but what other alternative do we have?' . . . The Central Bank answers to Parliament, and it is not clear what the government, which is split on this issue, can or will do if the bank holds firm." (Radio Free Europe/Radio Liberty, *Daily Report* 148, August 5, 1992). In Ukraine, similarly, the prime minister "accused . . . Parliament of interfering in the cabinet's work and said that if a clear delineation of powers between the government and Parliament was not established, his government would resign" (*Wall Street Journal, Europe*, July 6, 1992, p. 7).

10. To deal with this problem, Poland introduced a threshold in 1993 of 5 percent of total votes for single parties and 8 percent for coalitions. The purpose of the threshold was to reduce the number of political parties in Parliament.

11. The Russian Supreme Soviet announced at the end of the first week of April 1992 that all pensions would be revalued as of May 1, 1992. The revaluation involved recalculating, without the help of computers, the pension of each individual, based on his or her earnings record over the previous ten years indexed by a complex formula. In addition, some pensioners, by submitting additional documentary evidence, could count as pensionable service additional periods which had previously been ineligible. Those in charge of implementing the decision estimated that the task would realistically take at least six months.

Part II

POLICY DESIGN
AND IMPLEMENTATION

Six

A STRATEGY FOR REFORM

NICHOLAS BARR

THIS CHAPTER LINKS THE DISCUSSION of Part I with the rest of the book. It draws on the analysis of the previous four chapters to set out a strategy which shapes the recommendations for policy and for implementation in the remaining chapters. To encourage economic growth and increase individual freedom, many countries included economic and political liberalization in their early reforms. Those actions, the driving forces they unleashed, and their major outcomes were the subject of chapter 4. Some outcomes, such as greater efficiency and greater responsiveness of government to the electorate, are major purposes of the reforms. Other less-beneficial outcomes become constraints and were discussed in chapter 5.

The following strategy seeks to harness the beneficial outcomes of the driving forces while taking account of the constraints. A fundamental purpose of the transition has been to dismantle the apparatus of central planning and to move toward a market system. As a practical matter, however, such a move was associated during the early transition with a decline in output and mostly also with a fiscal crisis. The policy strategy which follows is connected in a very direct way with the effects of the output decline, on the one hand, and the move toward a market system, on the other. It seeks to (a) maintain macroeconomic balance. Economic liberalization involves policies both (b) to build markets and (c) to regulate markets (the latter to address market imperfections and for equity reasons). Achieving these aims requires the capacity (d) to implement the reforms, taking account not only of fiscal and other economic constraints, but also of the constraints on political and institutional capacity discussed in chapter 5. Maintaining macroeconomic balance concerns stabilization; liberalizing the economy relates directly to restructuring.[1]

Maintaining Macroeconomic Balance

Maintaining macroeconomic balance, which involves both monetary and fiscal policy (see the glossary), is especially difficult during a time of declining output. Monetary policy, which, for the most part, lies outside the scope of this book, includes control of the money supply and regulation of the banking system in the interests of monetary stability and consumer protection. Fiscal policy has two prime concerns. It is concerned first with the balance of government revenue and expenditure. Attention should focus particularly on those areas which could yield substantial savings, rather than on changes which yield only marginal savings. A major purpose of this aspect of fiscal policy is to assist stabilization. The second concern of fiscal policy is to contain the size of the public sector, both to allow private sector growth and— by facilitating private investment—to assist economic growth.

Mobilizing Resources

Policies to mobilize resources include improving the system for collecting taxes and developing ways to increase the participation of the private sector. Often the way to accomplish this is to privatize, but even in areas where the state maintains a major role, it is possible to mobilize private resources. Charges can be levied in some areas, such as copayments for some types of medical care, and incentives can be used, mainly through the tax system, to encourage certain types of private activity.

Containing Expenditure

Containing expenditure a direct response to the fiscal crisis. The starting point is the notion of targeting, which has two aspects:

- Horizontal efficiency implies that benefits should go to *all* who need them, which means that there should be no significant gaps in coverage. This is mainly a matter of poverty relief.
- Vertical efficiency has two implications. First, benefits should go *only* (or largely) to those who need them (there should be no excessive leakage of benefits). In the context of poverty relief, this implies that benefits should be aimed at the poor rather than the entire population. Second, in the contexts of health care or education and training, treatment or education should be relevant and effective for the individual concerned.

Cost containment thus involves policies to increase vertical efficiency and arrangements to discourage exploding costs, important examples being pensions and health finance. Cost containment is particularly necessary during a time of fiscal difficulty—all the more because of the political and administrative constraints on raising revenue—to prevent taxes from being so high that they create significant incentives against work effort. The role of competitive labor markets is important in this context: wages should contribute as little as possible to inflationary pressures, and payroll taxes should not create incentives which hinder additional employment. Competitive labor markets also contribute more broadly to a country's macroeconomic health. They assist the performance of exports and increase the capacity of the economy to adjust to changing external conditions.

Building Markets

Building markets, the major vehicle for pursuing economic growth, involves the liberalization of (a) demand, (b) supply, and (c) prices and wages.

Increasing Consumer Choice

Increasing consumer choice is important for political reasons, as an aspect of individual freedom, and also for reasons of efficiency. Greater choice leads to changes resulting from the pursuit of personal satisfaction, in furtherance of cultural identity, and from a desire to move away from the old system. In education, for instance, consumers exert pressure for new textbooks. Consumer choice also arises for income transfers, as the range of voluntary private savings and insurance options increases, and health care, as the scope for individual choice expands and more and better consumer information becomes available.

Employers also make choices. Indeed, employment-driven changes are a central part of the reforms. In a system with hard budget constraints and wages determined largely by market forces, managers take increasing account of labor productivity in their employment decisions. Changes in consumer demand and in derived employer demand for education, training, and retraining therefore lead to major changes in the roles of the recipients of education and training, service providers, government, and employers. In particular, increased emphasis is placed on the relevance of training.

Diversifying Supply

Diversifying supply also contributes to individual freedom and to efficiency. Large sections of the economy will be returned to the private sector. Examples in later chapters include the phased introduction of private pensions and sick pay schemes, allowing different types of hospitals and primary health care and allowing both public and private providers of education, training, and employment services. More generally, such liberalization allows a more effective mix of public and private services.

Improving Incentives

Improving incentives is a central component of restructuring and also a response to the fiscal crisis. Prices should be efficient. In cases like food and clothing, this means that consumers should generally pay market prices. As discussed in later chapters, it does not imply that patients should necessarily have to pay the full cost of drugs, nor that schoolchildren should pay market prices for schoolbooks. It does, however, mean that providers of medical services and managers of educational institutions should be aware of the full cost of these commodities, in the interests of efficiency and also because of the fiscal crisis. A second aspect of incentives is to have an appropriate structure of income transfers, such as a system of unemployment benefits which gives workers the incentive to rejoin the work force promptly.

Employers, too, need to face efficient prices. Thus the cost of family support, such as maternity leave and child care facilities, as well as the costs of administering it, should generally be borne not by the employer but by the taxpayer. Employment costs should be neutral with respect to employing men and women; imposing on employers the costs of family support gives a significant incentive against employing women.

Raising Labor Productivity

Raising labor productivity is a response to the inheritance of low growth in output and of misallocation. Policy has two aspects: workers' existing skills have to be put to more effective use, and their skills have to be increased. An implication of the first is that labor needs to be more mobile, both in terms of acquiring new skills and, to some extent, of moving to where their skills are in demand. Wages more closely related to market forces contribute to this objective; so does better job information. Effective redeployment of labor also includes policies which minimize impediments to labor mobility, through appropriate labor legislation and through broader policies such as improved operation of the housing market.

So far as increasing the skills of workers is concerned, education, training, retraining, and other forms of investment in human capital have major roles to play in imparting flexible skills which allow workers to change jobs more easily; and wages, again, give individuals the incentive to upgrade their skills.

Regulating Market Forces

Regulating market forces is the counterpart to building markets. Market forces, introduced for their efficiency advantages, need to be contained both to address market imperfections and to come to grips with important equity issues.

Introducing a Regulatory Structure

Regulation is central to the effective operation of a market system for the reasons set out in chapter 2. At its most basic, the regulatory structure needs to include legislation governing property rights, the conduct of private corporations, and the operation of institutions such as the banking system and financial markets which are essential to a modern market economy. Regulation is also necessary to address important market imperfections, particularly for consumer protection in complex areas such as financial markets and medical care, where individuals are typically imperfectly informed.

Policies to Address Unemployment

Two sorts of policies are involved in coping with unemployment. Income support, for instance in the form of unemployment benefits, is discussed in chapter 9. Active labor market policies, discussed in chapter 8, include services providing information designed mainly to help workers to move more quickly to a new job, investment in human capital through training or retraining, and possibly also short-term job creation.

Ensuring Access to Basic Goods and Services

Two separate types of activity are involved in guaranteeing access to basic goods and services. The first, strengthening poverty relief, is a response to the widening distribution of income. An implication of targeting is that, so far as possible, benefits should go to *all* who need them. Thus benefits should ideally go to all the poor. This involves improving and expanding unemployment benefits and poverty relief.

A second set of policies seeks to provide access to health care, education, and similar services. This involves ensuring the availability of commodities such as critical drugs and textbooks and designing policies such as systems of health finance which do not exclude uninsurable risks. It also includes any necessary redistribution by central government to assist poorer parts of the country.

The joint pursuit of poverty relief and cost containment at a time of fiscal crisis means that well-designed targeting is more than usually important. Targeting requires identification of the poor, which can in principle be achieved in two ways. It is possible, first, to look at the *income* or *consumption* of individuals and families. This approach, however, faces significant difficulties: it runs into problems of poverty measurement, is administratively costly, and, particularly if based on family rather than individual income, is often intrusive and liable to cause political resentment. In addition, means testing is likely to be ineffective for individuals with a significant amount of income in kind, such as home-grown food.

An alternative approach is to look for *indicators* of poverty, that is, the possession by the individual or family of one or more characteristics which are highly correlated with poverty. Examples of such characteristics are ill health, old age, and the presence of children in the family. The advantage of this approach is that many characteristics are easily observable, which facilitates administration. The disadvantage is that the indicator may be correlated only imperfectly with poverty. This leads to either or both of two targeting errors: gaps in coverage or leakage of benefits.

Determining the extent to which policy should seek to prevent leakage of benefits (the pursuit of vertical efficiency) is a complex matter. Ignoring the problem leads to increased expenditure, violating fiscal constraints. Targeting too strictly, however, produces gaps in coverage. In addition, middle-class participation in at least some benefits helps to maintain political support for the relevant programs (World Bank 1990b, p. 92). A possible line of attack is to devise schemes based on self-targeting. The idea is to construct benefits in such a way that only the genuinely poor will apply for them. An example in the famine literature is to offer work at a subsistence wage, for which only the genuinely indigent will apply (Ravallion 1991). The issue of targeted poverty relief is one of the main topics of chapter 10.

Providing an Enabling Environment for Insurance and Income Smoothing

Alongside poverty relief, income transfers also seek to enable individuals both to insure themselves against the risk of income loss (for instance, because of unemployment) and to reallocate consumption over their life cycle. In principle, these could be done through private insurance, personal savings, and private pensions. However, the private market cannot supply insurance against some important risks, giving the state a role in ensuring that protection against such risks is available, and capital market imperfections and other problems mean that state intervention is required for pensions at least to regulate activity in the private sector.

Implementing the Reforms

Maintaining macroeconomic balance, building markets, and regulating markets all relate to policy design. Effective reform, however, also requires administrative capacity and the ability to manage the political dimension of reform. All three sets of skills—policy analysis, administrative expertise, and political experience—are needed simultaneously.

Building Institutional Capacity

Policies which build institutional capacity are a response to the constraints discussed in chapter 5 and also a necessary element of the design and implementation of the regulatory structures just discussed. The problems are acute. First, the task of central administration, though much smaller, becomes more difficult because complex new systems are needed to provide cash benefits, health care, education, and training in a period of constrained resources. Second, institutional capacity is low: skills in short supply include those involved in designing a legislative program and in drafting individual laws. Policymakers are not familiar with basic economic concepts such as opportunity cost and the importance of incentives and often do not understand how markets work, including the need for an effective regulatory structure. Low pay and poor morale have driven many of the best people out of public service.

To cope with these difficulties, administrative needs should be given greater weight:

- Policy design should be as simple as possible.
- Individuals with implementation skills should be involved from the start of the process of designing policy and drafting legislation and regulations.
- Greater emphasis should be placed on monitoring the outcomes of policies.

Building Political Capacity

Increased political skills are necessary to enhance the ability of democratic political systems to meet the wishes of the electorate, while simultaneously putting into place policies which assist the process of reform. These include implementing cost-containing policies during the fiscal crisis, coping with the political implications of rising unemployment, and accommodating disappointed expectations that living standards would increase rapidly. To be effective, policies should have two important components: top-down communications and feedback. Top-down communication needs to be improved so that politicians can be more effective in persuading the electorate of the correctness of their policies and can, where necessary, defuse unrealistic expectations. Of equal importance, better feedback is vital to inform politicians about what the electorate wants and about the main sources of support for and opposition to their policy proposals.

The strategy thus involves maintaining macroeconomic balance, building markets, regulating markets, and implementing the reforms. Their detailed application has twelve elements:

- Mobilizing resources
- Containing expenditure
- Increasing consumer choice
- Diversifying supply
- Improving incentives
- Raising labor productivity
- Introducing a regulatory structure
- Addressing unemployment
- Ensuring access to basic goods and services
- Providing an enabling environment for insurance and income smoothing
- Building institutional capacity
- Building political capacity.

These twelve elements have their roots in the decisions already made and in the nature of the transition itself. They form a mosaic which emerges in different patterns in each of the following chapters.

Note

1. Balcerowicz (1993a) argues that the transition has three components: stabilization, mainly through macroeconomic policy; liberalization (for example, of ownership options and prices and wages) to allow an effective market mechanism; and institutional restructuring. These three aspects broadly correspond to the reform strategy presented here.

Seven

LABOR MARKETS: WAGES AND EMPLOYMENT

RICHARD JACKMAN • MICHAL RUTKOWSKI

THE MOVE TO A MARKET SYSTEM IS CENTRAL TO THE REFORMS in Central and Eastern Europe. This is as true in labor markets as in other parts of the economy. Effective labor markets are essential to assist the movement of workers into jobs where they are most productive. This chapter discusses (a) methods of setting wages which give appropriate incentives to work effort and labor mobility while minimizing inflationary pressures and (b) ways of maintaining employment which nevertheless facilitate the movement of workers into more productive jobs.

The main message is that free enterprise labor markets need the economic infrastructure of a market economy—private ownership, competition, capital markets, labor mobility—if they are to function efficiently. The priority for policy is to put this infrastructure into place; until this can be done, governments retain important responsibilities for limiting wage inflation and for containing the growth in unemployment. Such interventionist policies should be phased out over the medium term as the growth of markets and the elimination of constraints on labor mobility remove the need for them.

Even where labor markets work well, however, unemployment remains a problem. In Central and Eastern Europe, at least in the short run, the problem is likely to be particularly acute. Unemployment is the subject of chapter 8, which addresses the failures of labor markets and recommends policies to assist labor mobility and to reduce the extent and duration of unemployment.

The Inheritance

There is a widespread image of labor allocation in countries before the transition as having been similar to the planned allocation of energy, raw materials, or goods. Workers are often assumed to have been allocated by planners to a job, enterprise, or region and obliged to stay there. This image is rooted in the Stalinist period, when workers were forbidden to quit their jobs, and specialists and skilled workers could be forcibly transferred from one enterprise to another. After the end of the Stalinist period in the mid-1950s, however, the principles of setting employment and, to a lesser extent, wages changed in many respects, becoming more like those of market economies. The labor market became more of a market than other markets. Except for a few jobs, which continued to be centrally allocated, workers were free to choose their job, skill, or profession, as well as the region where they worked, and were free to quit their jobs. Even in the U.S.S.R., where restrictions remained more widespread, there is little evidence of effective compulsion either on an employer to accept a particular worker or on a worker to accept a particular job. Employers were legally permitted to create redundancies, and, although state employment services had existed in Poland since the late 1950s, in the U.S.S.R. since the 1960s, and in many other countries, they did not necessarily exercise a monopoly over the allocation of labor (for details of the former Soviet system, see Marnie 1992).

Restrictions on wages were greater. Some countries, such as Bulgaria, Czechoslovakia, and the U.S.S.R., developed a system based on "scientific" wage differentials, grading all jobs according to output levels, hours worked, regional coefficients, and so forth. Other countries, such as Hungary and Poland, replaced a wage system based on a central tariff structure with upper and lower tariff limits negotiated by the government and trade unions at the industry level. The individual wage structure remained complicated, involving bonuses, premiums, and regional and special coefficients (for example, for arduous work) paid on top of the basic wage. When the system was unable to contain wage pressure, various types of incomes policy, including a tax-based incomes policy, were applied.

To avoid confusion between the different periods, it is useful to distinguish the traditional (early or Stalinist) labor market and the reformed market (late or post-Stalinist). Five general features of the early labor market were substantially reformed after the mid-1950s:

• *General admonitions to work.* Workers not only were guaranteed the right to work but had a positive obligation to do so (staying outside the labor force was considered "parasitism"). For that reason, the countries of Central and Eastern Europe did not have a system of unemployment benefits. The principle was "he who does not work, neither shall he eat." Absenteeism and tardiness were often subject to criminal sanctions.

• *Mandatory assignment of graduates.* In some countries, the state had a right to assign the graduates of technical schools and universities to a specific location for a few years. The state's right to assign students geographically (for instance, to remote areas) was viewed as a means by which students could repay society for their free education. Noncompliance was always high, however, even in the U.S.S.R., where the system was harshest.

• *Forced labor.* The U.S.S.R.—and also some other countries—had a record of using concentration camp labor for construction, mining, and forestry. The economic benefits of forced labor may have been significant during the initial drive to collectivization and industrialization.

• *Mobilization campaigns.* Some organizations, in particular mass youth organizations, frequently engaged in the semi-voluntary recruitment of workers and students to carry out special tasks, such as construction or agricultural projects, or to help collect the harvest. These arrangements substituted for wage differentials when labor had to be mobilized rapidly.

• *Strict control of migration to urban areas.* A system of internal passports or permanent resident permits was introduced. Initially it was applied widely, but later only in some cities. The major objective was to restrain the migration of peasants from rural to urban areas.

These rules were all lifted or substantially relaxed in the 1970s and 1980s, though at different paces in different countries, Hungary and Poland being ahead of the others. Thus the labor market just prior to the transition was very different from the traditional one; it was, in a sense, eclectic, because it combined features adopted from different origins. Below we review the major aspects of the late, post-Stalinist system.

Strengths

The old system had significant strengths: it offered job security, essential benefits were guaranteed through employment, workers were involved in enterprise matters, and labor force participation was high.

First, by the post-Stalinist period, workers in most countries had job security and were free to leave one job in favor of another and to relocate geographically. Although their mobility was constrained somewhat by housing and endemic administrative restrictions, they could move and were more or less guaranteed employment and job security. The combination of these three possibilities was hardly achievable in Western countries. Contrary to common belief, labor mobility in Eastern European countries and the U.S.S.R., measured in turnover rates and number of moves between regions, was not much below that of Western Europe (see tables 7-1 and 7-2; Marnie 1992).[1]

Second, essential social benefits were guaranteed through employment. In a sense, everybody was a civil servant, knowing that within a substantial range his or her benefits were independent of performance. All family members, whether working or not, were automatically covered. The system was broadranging and did not require active pursuit of security, for example, by buying insurance. Everyone enjoyed a feeling of stability and certainty.

Third, workers were more involved in enterprise matters than is commonly realized. Almost everybody employed in a factory—assembly line worker, janitor, engineer, and supervisor—belonged to the same national trade union organized on an industrial basis. Union membership, though not compulsory, was very high because it gave preferential access to some social benefits. During an initial period based on a Taylorist approach to work organization, workers were expected to perform very narrowly defined functions.[2] Once this period had passed and workers' councils and unions were more active, however, employees became much more involved in the enterprise, especially in Hungary and Poland. This, however, was only partially successful in

Table 7-1. *Employee Turnover in Various Countries and Years, 1987–90*
(percentage of total employment)

| Region and country | Hired | | Quit | | Reason for leaving | | | | |
	All	Industry	All	Industry	Contract ended	Resigned	Dismissed	Transferred	Other
Central and Eastern Europe									
Czechoslovakia, 1989a	19.1	19.5	19.6	19.4	—	9.1	—	0.7	9.8
Hungary, 1989	20.2	—	22.6	—	1.3	16.3	0.5	1.9	2.6
Poland, 1989b	16.2	15.8	19.8	19.3	0.7	12.5	—	0.3	4.6
Romania, 1989c	—	—	—	10.5	0.5	5.5	4.5	—	—
Western Europe									
France, 1990d	32.2	22.3	31.8	22.9	15.9	8.0	2.7	1.7	3.5
Italy, 1987e	—	5.4	—	8.0	—	—	—	—	—
United Kingdom, 1988f	—	23.2	—	22.6	—	—	—	—	—
Asia									
Japan, 1988	18.9	15.5	17.7	15.0	0.9	11.7	0.7	3.4	1.0

a. Excluding farming cooperatives and, for industry, including manual workers only.
b. Full-time workers in the state sector only; reasons for leaving are for 1988.
c. The categories "transferred" and "other" are included in resigned.
d. Establishments with 50 or more employees.
e. Establishments with 500 or more employees.
f. Calculated as twelve times the four quarter moving averages for August 1988.

Source: Boeri and Keese 1992.

Table 7-2. *Internal Migration in Various Countries and Years, 1987–90*
(percentage of total population)

Region, country, and year	Internal migration	Population (millions of persons)[a]
Central and Eastern Europe		
Czechoslovakia, 1989	2.4	7.6
Hungary, 1989	1.9	10.6
Poland, 1990	1.4	38.0
Western Europe		
Germany, Fed. Rep. of, 1987	1.1	61.1
Italy, 1987	0.5	57.3
Netherlands, 1989	1.7	14.8
Spain, 1987	0.5	57.3
Sweden, 1990	2.1	8.5
North America		
United States, 1988	2.7	237.4

Note: Internal migration includes migration across twelve regions for Czechoslovakia, twenty regions for Hungary, forty-nine counties for Poland, eleven *Länder* for Germany, twenty regions for Italy, twelve provinces for the Netherlands, twenty-two *comunidades* for Spain, twenty-four counties for Sweden, and fifty states for the United States.

a. At the beginning of the year, except Spain, which is mid-year.

Source: Boeri and Keese 1992.

increasing the voice of workers because managers and sector ministries resisted the reforms. The major consequence of workers' councils and union activity in practice was the rising number of industrial conflicts. At the same time, workers could increasingly influence their employers by threatening to leave the enterprise.

Fourth, labor force participation was high. Countries in Central and Eastern Europe achieved a higher overall rate of participation than the highly indus-trialized countries primarily as a result of the greater economic activity of women, as shown by table 7-3 for women forty to forty-four years old, who are mostly past their peak childrearing age but still below retirement. The high participation rates of females reflected the abundance of jobs created by rapid industrialization and, more recently, the demographic imbalance between men and women as a result of the Second World War. As noted in chapter 3, economic growth in Central and Eastern Europe typically was spurred by the extensive, and only to a lesser extent the intensive, use of inputs, including labor. High participation rates were associated with greater equality for women in education and training and, consequently, with better career prospects for women in many sectors than exist in a number of market economies. By enhancing the financial independence of women, labor force participation helped to raise the social and political status of women. Child

Table 7-3. *Rates of Participation in the Labor Force of Women Forty to Forty-four Years of Age in Various Countries, 1950–85*
(percent)

Region and country	1950	1960	1970	1980	1985
Central and Eastern Europe					
Bulgaria	78.6	83.4	88.5	92.5	93.3
Czechoslovakia	52.3	67.3	79.9	91.3	92.4
German Democratic Republic	61.9	72.7	79.1	83.6	86.1
Hungary	29.0	51.8	69.4	83.2	84.7
Poland	66.4	69.1	79.5	83.2	84.7
Romania	75.8	76.4	79.5	83.1	85.1
U.S.S.R.	66.8	77.9	93.2	96.9	96.8
Northern Europe	30.9	39.9	53.8	69.9	71.1
Western Europe	34.5	39.5	46.4	55.1	55.6
Southern Europe	22.4	25.3	29.7	35.7	37.1

Note: The countries covered are Austria, Belgium, France, Federal Republic of Germany, the Netherlands, Switzerland, Luxembourg, United Kingdom (Western Europe); Greece, Italy, Malta, Portugal, Spain (Southern Europe); Bulgaria, Czechoslovakia, German Democratic Republic, Hungary, Poland, Romania, U.S.S.R. (Central and Eastern Europe); and the Scandinavian countries (Northern Europe). Regional averages are unweighted. The classification of countries by regions in this table is different from the International Labour Organisation's original classification.
Source: Kornai 1992, p. 207.

care facilities and generous maternity leave allowed women to stay in the labor force even when they spent significant amounts of time at home. On the darker side, many women may have been forced to work by social pressures or low incomes. High participation may, in the long run, have contributed to lower birth rates (Kuniansky 1983). Given all the other burdens placed on women, their productivity in work may often have been low (Kornai 1992). As figure 7-1 shows, participation rates were higher in Central and Eastern Europe at each level of income than in capitalist countries.

Weaknesses

Alongside these strengths were major weaknesses (sometimes the opposite side of the same coin): labor productivity was low; wages bore no relation to productivity, with a variety of consequential ill effects; the approach to unemployment was deficient; the system of wage bargaining contributed to inflationary pressures; employment was excessively concentrated in industry and agriculture; and international migration was very limited.

LOW LABOR PRODUCTIVITY. Low labor productivity was in part a consequence of labor hoarding. Although there are major problems of measurement, particularly when trying to adjust the value of inputs and outputs for differences in

Figure 7-1. *Rates of Participation in the Labor Force and Level of Development, Selected Countries, 1980*

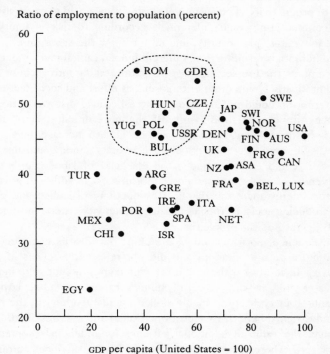

Ratio of employment to population (percent)

GDP per capita (United States = 100)

Note: In ascending order of GDP per capita, the names of countries given above in codes: EGY, Egypt; TUR, Turkey; MEX, Mexico; CHI, Chile; ROM, Romania; YUG, Yugoslavia; ARG, Argentina; POR, Portugal; GRE, Greece; POL, Poland; BUL, Bulgaria; HUN, Hungary; ISR, Israel; USSR, U.S.S.R.; SPA, Spain; IRE, Ireland; CZE, Czechoslovakia; ITA, Italy; GDR, German Democratic Republic; JAP, Japan; UK, United Kingdom; NZ, New Zealand; DEN, Denmark; ASA, Austria; NET, Netherlands; FRA, France; SWI, Switzerland; NOR, Norway; BEL, Belgium; LUX, Luxembourg; FRG, Federal Republic of Germany before reunification; FIN, Finland; SWE, Sweden; AUS, Australia; CAN, Canada; and USA, United States.

All the formerly socialist countries are in the ringed area above the capitalist countries and are at a comparable level of development. In order of GDP, the socialist countries included are Romania, former Yugoslavia, Poland, Bulgaria, Hungary, U.S.S.R., and former German Democratic Republic.

Source: Kornai 1992, fig. 10.1, p. 209. Used by permission.

quality across countries, studies comparing estimates of per capita gross domestic product (GDP) suggest that for the mid-1980s labor productivity in Central and Eastern Europe was about one-third of that in a middle-income country in the Organization for Economic Cooperation and Development (Boeri and Keese 1992).

WAGES BEARING NO RELATION TO PRODUCTIVITY. The second, and arguably the key, adverse inheritance affecting the labor market was that wages bore no relation to productivity. The system of setting wages paid considerable attention to compensating wage differentials according to risk of injury, fringe benefits, job status, job security, and so forth. At the same time, it did *not* capture differentials resulting from different human capital investment.

Governments tried to set the wage structure, taking into account notional marginal productivity and other criteria such as effort and social factors. Even the first criterion could not be implemented because distorted prices made productivity impossible to measure. If the government (sector ministry) noticed a shortage of, say, carpenters and decided to increase their wage rate (which was far from easy because of collective agreements and institutional rigidities), it could easily misjudge the elasticity of labor supply and the responsiveness of workers to training opportunities. In addition, government tended to underestimate the marginal productivity of skilled and educated labor. Introducing social factors alongside productivity criteria in wage setting obviously aggravated the problem.

In consequence, notwithstanding the effort put into its construction, the wage structure became random and did not reflect differences in skills or productivity. In contrast with the West, high incomes did not go to individuals with greater skill in the sense of human capital—doctors, lawyers, or accountants—but typically to highly skilled manual workers in the material, largely industrial, sphere. If two jobs, one skilled and one unskilled, have the same wage, the value of additional earnings for individuals contemplating training is zero, whereas the costs are positive (books, sacrificed earnings, and so forth). Thus workers had no incentive to invest in education. Governments tried to counteract such distortions by making education free, but this, in turn, meant that arbitrary wage differentials or other factors could lead to excessive training in some areas. The data presented in table 7-4 and figures 7-2 and 7-3 show that the spread of earnings between different high-paying jobs was much wider in the West than in Central and Eastern Europe. With the exception of Hungary, the overall differentials between skilled and unskilled wages were not that large in Central and Eastern Europe. Wage dispersion across industrial sectors also appears to have been lower in Central and Eastern Europe than in the West. The International Labour Organisation's *Year Book of Labour Statistics* (various years) shows wage dispersions in Czechoslovakia, Poland, and the U.S.S.R. which were lower than in any market economy except Sweden. Rather surprisingly, it also shows a significantly higher dispersion in Hungary. But the key element lacking in the wage structure of Central and Eastern Europe was its failure to take account of investment in human capital.

The absence of any connection between wages and human capital had further ill effects. First, the system did not encourage workers to acquire general skills and produced many narrow specialists with nontransferable

Table 7-4. *Wage Relativities, by Occupation and Gender in Various Countries and Years, 1988–90*
(percent)

Region and country	Nonmanual-to-manual workers	Female-to-male workers All	Manual
Central and Eastern Europe			
Bulgaria, 1990	107	74	—
Czechoslovakia, 1990	111	65[a]	—[a]
Hungary, 1989	162	64	65
Poland, 1990	119	76	—
Romania, 1989	118	—	—
Western Europe			
Belgium, 1988	175[b]	—	75
Germany, Fed. Rep. of, 1989	147	—	69
United Kingdom, 1990	140	60	60

Note: For all countries, wages refer to average monthly earnings except average annual earnings for Germany and average weekly earnings for full-time employees for the United Kingdom.

a. For 1988.

b. Estimated from data on hourly wages and average weekly hours of manual workers and average monthly earnings of nonmanual workers.

Source: Boeri and Keese 1992.

skills. A much higher proportion of secondary students attended vocational schools in Central and Eastern Europe, though probably not in the U.S.S.R., than in the West, even compared with, for example, the western part of Germany with its traditional emphasis on vocational training (see table 7-5). Moreover, by the late 1980s the overall share of skilled and educated workers in the labor force was substantially lower than in the West, as were enrollment ratios, which had been at the same level as in the West in the mid-1960s (see tables 7-6 and 7-7). The structure of skills was different, too. The lack of appropriate wage differentials hindered the movement of labor toward the services sector. Decisions about investing in human capital were detached from the real costs of acquiring skills. Free education apparently did not contribute to the social advancement of the rural or poorer population.

Another ill effect was that there was inadequate variation in wages both by skill and by region. One result was that job changes frequently did not improve the allocation of labor. If wages did not reward skills sufficiently, it might, for example, be rational for a skilled worker to move from a job where his or her skill was required to one which required no particular skill. Several studies point to a low dispersion of wages and stability over time (see, for example, Boeri and Keese 1992). The structure of wages with respect to educational qualifications was also compressed, although not as much as com-

(Text continues on page 133.)

Figure 7-2. *Relative Earnings by Occupation, Selected Industries and Countries, 1990*
(earnings of laborers in the chemical industry = 100)

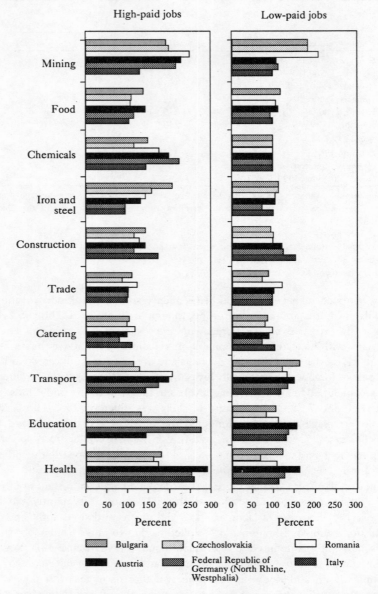

Source: Boeri and Keese 1992, chart 1, which used data from International Labour Organisation 1991.

Figure 7-3. *Occupational Earnings Differentials, Selected Industries and Countries, 1990*

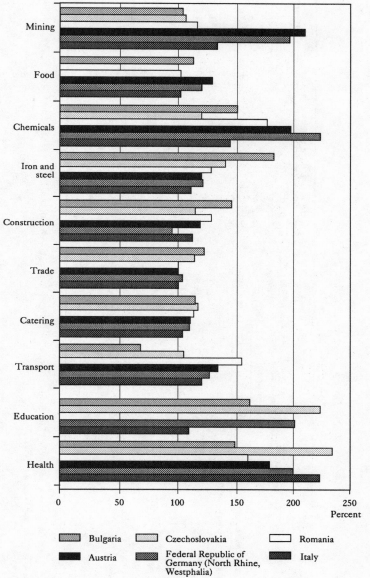

Note: Differentials measure earnings in high-paid jobs relative to low-paid jobs.

Source: Boeri and Keese 1992, chart 1, which used data from International Labour Organisation 1991.

Table 7-5.   *Type of Education Prevalent in Various Countries and Years, 1987–89*
(percent)

Region and country	Secondary enrollment, 1989		Higher education graduates [b]				
	Vocational [a]	General	Engineering, industrial trades, and architecture	Natural sciences and mathematics	Medicine	Education	Other
Central and Eastern Europe							
Bulgaria	60.3	39.7	28.1	4.7	7.5	29.4	30.3
Czechoslovakia	53.8	46.2	38.9	2.6	6.6	17.8	34.0
Hungary	76.1	23.9	21.7	4.2	7.3	35.3	31.5
Poland	77.4	22.6	16.9	2.9	15.6	29.5	35.1
Western Europe							
Austria	28.5	71.5	12.5	8.5	14.3	14.4	50.3
Germany, Fed. Rep. of	36.3	63.7	20.9	7.1	24.3	7.0	40.6
Italy	40.6	59.4	11.5	9.6	25.8	2.8	50.3
Netherlands	44.3	55.7	16.4	4.1	11.7	14.5	53.3
Sweden	35.6	64.4	29.5	5.5	23.1	16.4	25.5
United Kingdom	9.8	90.2	14.3	13.9	16.8	8.3	46.8

a. Includes teacher training.
b. For Germany, Netherlands, and United Kingdom, figures are for 1987; for Austria, Czechoslovakia, Italy, and Poland, figures are for 1988; and for Bulgaria, Hungary, and Sweden, figures are for 1989.
Source: Boeri and Keese 1992.

Table 7-6. *Level of Education of the Labor Force in Various Countries and Years, 1988–90*
(percentage of total labor force)

Region, country and year	Basic or less	Vocational	Secondary	Higher
Central and Eastern Europe				
Bulgaria, 1990	44.6	15.8	30.0	9.6
Czechoslovakia, 1989	26.0	21.0	43.8	9.2
Hungary, 1990	38.4	23.1	26.9	11.6
Poland, 1988	34.2	29.5	27.9	8.4
Romania, 1990	35.8	31.4	24.0	8.8
Western Europe				
Austria, 1990	28.8	57.8	6.3	7.1
France, 1989	35.3	..	46.0	14.6
Greece, 1989	52.6	..	35.3	11.4
Ireland, 1989	26.8	..	55.5	17.5
Italy, 1990	26.6	..	66.2	7.2
Netherlands, 1989	12.6	..	61.3	19.7
Spain, 1990	48.4	..	46.1	5.5

.. Not reported.

Sources: Boeri and Keese 1992 for Czechoslovakia and Poland; Lado and others 1991 for Hungary; Raboaca 1991 for Romania; and data base LABEDUC of the International Labour Organisation, as reported in Imbert 1991, for all OECD countries.

monly believed. As far as the overall distribution of earnings is concerned, the average earnings of workers in the ninth decile of the earnings distribution in Poland was 2.4 times that of the first decile in 1989; the comparable figure for the United Kingdom in 1990 was 3.3 (Boeri and Keese 1992). The distribution of wage and non-wage income, both in the 1970s and in the 1980s, also appears to have been more equal in Central and Eastern Europe than in the West (see Wiles 1974; Atkinson and Micklewright 1992).

A further ill effect of wage controls was excessive reliance on fringe benefits. Historically, wage setting was administrative and highly centralized. Because enterprises were generally unable to raise wages, they made increasing use of fringe benefits to recruit workers and reward effort. Over time, more decentralized systems evolved, and by the early 1980s, some countries introduced tax-based incomes policies. Various forms of this type of policy were put in place; all imposed high taxes on wage growth above a target level. This again provided an incentive to increase fringe benefits, which were not taxed.

A DEFICIENT APPROACH TO UNEMPLOYMENT. The third weakness of the old system was that its approach to unemployment was deficient. The root of the problem is that government denied its existence, and, as a result, policies to

Table 7-7.   *Enrollment as a Percentage of Age Group, by Level of Education in Various Countries and Years, 1965–89*

Country, age group, and year	Secondary education			Higher education
	All	Vocational[a]	General	
Poland[b]				
1965	69	71	29	18
1989	81	77	23	20
Former Czechoslovakia[c]				
1965	29	75	25	14
1989	87	54	46	18
Hungary[c]				
1965	—	64	36	13
1989	76	76	24	15
Greece[d]				
1965	49	18	82	10
1986	97	14	86	28
Germany, Fed. Rep. of[e]				
1965	—	58	42	14
1988	99	36	64	32
Hong Kong[e]				
1965	29	6	94	5
1987	74	10	90	—
Korea, Rep. of[e]				
1965	35	16	84	6
1989	86	16	84	37
Spain[e]				
1965	38	31	69	6
1987	105	27	73	32
Sweden[e]				
1965	62	39	61	13
1988	90	36	64	31

— Not available.
a. Includes teacher training.
b. Ages 15–18 for secondary; 20–24 for higher.
c. Ages 14–17 for secondary; 20–24 for higher.
d. Ages 12–17 for secondary; 20–24 for higher.
e. Ages 11–17 for secondary; 20–24 for higher.
Source: UNESCO 1991.

address it were inadequate. The resulting ill effects were twofold: hidden unemployment and lack of assistance for the genuinely unemployed.

Hidden unemployment is a major adverse inheritance. The most important strength of the old system—the lack of open unemployment—was possible only because it was replaced by hidden unemployment, which was caused by "over-full employment planning," distorted relative prices, soft budget constraints, and shortages (see Gora 1993 for details). Hidden unemployment increased after the end of the extensive period of industrialization, when labor reserves were already absorbed, and reached an estimated 15 percent of the labor force in Czechoslovakia in 1979 and more than 25 percent in Poland in 1987 (Brada 1989; Rutkowski 1990). The direct costs of hidden unemployment are the output lost by not employing workers in enterprises where they could contribute positively to output and the low productivity, poor labor discipline, and excessive degree of capricious labor mobility associated with a regime of permanent excess demand.

Lack of assistance to the unemployed was a consequence of job security. Although unemployment was low, some seasonal, frictional, and structural unemployment did occur (Bornstein 1978). Unemployment benefits were nonexistent; moreover, individuals who became unemployed were not automatically covered by other forms of income support since social assistance was limited to certain vulnerable categories. The so-called job placement offices primarily mobilized labor reserves, mainly moving workers from agriculture to manufacturing.

INFLATIONARY PRESSURES. The fourth adverse inheritance—problems with inflation—arose because no countervailing forces contained the growth of wages. The system of wage bargaining and wage setting was liberalized and made dependent on managers' decisions, subject to minimum and maximum tariff rates negotiated between branch-level trade unions and the relevant ministry. Although the myriad collective agreements at the industry level initially made it difficult for enterprises to set their own wages, these constraints decreased over time as the autonomy of enterprises increased. Enterprise-level wage setting was a source of inflationary pressure which led in the 1980s to the introduction of tax-based incomes policies. Because of numerous exemptions, these policies were rarely binding: wage-push inflationary pressures were not contained, and the problem was exacerbated by the absence of countervailing pressure from profit-earners. Powerful trade unions widely perceived tax-based incomes policies as the enemy of justifiable differentials among industries, and they are now exerting strong pressure to restore industry-level bargaining. Since industry-level bargaining has traditionally played a decisive role, it may not be easy to switch either into decentralized (enterprise level) or centralized (nationwide) bargaining, which would be desirable from the viewpoint of containing wage pressure.

The lack of employers' organizations made it impossible to have proper collective bargaining. Prior to the transition, countries in Central and Eastern Europe did not have a bargaining system based on tripartite agreements, mainly because no clear distinction was made between the state and the employer. State involvement in wage bargaining was thus high, with even minor industrial conflicts becoming "state" issues. In one or two countries, most notably Poland, trade unions were powerful and, together with the lack of employers' organizations and of bargaining at the industry level, aggravated the pressure on wages.

[135]

DISTORTED PATTERNS OF EMPLOYMENT. Employment was concentrated excessively in industry and agriculture. As a result of the slow growth of productivity and policies to promote heavy industry, the share of total employment in industry and agriculture remained high in Central and Eastern Europe (table 7-8 and figure 7-4), whereas it declined considerably over the last two decades in the West. The inherited structure will exert a negative impact on labor productivity for a long time.

LIMITED INTERNATIONAL MIGRATION. The final weakness of the old system was that international migration was very limited. Strong arguments suggest that surplus labor in one country can be productively employed in jobs which the population in another country is reluctant to undertake. Emigration and immigration in Central and Eastern Europe were formerly prohibited and then severely restrained. Since liberalization, restrictions on migration have been replaced by restrictions imposed by potential host countries.

Perennial Problems

The old distribution of income was widely perceived to be unjust, in the sense that it was not related to effort, to individual productivity, or to the quantity and quality of work performed. Despite wage compression, incomes and welfare differed across regions and sectors, but not in ways which reflected the quantity or quality of an individual's work. Economic reform will change matters, but income differentials will still not be related to *individual* effort or the *individual* productivity of the worker. Income differentials are not related to individual productivity in market economies, because capital intensities and rates of technological progress are different in different industries and regions. The feeling of a weak association between effort, productivity, and wages will remain and, indeed, be augmented by a rising share of income derived from sources other than work. Much time will undoubtedly pass before popular understanding of the rationale of wage and income differentials develops.

A widespread black market for labor developed under the old system, driven by collusion between employers and employees to evade taxation. This is not inevitable, but it will remain for a long time because of cultural patterns (cheating the state is not regarded as shameful) and inefficient mechanisms to control it. The endemic problem of the lack of rule of law, which emerged repeatedly in chapter 5, is as much a constraint on the design and effective implementation of labor market policies as it is elsewhere.

The Forces Driving Change

Fundamental political and economic forces were instrumental in efforts to change the labor market. The rush to democracy, liberalization, and stabiliza-

Figure 7-4. *Distribution of Employment by Sector, Various Countries, 1980–91*

Percentage of total employment

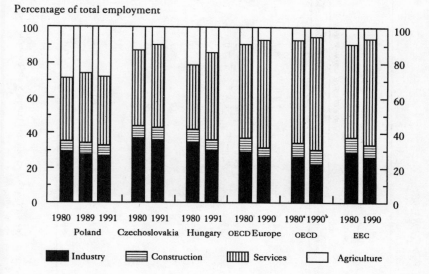

a. 1979 for Denmark and Switzerland; 1981 for Greece. Industry includes Japan's hotels sector.

b. 1989 for Greece and Iceland. Industry includes Japan's hotel sector.

Note: OECD, Organization for Economic Cooperation and Development; EEC, European Economic Community.

Source: OECD data, collected from various government sources.

tion created new incentives as well as new constraints which changed the economic landscape in important ways.

Political Forces

The major political force initially driving labor market reforms was a desire to make the economic system more democratic. However, there are two very different interpretations of what a democratic labor market should look like. The first view emphasizes the importance of giving employees participatory rights in the management of enterprises. The second argues that the introduction of market forces per se would produce a more just system. The reform programs put forward in the early transition endorsed the latter approach and advocated policies of mass privatization which would transfer the control of enterprises from the state to capital markets. In practice, however, privatization proceeded slowly, and workers' councils and trade unions found themselves playing an important role in managing enterprises. In countries such as Poland, and especially the Russian Federation, privatization legislation retains a significant amount of worker participation in the ownership of enterprises.

Table 7-8. *Industrial Employment by Occupation in Various Countries and Years, 1985–90*
(percentage of employment)

Region, country, and year	Professional, technical	Administrative, clerical	Sales, service	Production, transport	Other, not defined	Share of production workers in total employment[a]	Share of total employment			
							Agriculture	Industry	Construction	Services
Central and Eastern Europe										
Bulgaria, 1985	14.9	2.9	2.8	79.3	0.1	54.5	16.5	37.9	8.7	36.9
Czechoslovakia, 1990	21.0	–	–	75.6	3.4	50.7	10.1	36.5	9.3	44.1
Hungary, 1989	9.5	12.0	4.8	73.3	0.3	49.6	17.0	32.5	5.2	45.3
Poland, 1988	16.0	6.0	0.4	75.3	2.2	53.4	27.8	28.2	7.9	36.0
Western Europe										
Austria, 1989	7.3	16.6	3.7	72.1	0.2	38.3	7.8	28.9	8.5	54.8
Germany, Fed. Rep. of, 1989	10.6	19.4	6.4	60.3	3.3	34.0	3.7	33.5	6.7	56.1
Netherlands, 1990	13.8	18.9	9.9	55.7	1.8	25.7	4.6	19.7	6.5	69.1
Sweden, 1985	18.7	9.9	6.0	59.0	6.4	26.7	5.0	22.7	6.0	66.4
North America										
United States, 1990	12.2	23.7	5.3	58.4	0.5	27.5	2.8	19.9	6.5	70.7
Asia										
Japan, 1990	4.8	18.3	5.5	71.3	0.1	38.1	7.2	24.6	9.4	58.7

– Not available.

a. Excluding agriculture and agricultural occupations.

Sources: Boeri and Keese 1992. Data for Bulgaria, Poland, and Sweden are from the census; data for Czechoslovakia and Hungary are from the microcensus; and data for Austria, Federal Republic of Germany, Japan, Netherlands, and the United States are from labor force surveys.

Several reasons explain the initial preference for making enterprise management responsible to capital rather than to workers. First, there is a pressure to throw out everything connected with the old system. The transition is interpreted as a transition to capitalism and, as discussed in chapter 4, capitalism is mistakenly interpreted to mean unfettered private markets. This interpretation implies that any type of worker participation violates the employer's right to choose the most efficient options and that employer freedom is essential to creating a system in which workers, although deprived of the right to participate in decisionmaking, would feel justly appreciated and remunerated. Seen from this perspective, the concept of worker participation appears "socialist" and hence lacks credibility.

A second reason for emphasizing market forces is the weakness of trade unions, not least because of their prominence under the old system. Workers tolerated the old system because it seemed egalitarian (relative wage differentials were small) and because it was promulgated and supervised by the trade unions within a totalitarian framework. In the new system, trade unions occupy an ambiguous position because their actions supporting workers can easily be misinterpreted as supporting the old system. In exceptional cases, for instance Poland, where independent trade unions became powerful under the old regime, trade unions either function in ways similar to political parties (and hence are subject to the pressures toward market forces) or are reluctant to oppose the parties coming out of their camp.

Policymakers have pressured workers not to pursue participatory rights, in part because the stabilization program requires a sharp decline in real wages which might not be acceptable if worker participation were significant. Even in a non-participatory regime, workers will accept a dramatic decline in real incomes only when the drop is regarded as temporary and they expect to regain their former incomes later on (Rutkowski 1991).

The drive toward democracy, if interpreted as getting rid of *all* features of the old system, may lead to the excessive swing of the pendulum discussed in chapter 4.

Economic Forces

The fundamental economic forces driving changes in the labor market are the liberalization of demand, liberalization of supply, liberalization of prices and wages, and macroeconomic stabilization. These changes create new incentives and new constraints, to which, as described shortly, different groups in the economy adjust with different speeds.

Economic disruption occurred in many ways, of which two should be highlighted: inflationary pressures and supply dislocations. In all countries, price liberalization led not to a simple once-and-for-all adjustment of relative prices, but to a sharp jump in the overall level of prices, reflecting the end of the overvaluation of the domestic currency and the traditional excess demand for goods at previously fixed prices (the "monetary overhang"). The increase in the price index using conventional methods overestimates the actual rise in prices because it takes no account of the increased availability of goods and the end of black markets. Although the precise magnitude of the problem is in dispute, the large jump in prices has clearly made all of the transitional economies extremely vulnerable to very rapid inflation.

The liberalization of supply and demand led almost immediately to the collapse of trading relationships between enterprises which had been administered through the planning system. Supplies of raw material inputs and semi-manufactured goods prescribed by the plan were disrupted. The problem arose within countries, between countries (given the collapse of the Council for Mutual Economic Assistance), and among the former Soviet republics. Inputs could, of course, be purchased from the West, but normally only at much higher prices than previously. This dislocation of supplies obliged firms to reduce production and to cut real wages.

Different groups responded to these disruptions in very different ways. Households adjusted most quickly. They reacted swiftly to price increases and adjusted their work effort in the light of changes in real wages and the emergence of unemployment. The main effects were twofold: household demand for goods declined, with adverse effects on employers' demand for labor; and households increased their supply of labor and effort.

State enterprises responded to the driving forces, primarily the dislocation of input supplies and product markets, by reducing activity (Estrin, Schaffer, and Singh 1992). The fear of the imposition of tight budget constraints and bankruptcy laws, however misplaced, induced state enterprises initially to seek no more than viability (an operating surplus sufficient to service debt), not necessarily to maximize profits. In practice, in the early stages of the transition, given past injections of free capital from the budget and low and often negative real interest rates, financial viability was relatively easily attained. Once it was ensured, enterprises tended to stop exploring new cost-saving opportunities and continued to hoard surplus labor and to use it inefficiently; they also continued to waste energy and materials. The first signs of efficiency adjustment have been slow in coming. As the financial situation deteriorates, the main concern is often one of maintaining employment, in part because employers still see it as their responsibility (not least because in some countries the employer still offers the only access to social benefits) and in part because in some countries management is responsible to the representatives of the workers. In the face of falling revenues, state enterprises have often chosen to allow real wages to fall rather than to cut back employment. In the context of the transition, this approach may be appropriate, in that a faster rate of job loss in the state sector coupled with slow growth of employment in the private sector would produce even higher unemployment. Nevertheless, these decisions are being made for nonmarket reasons and indicate slow adjustment to market forces.

Private enterprises responded to developments faster than state enterprises, although less promptly than households. As they found new markets, their demand for labor increased, with a preference for better-educated workers. The effect on the labor market, however, has not been significant because, though growing fast, the private sector is still small. Newly established private firms have difficulty reaching the minimum size necessary for efficient manufacturing and hence tend to be concentrated in trade and service sectors. In a market economy most private sector growth comes from the growth of medium-size firms and from new firms founded by medium-size and large private firms. As Nuti (1992) points out, neither is available domestically in transitional economies. Plans for privatizing large-scale state enterprises have proceeded slowly.

The driving forces and the responses to them produce five key effects:

- The collapse in output and the resulting fiscal deficit make it necessary to restrict expenditure on labor markets and social protection, just when the need for them is rising sharply.
- Incomes are decompressed by wage liberalization and by the growing number of low incomes as unemployment and short-time working increase. This takes place in spite of incomes policies aimed at egalitarian indexation of wages and benefits, in part because the private sector is generally exempted from incomes policies.
- The private sector is growing rapidly, leading to increasing diversity of potential employers.
- Union coverage decreases, mainly because liberalization of supply leads to an enlargement of the private sector and increasing self-employment and to a general decline in employment in state enterprises.
- Unemployment increases for all the reasons discussed in chapter 4. During the early transition, jobs were lost primarily by older workers and women, and the rise in open unemployment was usually preceded by a period of unpaid vacation and reduced working hours. Notwithstanding the amount of labor hoarding which occurred prior to the reforms, the decline in employment over the first three years of the transition was much smaller than the decline in output (see table 7-9). Open unemployment, the main subject of chapter 8, is a new and major phenomenon in Central and Eastern Europe. Its size and speed of growth are one of the main challenges for labor market policies.

Policy

Within the labor market, the essential element of reform is the replacement of a system of wages and employment regulated to a greater or lesser extent by the state with one in which they are determined in the market by the forces of supply and demand.

Table 7-9. *Decline in GDP and Rate of Unemployment in Selected Countries of Central and Eastern Europe, 1992*
(percent)

Country	Decline in GDP compared with 1988	Unemployment rate	Proportion of unemployment rate to decline in GDP
Bulgaria	32	13	0.46
Czechoslovakia	15	7	0.47
German Democratic Republic	42	14	0.33
Hungary	16	9	0.56
Poland	19	11	0.58

Source: National economic statistics.

The liberalization of wages and employment might be expected to provide individual workers and employers with the incentives to allocate resources efficiently in the labor market. Market-clearing wage levels might, in principle, also be expected to be consistent with full employment of labor in conditions of macroeconomic stability. Provided labor is reasonably mobile, wages will also be consistent with horizontal equity (workers doing the same work can expect to receive the same pay). There is, however, no presumption that the distribution of income generated by market-determined wages will be socially or politically acceptable.

The Strategy

Although the objective of policy may have been to have wages and employment determined by the market, the reality during the first years of transition was very different. Even in countries like Poland, which introduced comprehensive price liberalization, wages remain subject to administrative control, with a penal tax on firms exceeding their permitted levels of wage increase. Incomes policies (generally tax based) have also been imposed in Bulgaria, the Czech Republic, Hungary, Romania, and intermittently in Russia and other countries of the former U.S.S.R. The employment decisions of state firms are also determined as much by political considerations as by market forces. In general, enterprises have been prepared to extend and take credit from one another irrespective of their financial situation in the belief that, in the final analysis, the government (or the state banking system) would underwrite their debts to prevent mass layoffs.

The immediate reason for continuing government intervention in decisions about wage setting and employment has been macroeconomic. As already noted, the countries are very vulnerable to accelerating inflation. The "shock therapists" (Bulgaria, the Czech Republic, and Poland) have made the control of inflation a priority and have used incomes policies to support restrictive demand management. A second objective of incomes policy has been to hold down wages to improve competitiveness and thereby to help maintain employment. Incomes policies have a number of allocative and distributional effects. Insofar as they affect wage setting, they maintain the compression of the wage distribution. This prevents, or at least slows down, the emergence of wage differentials as a guide to decisions about how to allocate resources in the labor market; it also cushions the fall in the living standards of low-paid workers. Likewise, delaying the imposition of financial discipline on enterprises slows down the rise in unemployment and thus may help contain one of the main causes of household poverty.

The very cautious approach to wage liberalization and to the adoption of hard budget constraints for state firms in the first years of reform reflects the inheritance from the planned economy. Efficiency in the labor market requires not only the liberalization of wage and employment decisions but also the institutional infrastructure of the free market. This means private ownership and competitive markets for products, capital, and labor, with labor mobile between employers, occupations, and localities. The competitive labor market is characterized by an absence of government intervention in wage setting or in employment decisions, but behind the scenes, as it were, the government has the responsibility for upholding private property rights and maintaining competition.

As already noted, the inheritance of the economies of Central and Eastern Europe is very different. The absence of private ownership means that enterprises do not face incentives to minimize costs in terms of wages paid or employment decisions. The concentration of industry means that price and wage liberalization offers opportunities to exploit monopoly power. The absence of capital markets means that there is no basis for efficient investment decisions either for firms or for workers. And there are many restrictions on labor mobility. One essential component of labor market policy, then, consists of structural policies designed to establish the institutions and legal framework necessary for an efficient labor market. In part, this depends on the progress of the reforms generally: developing the private sector by privatizing existing enterprises, enabling and encouraging new ventures to start, setting up capital markets, and opening up product markets to competition, for example, through trade liberalization. In terms more specifically of the labor market, it implies removing impediments to labor mobility such as residence permits or employment-linked social benefits, ending administrative restrictions (for example, on self-employment), and introducing legislation to define the rights of workers, employers, and trade unions and the role of government within the context of a market economy.

The policy strategy thus has two aspects:

- In the short run, the main concern is with the management of wages and employment in the absence of the infrastructure of a market economy (primarily an issue of macroeconomic stabilization).
- Over the longer term, the main objective is institutional reform.

Maintaining Macroeconomic Balance

Macroeconomic stabilization is usefully discussed in terms of wages and incomes policies (the price of labor) and employment policy (the quantity of labor).

WAGES AND INCOMES POLICY. Policies on wages and incomes are intended to restrain absolute wages, preferably without unduly affecting relative wages.

Containing absolute wages. All the countries in Central and Eastern Europe have retained wage controls to contain absolute wages, notwithstanding the disappointing record of incomes policies both in industrial economies and in Latin America (Kiguel and Liviatan 1992). There are several reasons why incomes policies might be more effective in the context of transitional economies than elsewhere. First, the inheritance of government wage regulation means that the administrative apparatus for incomes policy is already in place. Enterprise managers can be expected to set wages in accordance with the policy because that is what they have always done. Second, and also deriving from the inheritance, state enterprises in transition economies may be less resistant to wage demands than would be a private firm motivated by profit maximization (Hinds 1991; Lipton and Sachs 1990). In some countries, such as Poland, the managers of enterprises are to a significant extent responsible to the representatives of workers through workers' councils, which might be expected to weaken their opposition to wage claims. Third, the inflationary turbulence of the last years of communist rule and the enormous price leaps at the time of price liberalization may have created a situation in which inflation expectations are highly volatile and thus perhaps more easily influenced by policy.

Incomes policy has typically taken the form of regulating the rate of increase of the firm's wage bill, with punitive tax penalties levied on firms exceeding the norm.[3] This method of regulation basically continues policies adopted in the communist era (Bosworth 1991; Commander and Staehr 1991). These policies were not particularly successful under the previous regimes, no doubt because they were applied in conditions of generalized excess demand, rather than in conjunction with restrictive fiscal and monetary policies. It is not clear whether in the early transition these policies helped to contain inflation (or did so with less unemployment than would have occurred in their absence). In Poland, for example, wage growth was significantly below the rate permitted by policy during the first six months following the reforms but significantly higher thereafter. There is some evidence that Polish enterprises responded to an incomes policy related to their total wage bill by reducing employment in order to pay higher wages to the workers who remained, but as unemployment rose, the wage ceiling was redefined in terms of the average wage per employee. Nor is it clear whether incomes policies contributed to the sharp initial fall in real wages which characterized labor markets in the early transition. Real wages have fallen everywhere as a result of the collapse in output and the associated fall in labor productivity. The degree of compliance with the excess wage tax has also been a cause of concern. In Poland, tax collections at the beginning of 1991 were only about half what was owed, while in Bulgaria only about 10 percent of the excess wage tax was apparently collected in 1993. By contrast, in Russia, the excess wage tax is collected along with and at the same rate as the profits tax, and compliance has therefore been quite high.

The effects on relative wages. Incomes policies have not prevented, and were not intended to prevent, any adjustment of *relative* wages in response to relative scarcities of different types of labor. Relative wages can adjust within the confines of the policy in at least four ways:

• Incomes policy operates at the level of the firm. Whether the control is based on the total wage bill or on the average wage per employee, there is scope within the firm to give larger wage increases to some workers than to others, within a given overall constraint. In conditions of rising unemployment, it is generally better for policy to relate to the average wage rather than to the total wage bill if the objective is to restrain wage inflation, since using the total wage bill encourages firms to lay off workers in order to raise the wages of those who remain.

• Various countries such as the Czech Republic and Hungary have exempted firms with particularly rapid growth in output, productivity, value added, or profits.

• Private firms, and sometimes small firms, have generally been exempt from the scope of the policy.

• In any event, firms facing labor shortages may choose to increase wages in excess of the limit and pay the tax liability thereby incurred.

In writing about Poland in 1990, the first year of the reform, Coricelli and Revenga (1992) note clearly disparate patterns of sectoral wage behavior and, in particular, a strong correlation between wage growth and profitability, on the one hand, and output performance, on the other. The Polish experience seems to suggest a rather limited impact of incomes policy on relative wage adjustment, and the evidence for Bulgaria and Hungary appears to point in the same direction. In most countries, in contrast, the minimum wage has been increased by at least the same rate as the permitted rate of wage growth for the economy as a whole and in countries such as Bulgaria in 1991 by considerably more. In principle this will protect the low-paid workers (or at any rate those in full-time work) from any reduction in their relative wage.

The longer term. A system of wage controls supported by punitive taxation cannot be consistent with the efficient workings of a market economy over the longer term. Current tax-based policies maintain one of the worst labor market features of the inheritance: low effort and poor workmanship on the part of a relatively well-educated, skilled, and potentially highly motivated work force. The question is how best to loosen the controls. One approach, adopted in Poland in January 1992, is to exempt private firms from the policy. To the extent that incomes policies were put in place to counterbalance a lack of financial discipline—the result of an implicit understanding that any losses incurred by the enterprise would be underwritten by the government—the lack of discipline is specific to state firms. This then implies that incomes restraint should not be imposed on private firms. Exempting private firms not only encourages the process of privatization but also, rather conveniently, means that as firms become privatized, incomes policy automatically fades away.

Arguments can be made, however, for extending equal treatment to state and private firms on grounds of equity. Doing so, it is argued, would prevent a brain drain in which the best workers move to the private sector irrespective of relative productivity and would help newly privatized enterprises to contain wage pressure and generate a return on capital (Lane 1992). Evidence from highly industrialized economies suggests that private ownership and reasonably competitive markets are not necessarily sufficient to contain wage pressures. It may also be necessary either to take measures to eradicate the sources of inflationary pressure by weakening the trade unions, restricting unemployment benefits, and the like or, alternatively, to involve unions and employers with government in setting wages.

EMPLOYMENT POLICY. The key questions for employment policy in the short run are (a) whether government should allow unemployment to rise sharply and, if not, (b) what policies they might adopt to contain unemployment. This section addresses mainly the former question, discussing in particular the issues surrounding continuing government support for enterprises. The question of how best to respond to unemployment as it emerges is postponed until the next chapter.

In the early transition, the collapse in output was associated with a less-than-proportionate fall in employment in the short term. But attempts by enterprises to maintain employment have been leading, in many cases, to substantial and growing financial deficits. Governments have tolerated continued lending by the banking sector to financially insolvent enterprises and have delayed the introduction of bankruptcy laws. The indiscriminate imposition of hard budget constraints and the bankrupting of insolvent enterprises would lead to massive losses of jobs. Although growing rapidly, the private sector was initially too small to provide new employment on the scale required.

In contrast with the situation with wage controls, where policy has been explicit, employment policy has thus typically taken the form of an uneasy compromise between (a) the explicit objective of ending financial support to loss-making state enterprises and (b) the political pressure to avoid mass layoffs and large-scale unemployment. Much expert advice, both private and official, has argued for imposing hard budget constraints on firms, with high levels of unemployment, should they emerge, being dealt with by a mix of active labor market policies and the social safety net.

There are arguments for allowing unemployment to rise sharply (Gomulka 1991). Unemployment, it can be argued, is helpful in the medium term for the growth of the private sector. High rates of unemployment might be expected to facilitate the starting up of new enterprises by ensuring an abundant supply of labor. High unemployment will also hold down wages and strengthen the hand of employers in enforcing discipline and work effort.

Some unemployment is no doubt inevitable and will have to be accepted as part of the price of a market economy. Most private sector firms recruit their workers directly from state enterprises, however, and it is by no means clear that unemployment plays a useful role in the reallocation of labor. Very high levels of unemployment create economic waste, social distress, and severe personal hardship. They are likely to be associated with a high incidence of long-term unemployment which, evidence from OECD countries suggests, has corrosive economic and social effects. Evidence from the OECD countries also suggests that, beyond a certain point, an increase in unemployment, and in particular an increase in long-term unemployment, has very little effect in holding down wage pressures (Layard, Nickell, and Jackman 1991, chap. 9). The costs of unemployment must be measured not only in terms of lost output and fiscal pressures but also in terms of the stress placed on families and society, which brings with it the potential for generating a political backlash against the whole reform program. At some point, the costs of unemployment become so severe as to justify intervention. This point cannot be quantified with any precision, not least because it will obviously differ according to the history and circumstances of each country; but unemployment rates well in excess of 20 percent, as experienced in many industrial towns and cities and sometimes even whole regions within Central and Eastern Europe, go far beyond levels which, on the experience of OECD countries, could be regarded as necessary for the efficient working of a market economy. To slow the growth of unemployment, government has two lines of attack: it can continue to support loss-making state enterprises, or it can introduce active labor market policies, such as the provision of temporary jobs and the establishment of large-scale retraining schemes.

Active labor market policies. The usefulness of such policies as the main instrument for combating mass unemployment should not be overstated. As discussed in the next chapter, they can be very useful in particular situations; but there must be serious doubts about whether they can be applied on the scale required:

• The first doubt relates to the effectiveness of such policies in the context of a generalized collapse in output. With few jobs on offer, policies to promote labor mobility are likely to place few workers in new jobs and to encounter low take-up. Among the highly industrialized countries, only Sweden has attempted to provide training and temporary work schemes on the scale required to cater to everyone who becomes unemployed. But Sweden has not had to confront the levels of unemployment experienced in parts of Central and Eastern Europe, and it would not necessarily be feasible to provide temporary jobs, public works, or retraining at the local level for 20 or 30 percent of the work force.

• Even on the more limited scale actually implemented in Sweden, active policies are likely to be beyond the resources of countries in Central and Eastern Europe. The policies are too costly and involve too much administration (Sweden spends around 1.4 percent of GDP on active policies and has one labor office official for every nine persons unemployed). Given very limited resources, active labor market policies will need to be applied selectively and targeted toward particular groups of individuals, such as the long-term unemployed, who have specific needs.

Policies to assist enterprises. This is an alternative way of trying to maintain employment. Governments are already pursuing such policies, through implicit or explicit subsidy, delay or non-enforcement of bankruptcy laws, and so forth, and are likely to remain under political pressure to continue doing so. Not only will producer interests argue for continued support for individual enterprises, but public opinion is likely to support so-called common sense arguments that it is better to pay people for producing something than for producing nothing and that unemployment is bad for individuals. All these arguments have particular force where a commercially insolvent firm is the main source of employment in a locality.

Nonetheless, existing large-scale and indiscriminate policies of enterprise support insulate enterprises from commercial pressures and the need to adjust to the market economy. A total and immediate cessation of such policies, however, would add intolerably to unemployment and economic dislocation in the short run. What is needed is to replace policies of *generalized support* with those of *selective support*, targeted toward firms which have some prospect of economic viability and which operate in areas of acute economic distress. In assessing whether particular firms might qualify for support, several questions must be considered:

• Is value added positive? This is an empirical question of whether, evaluated at market prices, the value of the output produced exceeds the value of the inputs used. Since, in most enterprises, employment is already excessive, quite substantial reductions in employment may perhaps be achieved with no loss of output. But if an enterprise is on the point of closure, a loss of output is inevitable, and the relevant issue is the average value added in the industrial sector. An attempt to measure value added at world prices has been made both for the (non-Soviet) countries of Central and Eastern Europe (Hare and Hughes 1992) and for those of the former U.S.S.R. (Senik-Leygonie and Hughes 1992). Although some sectors have negative value added (up to almost one-quarter in some countries), most countries have positive value added in most sectors except food processing. In general, state enterprises make a positive contribution to economic welfare.

• Is the fiscal cost sustainable? Of more immediate concern for many governments is the question of whether there would be a budgetary saving from closing down a loss-making enterprise if most of the workers are likely to remain unemployed? Some savings will arise because the net of tax wages of persons in work is likely to be higher than the unemployment benefits paid (though maybe not much higher, given that many workers have part-time jobs or are obliged to take periods of unpaid leave). In contrast, a firm with positive value added will produce a net cash income. To think in terms of illustrative numbers, one might have value added per worker in an enterprise of $40, a wage of, say, $70, and unemployment benefits of $50. Supporting the enterprise is then both cheaper to the government (costing $30 per worker as against $50 per worker for benefits) and economically more efficient in that the value added in the enterprise is positive.[4]

• Can public authorities deliver benefits effectively? This question asks whether local authorities or other public agencies have the capacity to take over the social services and other benefits formerly provided by enterprises. Again, enterprises often continue to provide benefits to former workers, but if the enterprise itself closes down, other arrangements are needed.

• Is insolvency only transitional? A further consideration concerns longer-run industrial policy. There may be no point in keeping alive enterprises which are doomed to extinction, but in many cases enterprises face financial problems attributable as much to temporary features of the transition, such as the collapse in output and its effect on demand, supply disruptions, and so forth, as to a lack of longer-term viability. Notwithstanding their obsolete technology and antiquated capital, at appropriately competitive levels of the exchange rate and labor costs, most countries in Central and Eastern Europe can expect to continue production across a wide range of industrial sectors.

In principle, then, an argument can be made for supporting some enterprises. This support, however, should be exceptional, temporary, and offered only in very specific conditions:

- Where the local unemployment rate is very high relative to the national average
- Where external constraints, such as housing rigidities or lack of the language skills necessary for migration, limit the mobility of labor
- Where high unemployment clearly creates major social and political tensions (an extreme example is the former Yugoslavia, where the large pool of unemployed youth exacerbated the tensions leading to the war)
- Finally, where support for enterprises is transitional and transparent.

The most difficult task of industrial support is to avoid creating perverse incentives at the level of the enterprise. A general employment subsidy (proposed for eastern Germany by Akerlof and others 1991) would be relatively non-distortionary but enormously expensive. Given the resource constraints facing most countries in Central and Eastern Europe, any scheme of industrial support will need to be selective, for instance, enterprises in areas of high unemployment or company towns, or enterprises with some chance of surviving without subsidies in the longer term. In practice, this may mean that support has to be discretionary. Most important, support should be for a limited period (perhaps three years) and should stop automatically at the end of that time, so as to ensure that the enterprise uses the support to restructure its activities rather than to perpetuate inefficiency (as has often been the case with industrial policy in the West). The level of support should be a predetermined amount (not a policy of writing off all losses) and conditional on employment within the enterprise. To ensure compliance with these conditions, the firm may be required to file for bankruptcy and to hand over management to an official administrator while receiving support. Finally, any support should be explicit to avoid confusion between (a) policies, funded mainly by taxes, whose objective is to mitigate the social and financial costs of unemployment and (b) loans made to firms by banks on the basis of purely commercial criteria. This is another area where government has to make difficult choices which have a major political dimension.

Building Markets

We turn now from short-run economic management to longer-term issues of institutional reform. The basic prerequisite for economic efficiency is a well-functioning market, but it is by no means sufficient. The labor market is prone to many of the market failures discussed in chapter 2, and in many instances such problems justify some form of government intervention. Of course, market failure does not always or automatically justify intervention; governments often do worse than an imperfect market, and we give some examples of this below. The extent of intervention differs considerably across the highly industrialized countries, partly for reasons of culture or political ideology and partly because evidence about the relative efficiency of different policies is far from conclusive. Whatever the degree of intervention, however, the labor market remains at root a market in all market economies, and this implies a general acceptance of the idea stressed in chapter 1 that the prime purpose of wage differentials is to guide the allocation of labor in the interests of efficiency rather than to achieve an equitable distribution of income.

In developing policy for the longer term, therefore, a first objective is to create the institutions of the free market as a precondition for liberalizing wages and employment. The most important of these institutions are the private ownership of the means of production, free access to all markets, and a legal framework to establish the rights of employers and workers and to define the role of government in the labor market.

DIVERSIFYING SUPPLY. Efficiency in a free market requires competition among a large, or potentially large, number of profit-seeking individuals and enterprises. It has, in practice, proved relatively easy to privatize significant areas of the economy, such as shops and local services, where enterprises are typically small and capital requirements low. Self-employment has also grown in these areas. In some countries, the private sector has expanded significantly in agriculture. But generally the privatization of large-scale manufacturing enterprises has proceeded very slowly. By late 1993, only Russia and the Czech Republic had actually started to implement a scheme of mass privatization of manufacturing enterprises, and Poland and Bulgaria had enacted legislation to privatize all state enterprises (see Estrin 1994).

Even in the absence of private ownership, enterprises in the state sector apparently are beginning to behave more like private firms. The collapse of the central planning system has devolved managerial responsibility to the enterprise, and price liberalization plus the imposition (or threatened imposition) of hard budget constraints has obliged firms to adjust to more market-like behavior. Even without changes in formal ownership, effective control of the enterprise has passed from the state to the firm's managers and workers.

INCREASING CONSUMER CHOICE. In a competitive market, workers are in principle free to choose their employer and type of work, just as the consumer is free to choose between the different goods and services on sale. In the early years of transition, potential employers became visibly more diverse as shops and other small enterprises were privatized and numerous service sector activities started up.

Although the range and variety of employment activities have already increased and can be expected to increase further, employment opportunities for the average worker have not expanded commensurately. In a depressed labor market with high unemployment, few firms will be taking on new recruits, and workers, far from being in a position to choose among the jobs that exist, may in practice have no alternative but to stay in the job they already have. In the longer term, the extent of choice over jobs seems to be much more restricted in market economies than is the choice of goods. Under perfect competition, there are many buyers and many sellers and no one has any market power. Under imperfect competition, sellers set prices above marginal costs and thereby create buyers' markets. In product markets, producers struggle to sell, while consumers have an abundance of choices. Labor markets, too, are buyers' markets. Wages tend to exceed marginal products for reasons of "insider" power and of efficiency wages (see Layard, Nickell, and Jackman 1991). As a result, workers, who sell labor, look hard for work, while employers, who buy it, can pick and choose. In a market economy, there are many potential employers, but workers have difficulty getting any particular job. In the medium term in Central and Eastern Europe, even if economic recovery, liberalization, and restructuring create conditions similar to those prevailing in market economies and even if the quality of the jobs themselves improves, any individual worker may find it just as difficult to secure a particular type of job than under the excess demand regime in place before the reform.

IMPROVING INCENTIVES. In the context of the labor market, improving incentives means giving both workers and employers the incentive to use scarce labor resources most productively. In theory, under the assumptions of private property, perfect competition in product and factor markets, full information, and an absence of externalities, market-clearing wages will allocate labor efficiently in a static sense and also, intertemporally, will provide the right incentives for investment in education, training, and job mobility. In competitive markets, profit-maximizing firms pay wages reflecting the value at the margin of what they produce. On the supply side, market-determined wages should therefore give workers appropriate incentives to put forth effort and acquire skills, to take responsibility, and to move between enterprises, occupations, and localities. More generally, a competitive market should give people the right incentives to set up businesses, to become self-employed, and to take risks. On the demand side, market-determined wages give firms criteria to choose between more or less labor-intensive techniques of production and an incentive to take account of relative scarcities of different types of labor in their employment and investment decisions.

Two sets of incentive issues are particularly relevant in the labor market: (a) the effects of minimum wages and income transfers and (b) incentives affecting employment. So far as minimum wages and income transfers are concerned, the fundamental incentive issue is the conflict between economic efficiency and income distribution. A central purpose of the reforms is to have wages determined in the market for efficiency reasons. One of the results has been a widening distribution of earnings. At a time of falling real wages in many countries, however, this can greatly worsen poverty. Intervention can then take one of two forms:

- Government may intervene in the process of wage setting, most obviously by imposing a minimum wage floor or
- It may introduce a system of income transfers.

The provision of a social safety net will of itself put a floor beneath the wage distribution and prevent wages from being set at market-clearing levels. Both forms of intervention are common in industrial market economies (although more so in Europe than in the United States); they have not been a feature of the newly industrializing economies of East Asia or Latin America.

Incomes policies, and in particular the relative protection of the minimum wage, were significant in the first years of the transition. Yet to maintain a government-determined minimum wage clearly violates the principle of having wages determined by the market. The general presumption must be that minimum wages can price low-skilled or young workers out of a job and hence cause unemployment. In addition, in some countries such as Belarus and Ukraine, the minimum wage has become a sort of numeraire for the whole wage structure, thus inhibiting the flexibility of relative wages. In the highly industrialized countries, minimum wage legislation tends to be seen more as a means of protecting unorganized or casual labor rather than as a significant instrument of poverty relief. These considerations suggest that, as the economy starts to recover, the minimum wage should be abolished or at least substantially reduced.[5]

Unemployment benefits also place a floor beneath wages and can again price young or unskilled workers out of a job. Additionally, the tax costs of financing benefits weaken incentives when numerous other budgetary demands and a diminished tax base mean that tax rates are already high. Such costs may be limited in two ways, consistent with the commitment to ensuring that no one falls below a minimum poverty line. Government could seek to introduce a system of well-targeted benefits, either for all workers or for particular groups (for example, young persons or married women). It could also require the unemployed to search for a job or to participate in a public works scheme as a condition for receipt of benefits so as to reduce the duration of unemployment and deter fraud. Even so, unless the poverty line is very low indeed, poverty relief will entail significant efficiency costs in two ways. First, the tax burden must be increased to pay for benefits. Second, incentive effects exist for the low-paid: in many economies, the unemployment benefit is higher than the minimum wage, and many people work part-time or are made to take unpaid holidays and thus may receive less than the minimum wage. For both reasons, unemployment benefits can be higher than the realistic alternative earnings.

If the minimum wage is not abolished, the question arises of its relationship to the poverty line. Given the adverse conditions of the labor market, an argument can be made for allowing the minimum wage to drift downward to reduce its harmful effects on employment. However, it can hardly fall much below the poverty line. First, such a move would create adverse incentives at the lower end of the wage distribution. Second, it would be ethically and politically difficult to argue that individuals who work should accept a lower standard of living than the unemployed or pensioners. Third, in the short run, it is not financially or administratively easy to supplement an inadequate minimum wage with means-tested support to bring low-paid workers up to the level of the social safety net.

If the government sets both the minimum wage and the poverty line, it will probably need to set them at about the same level, taking into account the consequences for both the budget and the labor market. It is essential to avoid an automatic link between the minimum wage and the poverty line. Otherwise, an increase in the minimum wage would lead to a commensurate increase in many cash benefits, which could lead to unsustainable budgetary costs.

Alongside their effects on labor supply, incentives also affect employment. The effect on employment of a high minimum wage has already been discussed. The employment status of women can create similar problems. In contrast with the West, the enterprise rather than the state has been responsible for providing support services ranging from maternity leave to preschools. Women on maternity leave in the countries of Central and Eastern Europe have been counted as employees, whereas in Western countries, they are recorded as "out of the labor force." In the transition, enterprises are becoming more and more reluctant to employ women with young children, because they may have to pay wages and child care costs for workers who may be at home much of the time on maternity leave. These considerations argue for divesting social benefits from the enterprise to the state.

RAISING LABOR PRODUCTIVITY. In the short run, two sets of policy to increase the productivity of labor stand out: (a) those aimed at adjusting effort and hours and (b) those aimed at increasing labor mobility (that is, reallocating labor to more productive uses). The scope for increasing work effort and hours worked is considerable. Under central planning, total factor productivity was low, but raising it encounters problems in the short run. First, relevant managerial experience is absent. Efficiency in the use of resources was never the main objective of firms under central planning, and the techniques for achieving it are thus basically unknown. Second, in conditions of depressed demand, raising labor productivity is seen as destroying jobs, and managers are under considerable pressure from their work force to permit productivity to fall to protect jobs.

Mobility of labor prior to the reforms was substantial. This might suggest that few barriers exist to the movement of labor between enterprises, occupations, and regions during the transition. In particular, the worry that linking the provision of social benefits to the employer would inhibit mobility may perhaps have been exaggerated. But it may also be true that in the past, much of the mobility was a "churning" around of people with little overall change in the structure of employment. Experience in highly industrialized economies suggests that manual workers are particularly resistant to moving from manufacturing to service jobs. Similarly, a large-scale movement from industrial towns to major cities, where service activities inevitably concentrate, would put pressure on housing. These difficulties are aggravated by the absence of a market in housing and by the unresolved questions of property restitution, which mean that people are reluctant to leave their homes. Clearly, any remaining administrative controls on a household's place of residence should be removed, but this is unlikely to have more than a marginal effect on labor mobility. Occupational mobility, involving the acquisition of new skills and the retraining of workers, seems also likely to run into bottlenecks if everyone wants to learn the same thing (for example, management or accountancy). In the long run, the most important factor is the education provided in schools and colleges.

Regulating Market Forces

An important element of labor market policy consists of determining the appropriate longer-term role of government and establishing the required types of labor market institutions. On the macroeconomic side, the important choice is between centralized and decentralized bargaining. If government chooses to take the "corporatist" route, it will be necessary to build up the appropriate institutions representing employers and the trade union movement. On the microeconomic side, the role of government in the labor market goes beyond the fundamental responsibility for upholding private property and maintaining competition. Many countries have laws concerning minimum wages, working conditions, unfair dismissal, the right to strike, and other aspects of the employment contract. Numerous regulations and codes of practice, including those of the International Labour Organisation to which many OECD and Central and Eastern European nations are signatories, affect the conditions of wages and employment. They seek, for the most part, to protect vulnerable groups such as children, to outlaw discrimination by race or gender, and to protect workers' rights.

Although intervention is widespread, its scope and nature vary considerably. With regard to wage bargaining, for example, several of the smaller economies in Central and Eastern Europe have set up centralized wage bargaining arrangements along the lines of the Scandinavian countries or Austria, where wages are set by agreement among the social partners (employers, trade unions, and government) rather than by the market. Empirical research (see, for instance, Bruno and Sachs 1985, chap. 11) has often noted that the more regulated labor markets of the corporatist economies have been able to achieve higher and more stable employment rates than economies with decentralized wage bargaining.

There is not space here to review at any length the experience of different wage bargaining systems in the OECD countries (see Calmfors and Driffill 1988). However, three conclusions of particular relevance to the economies of Central and Eastern Europe stand out:

- Recent crises in the Nordic countries notwithstanding, the corporatist economies have been conspicuously successful in holding down unemployment, though arguably at the expense of economic growth (Jackman 1990; Layard, Nickell, and Jackman 1991).
- Centralized bargaining systems are inherently unstable. They hold together only in relatively small countries and even then only with difficulty (see, for instance, Calmfors and Forslund 1990 on Sweden).
- Where bargaining is decentralized, unemployment is higher when trade unions are powerful and is lower when markets are competitive (Calmfors and Driffill 1988; Layard, Nickell, and Jackman 1991).

Policymakers in these countries have to decide whether it is feasible or desirable to set up corporatist tripartite systems for determining wages. In countries such as the Czech Republic and Hungary, a tripartite centralized system of wage bargaining has been set up along the lines of the corporatist economies of the Nordic countries and Austria. Experience in the highly industrialized countries suggests that centralized bargaining may well be successful in the smaller economies (such as the Baltics, the Czech Republic, Hungary, and perhaps Belarus), but that it is less likely to succeed in the larger countries (Poland, the Russian Federation, or the Ukraine), particularly those with a history of union rivalry or other social divisions. In the medium term, however, maintaining centralized wage bargaining may not be desirable, given the highly distorted inherited wage structure and the objective of using market-determined wage differentials as one of the instruments for restructuring the economy.

Implementing the Reforms

The institutional capacity required for a corporatist approach to economic policy includes some form of centralized confederation or association of trade unions and employers which is competent to make decisions on behalf of individual unions and employers. The central organization must therefore be both representative and encompassing, so that decisions made centrally evidently reflect the wishes of the majority. In the absence of a consensus, it may be difficult to hold even a representative association together, and this limits the scope of its activity.

On the microeconomic side, matters are more straightforward, at least in principle. Employment will be governed by contracts between individual workers and their employers. These contracts will be legally enforceable and thus require the backing of a legal system. Employment tribunals will need to be set up to adjudicate disputes, unions will want to provide specialist legal advice to their members, and the government will want to frame legislation governing employment contracts (for example, to ensure that they protect workers' rights as defined by the International Labour Organisation charter).

Conclusions: Priorities and Sequencing

Labor market policies can be divided into those which require immediate attention and those which will become more important once restructuring is firmly under way.

Short-term Policies

The first phase (up to two years) should concentrate on alleviating the worst adverse effects of stabilization and should lay the foundations for restructuring the labor market.

Tax-based incomes policies should be maintained, at least in the state sector. The overriding objective in the early transition is to control inflation while avoiding a catastrophic collapse of output and employment. Tax-based incomes policies contribute to this objective by containing wage pressures, even though they may inhibit the adjustment of relative wages.

Palliative measures are necessary to alleviate labor market strains. To the extent possible, the brunt of the initial adjustment should be borne by the wage bill rather than the size of employment. This would be assisted by:

• Reducing working hours
• Introducing work-sharing
• Subcontracting workers to other enterprises
• Introducing more flexible working schedules for women.

Regulations which aggravate labor rigidities should be relaxed and then abolished. In particular, rules for hiring and dismissal should be made more flexible to improve the effectiveness of internal labor markets in enterprises.

Continuing to provide selective financial support for some loss-making enterprises should not be ruled out. Such a policy has some advantages because it would slow the adjustment in depressed areas. It softens the shock of the transition by preparing workers psychologically for possible layoffs in the future and may therefore induce some workers to quit. It may assist the political viability of the reforms, and it enhances the initial ability of the private sector to create new jobs, as well as the effectiveness of training institutions, both of which are limited. Despite the danger that support may hinder rapid closure of structurally inefficient plants (as happened in Western Europe, see Houseman

1988), the current financial weakness of firms is not necessarily an indication of their viability in the future (Svejnar 1992).

Nevertheless, such support needs to be designed with great care. It should be:

- Explicit and designed to minimize distortions
- Limited to particular localities where labor mobility is particularly low (due, for instance, to housing constraints, language problems, or enterprise- or region-specific skills)
- Paid only when palliative measures are insufficient to keep the unemployment rate in the short run below a politically tolerable level
- Paid only for a limited period and stopped automatically at the end of the period.

Medium-term Policies

Once labor market restructuring has started, attention should shift toward policies concerned with building viable labor market institutions. The most important are institutions to deal with unemployment, although they should be supplemented by arrangements which improve the use of resources within enterprises and by an appropriate collective bargaining scheme.

Policies against open unemployment should become a priority, with particular emphasis on policies against long-term unemployment. The already high proportion of long-term unemployed in the labor force may soon lead to the serious problems of deskilling, demotivation, and marginalization suggested by the OECD experience (see Layard, Nickell, and Jackman 1991, chap. 9).

Action is needed to improve the use of resources within enterprises. Such action is a proper concern of policymakers. The initial emphasis on measures to reallocate labor across enterprises and close down seemingly unprofitable organizations should be supported by measures to increase the efficiency of resource use within enterprises:

- On the supply side, policies should focus on education and training.
- Incentive structures within enterprises should be improved. The government should resist any temptation to apply the basic wage scale developed for the civil service to state-owned enterprises.
- The minimum wage should be adjusted by discretionary action.
- Fringe benefits incompatible with the incentive structure of market economies should be discontinued. This should not be difficult once the tax system is reformed, when workers will likely prefer cash payments.

Collective bargaining reforms should not be delayed beyond the end of the first labor market shock. The development of a stable bargaining system is vital:

- Large countries will generally find decentralized bargaining more useful than centralized bargaining.

• Small, open, and highly unionized economies should move toward more centralized bargaining, so that the benefits of wage moderation can be shared by all participants of the bargaining process.

Notes

1. The tables give data for the late 1980s to show the situation just prior to the major changes taking place in most of Central and Eastern Europe in 1990 and 1991.

2. Taylorism was a so-called scientific approach to management, in which workers were simply treated like another machine.

3. Such incomes policies can take two forms. They can impose a maximum rate of growth of x percent either on the *total* wage bill or on the *average* wage. Under the first approach, if some workers are laid off, those who remain can have a wage increase of more than x percent; the second approach does not have this effect. Thus the first approach gives a stronger incentive to shed labor than the second.

4. One might ask whether a better solution might not be to reduce the wage. Evidently, if the benefit is 50, and plausibly the disutility of work might be 20, the individual firm cannot reduce the wage below 70, but this obviously raises questions about how benefits are set.

5. One reason for retaining a minimum wage in the short run is the lack of capacity to administer income-tested poverty relief on a large scale. Targeting is therefore likely to be more effective if it is based on indicators of poverty such as unemployment or family size. The minimum wage is one way of targeting assistance to the working poor. This argument for a minimum wage becomes less compelling once the capacity exists to administer income-tested assistance to the working poor.

Eight

LABOR MARKETS: UNEMPLOYMENT

DAVID FRETWELL • RICHARD JACKMAN

SOME UNEMPLOYMENT IS AN INEVITABLE FEATURE OF A FREE MARKET ECONOMY, and the objective of policy in Central and Eastern Europe can no longer be, as it was under communism, to prevent *any* unemployment. Rather, it should be to moderate the rise in unemployment during the process of transition and to maintain it at reasonably low levels thereafter. Full employment in a market economy depends above all on a well-functioning labor market, with most jobs being provided by private firms producing market output. Over the longer term, a recovery of employment in Central and Eastern Europe will likewise depend on the growth of the private sector and on putting into place the infrastructure of an efficient labor market. In the meanwhile, labor markets in Central and Eastern Europe have been hit by enormous shocks and are subject to rigidities and various forms of market imperfection which may lead to unemployment rates in excess of what is economically efficient or politically tolerable.

This chapter examines the role of active labor market policies. Two questions precede the main discussion of the chapter: What is the rationale for this kind of policy? What are the major types of such policies?

Active labor market policies are aimed at helping the unemployed return to work (as against the passive policy of paying unemployment benefits). The starting point is to ask why such measures are necessary at all. Their rationale is at root the same as that for unemployment benefits—namely, to protect households from fluctuations in their income more violent than policymakers regard as acceptable. In the absence of private insurance, government has a role in supporting the living standards of individuals whose employment opportunities have collapsed. This insurance can operate through cash benefits, but it can also take the form of cushioning labor market shocks or of helping people move into new jobs.

The balance between active and passive measures, given a commitment to provide *some* help for the unemployed, will depend in part on their relative budgetary costs and in part on their economic effects—both positive and negative—on the workings of the labor market. Active measures, such as job counseling, training schemes, and temporary work, tend to be more costly than cash benefits but may decrease the duration of unemployment and increase productivity. Cash benefits involve less government interference but blunt the incentive to seek work and may be subject to abuse by the fraudulent or idle. What, then, determines the balance between support in cash and support in kind in the labor market?

Cash benefits allow the individual to choose between different types of jobs and to choose whether to acquire new skills or move to a different locality. If the consumer (worker) is the best judge of his or her best interests, he or she may well do better than a government agency in determining the best course of action. The unconditional availability of cash benefits may, however, deter people from taking jobs which pay wages close to the benefit level and may discourage them from devoting resources to seeking work or from incurring the costs of retraining or the risks of moving to a new locality. Cash benefits may also strengthen the underground economy, because persons working in undeclared activities may additionally be able to claim benefits. Of course, these problems can in part be countered by vigilant administration of the benefit system, but employment office staff in Central and Eastern Europe often have neither the resources nor the experience to prevent abuse. What has often happened instead is that entitlement to benefits has been restricted, for example, to those with recent work experience, or confined to a limited duration of unemployment. Although these administrative restrictions may cut abuse, many people, in particular the long-term unemployed, fall through the net.

Active policies have major advantages in that they attempt to counteract market imperfections, including the inefficiencies associated with the payment of benefits. Active policies help the unemployed to find work by means of counseling and bringing together information on vacancies. They also encourage labor mobility by financing training and relocation. They may maintain an individual's work experience by providing a temporary job during a downturn, thereby improving his or her chances of getting back into work once the recovery comes. And lastly, active policies may be an effective way of stopping abuse: benefit claimants who already have undeclared jobs, or who are not seriously looking for work, will stop claiming benefits if payment is made conditional on participation in some temporary work or training scheme.

But active labor market policies are costly and subject to many of the problems of government failure discussed earlier. It is one thing to organize interviews or to finance training schemes or temporary jobs, but quite another to ensure that these activities truly benefit the unemployed. In principle, obviously, people should be trained only in skills for which there is, at least potentially, a market demand. In practice, however, the persons running the employment services are not necessarily the best judge of what skills may be needed or how new skills should be taught. They may not even have an incentive to do so. So even where active policies are desirable in principle, they may well not be cost-effective in practice.

Why, then, are such policies desirable at all? They are not, after all, used in East Asia. The answer is that the relevance of such programs depends on the context. Active programs are a response to declining industries—something unknown in East Asia, but all too well known in the West. Persistent unemployment in countries of the Organization for Economic Cooperation and Development (OECD) is now generally attributed to structural factors, themselves resulting from various rigidities in the labor market (OECD 1990c). Western governments are therefore urged to make their labor markets more flexible, in part by removing rules and regulations which impede adjustment and mobility and in part by shifting from passive to active labor market policies, such as job counseling, work experience, or longer-term training, which are designed to help workers relocate themselves from one sector of the labor market to another. An extensive survey of their experience argues that active labor market policies, if properly designed, clearly targeted, and efficiently managed, can reduce unemployment (OECD 1993). Discussion later in the chapter therefore refers to different programs in the highly industrialized countries.

Given the enormous structural imbalances and the great barriers to mobility, these arguments may apply with particular force in Central and Eastern Europe. An example may illustrate the scale of the problem. In Poland, where unemployment was at 13.5 percent and affected about 2.25 million persons in mid-1992, employment in the steel industry declined from 190,000 in 1988 to 123,000 at the end of 1991. It is estimated that restructuring will lead to another 60,000 redundancies, about half of them in one region. Although the scale of this decline is dramatic, it resembles those which have occurred in other countries, if over a longer period. In the European Union, for example, employment in steel declined from 784,000 to 434,000 between 1975 and 1985, and the competitive forces of market restructuring were substantially mitigated in their impact both by industrial policies involving continuing support for the steel industry and by labor market policies assisting labor mobility.

The main message of this chapter, therefore, is that active policy measures do have an important role but, because they are costly, intervention will have to be selective where there is no immediate market return. The need is to target measures on areas where they can be most cost-effective. To illustrate the importance of context, the return to spending on retraining is likely to depend not only on the existence of gaps in the provision of private training but also on the existence of job openings. With rising unemployment, the cost-effectiveness of expanding training schemes indiscriminately may be quite low. In these circumstances, public employment services can be effective in assisting the disadvantaged, including the long-term unemployed (Jacobson 1991) and can help maintain employment by supporting public works and temporary employment schemes.

Turning to the second question—the range of policy instruments—active labor market policies in the context of Central and Eastern Europe may usefully be divided into three types, each addressing a different element of unemployment. Although the causes of emerging unemployment vary in intensity over time and across countries, three main dimensions arise from different sources and require different policy responses:

• In the short to medium term, economy-wide unemployment is the result of aggregate shocks, such as the impact of macroeconomic stabilization and the sharp fall in aggregate demand, the collapse of central planning arrangements including the Council for Mutual Economic Assistance (see the glossary), and, for many economies, large increases in the price of imported energy and raw material inputs. These shocks cause the demand for labor to fall across the economy as a whole, although the impact is often more severe on some sectors than on others. The response to the short-term economy-wide shocks resulting from liberalization and stabilization is discussed in chapter 7, where the main emphasis is on the possible role of wage policy as a means of containing inflation and on policies to make support for enterprises more selective and linked more explicitly to employment objectives. This chapter discusses whether arguments can be made for other measures of direct job creation such as public works or temporary job schemes.

• In the medium term, structural unemployment results from sectoral imbalances caused by changes in relative prices, exposure to world trade, modern technology, and management practices. These shocks, which are fundamental to the whole process of transition, make it necessary to restructure the economy and to shift labor from one sector to another. Active policies for retraining and assisting labor mobility are crucial in this area.

• Even in the long run, the labor market will have to adjust from the excess demand regime characteristic of centrally planned economies to one in which unemployment is a permanent feature of the economic landscape. In most market economies, the public sector is the main agency responsible for running employment exchanges for unskilled and semi-skilled workers. Redeployment or job brokering policies bring together information on job vacancies and advice and expertise about employment opportunities.

These three types of active policy—information services, retraining, and job creation—are summarized in table 8-1.

The Inheritance

Although formal unemployment was not recognized in Central and Eastern Europe before the transition, some active labor policies existed and were used to ensure the availability of workers for the productive sector. Production targets were the dominant force in shaping policy for the development and deployment of labor. Although individuals could choose where to work, career opportunities were primarily subject to the needs of the enterprise within the confines of the overall plan.

Table 8-1. *Description and Assessment of Active Labor Market Policies*

Measure	Description	Assessment in OECD countries
Employment services	Placement, counseling, and vocational guidance; job search courses and intensive counseling for the disadvantaged; assistance with geographic mobility	Found to be effective in increasing job placements in United Kingdom, United States, and Netherlands
Training	Training programs focused on the adult unemployed or those at risk of losing their job; training takes place usually at training centers or in enterprises	Mixed evaluations; targeted programs, including those for plant closures, successful in Canada, Netherlands, and United States; general programs improved job prospects in Norway, Sweden, and United Kingdom but were ineffective in Federal Republic of Germany and United States
Direct job creation		
Subsidies for regular employment	Wage subsidies for recruiting or retaining particular workers (such as long-term unemployed)	Severe problems of dead-weight and substitution, implying very low net employment effect
Enterprise allowance	Grants or prepayment of benefits to allow the unemployed a capital sum to start up their own business	Most studies find quite high survival rate of enterprises, but deadweight of about 50 percent
Temporary public sector jobs (public works)	Temporary work in the public sector for the unemployed	Relatively few studies of this practice, which is now uncommon in OECD countries

Strengths

Job security minimized social conflict and stress related to job loss. Job content and wages were highly structured. Although this facilitated job mobility within and between similar enterprises (the Lenin Steel Works in Hungary had more than a 20 percent turnover annually), it did not necessarily assist occupational or geographical mobility.

So far as redeployment was concerned, labor offices in several countries assisted the movement of workers between firms and had a limited capacity to organize social and occupational rehabilitation services for disabled workers. The offices and statistical institutions also had the capacity to collect statistical information on employment. This network of offices provides a platform to help administer active labor policy and related income support programs during the transition.

An extensive training infrastructure was present in both public institutions and state enterprises. Training was partially financed in some countries by a payroll tax (for example, a 2 percent contribution in Poland). In addition, enterprises made considerable in-kind contributions to pre-service youth training and adult in-service training.

Weaknesses

The weaknesses of the old system involved unhelpful attitudes toward unemployment and job seeking, inadequate institutions for resolving political conflict, restricted occupational mobility, and active labor policies and institutions which were poorly adapted to the needs of the transition.

The inherited attitudes to unemployment and job seeking are poorly suited to the needs of a market system. Workers under the old system did not have to worry about finding employment. They were not accustomed to unemployment, nor to the continued fluctuations of and changes in labor demand which are common in market economies. Nor were they used to the varied patterns of work in a market economy, including contract employment, job sharing, and part-time work. Taking responsibility for finding work is likely to be particularly difficult for displaced workers, many of whom will expect the state—or, more specifically, the technically weak and understaffed public labor offices—to allocate them a job, as under the old system.

Inadequate mechanisms exist for resolving political conflict. The social partners—government, labor, and employers—are less useful than they might be in resolving conflict because their role was subverted under the old system. Although unions existed, they were part of the state or party apparatus and did not represent workers in the way that trade unions do in the West. Employers were also part of the state structure. There was no experience of building a consensual approach to labor market policy based on the roles of the social partners in representing their respective constituencies.

Occupational mobility was restricted. The system of wage setting did not encourage workers to acquire skills, and many workers have only narrow technical skills, often specific to industries now in decline. Both factors reduce occupational mobility.

Active labor policies were unable to address effectively the needs of the transition. The first deficiency was that before the transition, labor services were poorly funded. Given low or nonexistent unemployment, expenditure was obviously lower than in most Western economies, particularly in the case of active policies.[1] Even so, labor services were understaffed and their employees poorly trained. Hungary and Romania in effect disbanded some active labor services such as occupational counseling.[2] In addition, there was a lack of technology and labor market information to support cost-effective administration.

The second deficiency of existing arrangements was the narrow range of employment services. Unemployment benefits were limited or nonexistent; public labor offices provided minimal local matching of workers with vacancies, counseling was normally available only for selected individuals with significant social or physical handicaps, and no policies or technical expertise dealt with large-scale layoffs, because they did not occur. Access to public employment services was limited, no private services were provided, and no competition existed among service providers.

Policy toward training and retraining was narrow, did not include provisions to target the unemployed, and did not address the needs of a changing and increasingly technological era nor of private and small-scale enterprises. There was a lack of competition in the provision of training services, a lack of private training capacity, and little investment in programs providing adults with easy access to retraining for career advancement or other reasons.

A final inadequacy of the inherited arrangements was that community economic development was, and is, poorly understood. Local authorities had no experience in identifying areas where their communities had competitive advantages for specific types of investors or in marketing these advantages to potential investors. The system did not encourage individuals or local authorities to take initiatives and offered almost no experience in the use of human capital as an investment incentive, such as offering state-supported worker recruitment and training services as part of a package giving firms an incentive to invest in an economically depressed area.

The Forces Driving Change

Unemployment, democracy, decentralization, and supply/demand shocks resulting from stabilization and restructuring are the key political and economic forces of the transition. Above all, countries urgently need to develop new mechanisms to enable the labor force to respond to market signals.

Political Forces

The political dimension of unemployment has already been stressed. Citizens need to understand the nature of unemployment in a market economy, receive the necessary income support when they are unemployed, and have confidence that economic measures, including active labor policy, will alleviate and shorten periods of unemployment.

The move to democracy, combined with the need for consensus among the social partners, makes it important to establish a tripartite structure to incorporate government, labor, and employer organizations in discussions and to find out how the interests of different factions can be heard. The development of tripartite mechanisms is complicated by two sets of factors:

• Many trade unions exist whose legal status and number of members are sometimes not clear. Unions which played a key role in the revolution feel that they are now being squeezed out of the political process and that they must change their role and take the side of workers in discussions with a government which they helped to create.

• New organizations representing private entrepreneurs in small businesses are only now being created. Even where they exist, they are sometimes in competition with inherited organizations, such as chambers of commerce, which primarily represent state-owned enterprises.

Decentralization is an additional major political force. If taken to extremes, it can inhibit the development of national labor exchange mechanisms and fragment previously established occupational standards which promote worker mobility. Decentralization can also lead to retention at a local level of the majority of resources for active programs (70 to 90 percent in some countries of the former U.S.S.R.), largely ruling out redistribution between localities.

Economic Forces

The shape of unemployment is becoming clearer:

• Employment in the manufacturing sector in Central and Eastern Europe was around twice as high as in market economies. Manufacturing enterprises, on average, may therefore expect to shed up to 50 percent of their workers during the transition. Many industrial sectors and enterprises will have to make even larger adjustments, including complete closure in a significant minority of cases.

• The closure of enterprises, when translated to the local level, will create large-scale unemployment in some communities. This is a special concern where employment is concentrated in a single state enterprise. As table 8-2 shows, the variation in unemployment rates across regions within countries is greater than in many Western economies, reflecting in part the fact that the regions themselves are smaller and, more important, that there is greater concentration of production.

• Although unemployment in many countries initially occurred largely among professional and skilled workers, as the collapse deepened, the unemployment rates of unskilled workers started to rise. In most countries of Central and Eastern Europe, as in most market economies, the highest unemployment rates are now found among the unskilled.

• During the initial phase of restructuring, women tended to be more affected than men by unemployment: they were laid off more often, and their chance of being re-employed was lower. Over time, the first factor has become less important, since mass layoffs generally affect men no less than women. However, the probability of finding a new job remains lower for women.[3] One would therefore expect women to be a majority among the unemployed, yet, as table 8-3 shows, by the third year of the transition, this

Table 8-2. *Indicators of Regional Unemployment Differentials in Various Countries, 1991*

Country, number of regions, and time period	Average unemployment	Standard deviation[a]	Coefficient of variation[a]	Unemployment rate[b]		Index of mismatch[c]
				Top quarter	Bottom quarter	
Bulgaria (nine regions)						
First quarter	2.6	0.7	0.24	4.0	2.2	0.27
Second quarter	5.3	1.0	0.17	7.0	4.7	0.43
December	12.5	2.1	0.19	14.8	9.7	0.48
Czechoslovakia (twelve regions)						
January	1.5	0.7	0.47	2.6	0.9	0.31
June	3.8	3.0	0.79	6.8	1.9	1.55
December	6.6	4.0	0.60	12.6	2.7	2.82
Hungary (twenty regions)						
January	2.1	2.4	1.14	2.9	0.3	0.36
June	3.9	2.3	0.59	5.2	1.0	0.57
December	8.5	6.2	0.73	13.9	2.9	0.98
Poland (forty-nine regions)						
January	6.5	2.5	0.38	9.7	3.3	0.84
June	8.4	3.2	0.38	12.5	4.2	1.07
December	11.4	4.0	0.35	16.6	5.9	1.33
Romania (nine regions)						
First quarter	0.7	0.7	0.73	2.0	0.3	—
Second quarter	1.5	0.8	0.44	3.1	1.0	—
France (eleven regions)[d]	10.5	1.7	0.16	12.9	8.6	—
Spain (eleven regions)[d]	20.1	5.0	0.24	27.7	14.7	—
Sweden (five regions)[d]	1.9	0.6	0.30	2.6	1.1	—

– Not available.

a. Weighted by the labor force.

b. Unemployment rates for the bottom quarter of the labor force by region were calculated by ordering regions in terms of ascending unemployment rates, taking regions until the cumulative labor force passed one-quarter of the total, including the last region in the calculation with an appropriate fractional weight, and doing the same for the top quarter.

c. Computed as follows: $I = 1/2 \sum s_r |(U_r - V_r) - (U - V)|$, where r = the number of regions; r, U_r and V_r are, respectively, regional unemployment and vacancy rates; and U and V are the countries' average unemployment and vacancy rates.

d. 1987.

Sources: OECD 1992a, table 6.7, p. 257. For Czechoslovakia, Hungary, and Poland, data from national statistical offices; for Bulgaria and Romania, Commission of the European Communities 1992, except December for Bulgaria, which was provided by the Bulgarian Ministry of Labor; data on OECD countries, OECD 1989a.

is no longer generally the case. The explanation of this apparent paradox is that women drop out of the labor force. Two pressures are at work. The deterioration of services and fringe benefits provided by enterprises, such as kindergartens and maternity leave, discourages women from participating in the labor force; the decline in real family income works in the opposite direction. During the early years of the transition, the first effect was stronger. The medium-term outcome is not clear, although international comparison suggests that increasing numbers of women will leave the labor force.

• In all countries, youth unemployment rates are very high, sometimes three or four times the overall unemployment rate. As in market economies, in bad times firms protect their existing workers but no longer recruit new ones (see table 8-4).

• A conspicuous feature of labor markets in the transition has been the abnormally low outflow rates from unemployment. With the exception of the Czech Republic, the prospects for unemployed people to find a new job fairly rapidly are far below those of the industrial economies. These countries face an increasingly severe problem of long-term unemployment.

Policy

A primary objective of active labor policies is to facilitate restructuring and to anticipate, shorten, and alleviate unemployment to the extent that doing so is feasible and cost-effective. At the same time, active labor market services can increase labor productivity, particularly for the economically disadvantaged and unemployed.

The Strategy

The provision of active labor market policies in Central and Eastern Europe in terms both of expenditure and of the number of participants is set out in table 8-5. Although the table is not comprehensive, it shows clearly the difference between, on the one hand, the former Czechoslovakia, where active policies absorbed well over half of total spending on the unemployed, and, on the other, the remaining countries, where active policies absorbed only some 10 percent of expenditure. Given the exceptional achievement of the former Czechoslovakia, particularly the Czech Republic, in achieving very low rates of unemployment (2.5 percent at the end of 1992), this might indicate the efficacy of active policies. Czechoslovakia, alone of the countries in Central and Eastern Europe, attempted such policies on the scale required to provide work or training for all who became unemployed. In evaluating the Czech approach, however, three factors should be taken into account:

• In common with other countries in Central and Eastern Europe, the Czech Republic postponed the introduction of bankruptcy legislation until April 1993, thereby implicitly supporting the continuing operation of numer-

Table 8-3. *Registered Unemployment Rates of Females and Males in Various Countries, by Quarter, 1990–93*
(percent)

Year and quarter	Bulgaria		Czechoslovakia		Hungary		Poland		Romania	
	Females	Males	Females	Males	Females	Males	Females	Males	Females	Males
1990	2.0	1.1	1.0	0.9	0.8	1.0	6.6	5.6	—	—
1991										
First quarter	3.5	2.4	2.3	2.5	2.4	3.0	8.1	6.4	1.2	0.5
Second quarter	6.3	5.5	4.0	3.6	3.3	4.2	9.6	7.4	2.3	1.3
Third quarter	9.6	8.7	6.3	5.0	5.0	5.9	12.0	9.2	3.1	1.7
Fourth quarter	12.6	11.3	7.3	6.0	7.1	8.2	13.2	10.2	4.0	2.1
1992										
First quarter	14.2	13.2	6.8	6.2	8.8	10.8	13.8	10.7	5.6	3.3
Second quarter	14.8	14.7	5.8	5.1	9.9	12.2	14.6	11.1	7.5	4.4
Third quarter	17.3	16.4	5.7	4.7	11.5	13.7	15.6	11.8	9.6	5.9
Fourth quarter	19.0	18.7	5.4	4.8	12.5	14.6	15.5	11.9	10.5	6.3
1993										
First quarter	—	—	3.3a 11.9b	2.5a 12.1b	11.6	16.1	16.1	13.0	11.7	7.7
Second quarter	—	—	3.2a 12.5b	2.1a 12.4b	11.4	15.4	16.8	13.1	11.6	7.2

— Not available.
a. Czech Republic.
b. Slovak Republic.
Source: Commission of the European Communities 1993.

Table 8-4. *Unemployment Rates, by Age in Various Countries, 1991 and 1993*
(percent)

Age (years)	Czech Republic, 1991	Slovak Republic, 1991	Poland, 1993	Hungary, 1993
0–19	20.9	51.2	—	34.9
20–24	6.5	18.5	36.0[a]	15.0
25–29	6.2	15.1	28.1[b]	11.1
30–39	3.7	10.8	24.5[c]	10.2
40–49	3.2	9.8	—	—
50–54	2.3	5.6	9.5[d]	8.6
55–59	0.9	2.6	1.9	8.5

— Not available.
a. 18–24.
b. 25–34.
c. 35–44.
d. 45–54.
Sources: For Czech Republic and Slovak Republic, Ham, Svejnar, and Terrell 1993; for Poland, Poland, GUS 1993; for Hungary, KSH 1993b.

ous loss-making state-owned enterprises. This policy may assist in holding down the growth in unemployment in the short run, but does so at the expense of postponing essential economic adjustment. However, the idea that low unemployment in the Czech Republic can be attributed solely to the delayed implementation of bankruptcy laws seems implausible since bankruptcy remains virtually unknown throughout Central and Eastern Europe.

• Like Sweden, but probably more successfully than other countries in Central and Eastern Europe, the Czech Republic has a centralized wage bargaining system, in which trade unions are involved both in setting wage norms and in enforcing policy. Wage moderation may have helped to maintain employment in state and private enterprises, thus easing the task of labor market policy.

• Again like Sweden, but unlike at least some of the other countries in Central and Eastern Europe, the Czech Republic has maintained very tight conditions, especially since January 1992, on the receipt of unemployment benefits and limited to six months the duration for which people can obtain benefits.

These elements of policy are mutually reinforcing. The conditionality and limited duration of benefits put pressure on unemployed people to find work or to go onto schemes, while the availability of places enables the government to impose a restrictive regime of benefits. Employment in the economy as a whole is maintained by these measures, together with enterprise support and wage moderation. Should employment in state enterprises begin to begin to

Table 8-5. *Public Expenditure and Participant Inflows in Labor Market Programs in*
(participant inflows as a percentage of labor force; public expenditures as a percentage of GDP)

	Bulgaria Participant inflows		Czech Republic Public expenditures		Participant inflows		Hungary Public expenditures,	Participant inflows,
Program	1991	1992	1992	1992	1991	1992	1992	1992
Active measures	—	—	0.18	0.32	—	—	2.78	—
Public employment services and administration	—	—	0.09	0.09	—	—	0.16	—
Labor market training	0.5	0.5	0.01	0.01	—	0.3	0.09	1.2
Training for unemployed adults and those at risk	0.5	0.5	0.01	0.01	—	0.3	0.09	1.2
Training for employed adults	..	..	..	..	..	..	..	..
Youth measures	..	..	0.01	0.04	0.3	0.4	..	..
Measures for unemployed and disadvantaged youth	..	..	0.01	0.04	0.3	0.4		
Support of apprenticeships and related forms of general youth training	..	..	..	..	..	..	..	..
Subsidized employment	..	0.2	0.08	0.16	..	2.3	0.14	1.3
Subsidies of regular employment in the private sector	..	..	0.05	0.10	0.4	1.2	0.04	0.8
Support of unemployed persons starting enterprises	..	0.1	0.02	0.03	0.2	0.5	0.01	0.1
Direct job creation, public or nonprofit	..	..	0.01	0.03	—	0.6	0.08	0.4
Measures for the disabled	..	..	..	0.01	..	..	..	..
Vocational rehabilitation	..	..	..	..	..	..	..	..
Work for the disabled	..	..	..	0.01	..	..	..	..
Mobilizing labor supply[a]	0.6	0.7	0.08	0.19	—	2.5	0.15	2.1
Non-targeted training[b]	..	..	..	..	..	..	..	..
Work as a social objective[c]	..	..	0.01	0.04	—	0.6	0.08	0.4
Passive measures	—	—	0.23	0.19	—	—	2.39	—
Unemployment compensation	—	—	0.23	0.19	..	..	2.36	—
Early retirement for labor market reasons	—	—	..	..	—	—	0.03	—
Total	—	—	0.41	0.51	—	—	2.78	—

—Not available.
.. Nil or less than half of the last digit used.
a. Training for unemployed adults and those at risk, measures for unemployed and disadvantaged youth, subsidies of regular employment in the private sector, support of unemployed persons starting enterprises, and vocational rehabilitation for the disabled.

Central and Eastern European Countries, 1991–92

Poland		Romania				Slovak Republic		
Public expenditures, 1992	Participant inflows, 1992	Public expenditures 1991	1992	Participant inflows 1991	1992	Public expenditures, 1992	Participant inflows 1991	1992
0.32	—	0.02	0.18	—	—	1.37	—	—
0.02	—	0.01	0.16	—	—	0.09	—	—
0.02	0.4	0.01	0.01	..	0.3	0.10	0.3	1.0
0.02	0.4	0.01	0.01	..	0.3	0.10	0.3	1.0
..	..	..	..	..	..	..	..	..
0.16	2.3	..	0.01	0.1	..	..	..	..
		..	0.01	0.1	..			
0.16	2.3	..	..	..	..	..	..	..
0.08	0.9	..	..	..	..	1.16	0.8	3.1
0.04	0.6	..	..	..	..	1.01	0.5	2.3
0.01	..	..	..	..	..	0.01	—	—
0.02	0.2	..	..	..	..	0.14	0.3	0.9
0.04	—	..	..	..	..	0.01	..	..
..	..	..	..	..	..	..	..	..
0.04	—	..	..	..	..	0.01	..	..
0.07	1.0	0.01	0.01	0.1	0.3	1.12	0.8	3.2
0.16	2.3	..	..	..	..	..	..	..
0.07	—	..	..	..	..	0.15	0.3	0.9
2.51	—	0.30	0.79	—	—	0.65	—	—
1.73	—	0.30	0.79	—	—	0.65	—	—
0.78	—	..	..	—	—	..	—	—
2.83	—	0.32	0.96	—	—	2.01	—	—

b. Training for employed adults and support of apprenticeships and related forms of general training for youths.

c. Support of unemployed persons starting enterprises and work for the disabled.

Source: OECD 1993, table 2.B.2.

fall more rapidly as a result of the bankruptcy law, unemployment may rise above the levels with which active policies can cope, and the duration of benefits will have to be extended to care for increasing numbers of unplaced people. This would reduce the pressure on the unemployed to look for work, thereby aggravating the rise in unemployment.[4]

Poland, in contrast, has had to grapple with more rapidly rising unemployment. Approximately 14 percent of the labor force was unemployed in 1992, an additional 11 percent of the labor force was on short time, and the number of large-scale layoffs was significant. Poland has a comprehensive active labor policy framework but only limited institutional capacity to administer it: in 1992 active labor policy involved only about 8 percent of the unemployed, and staff/client ratios in labor offices were in the range of about 1:300. With unemployment such as that on the Polish scale, active policies and institutions cannot cope well, and, though the duration of unemployment benefits has been reduced, additional social assistance schemes have been developed (in Poland about 30 percent of individuals who have exhausted their entitlement to unemployment benefits qualify for social assistance).

In other countries, as in Poland, rapidly growing unemployment has meant that resources, both administrative and financial, have been stretched to the limit, and active programs have suffered as a result. Eastern Germany is alone in attacking the sharp decline in employment, from 10 million in 1989 to 6.5 million in 1992, using active labor market policies on a large scale. This has been possible because of the availability of resources and institutional capacity from western Germany. Schemes for retraining, for promoting occupational mobility, and for directly creating jobs (together with "passive" measures such as early retirement, part-time work, east-west commuting, and migration) kept the rise in unemployment to 1.2 million, with 43 percent of expenditures being devoted to active as against passive measures. In the countries of the former U.S.S.R., enterprise support has been maintained almost without regard to fiscal cost, and this has moderated the rise of unemployment, but at the expense of allowing inflation to run out of control.

Given this experience of active labor policies in Central and Eastern Europe, a strategy for addressing unemployment might usefully comprise the following elements:

• The starting point is to monitor the incidence of unemployment by sector (industry, region, occupation, age, and gender), and in particular the incidence of large-scale layoffs, so as to initiate policies to help the unemployed find work. In the short run, there are many more job seekers than jobs (the ratio was about 200 to 1 at the public labor exchange in Romania in late 1992). With labor demand everywhere depressed, policy for the early years of the transition may need to concentrate on measures for creating jobs, particularly in areas or sectors where unemployment is exceptionally high (for example, in company towns or among young people); such policies might be allied to linking the payment of unemployment benefits to participation in public works or training. These topics are the main subject of the section on regulating market forces.

• At the same time, although with a more medium-term perspective, the infrastructure of a labor market can be put in place. As demand for labor begins to grow and as new private firms take root, policy should facilitate the movement of labor between jobs, regions, and occupations, including encouragement of new entrepreneurs and self-employment. The institutional arrangements for employment services and training or retraining programs should be refined to assist the efficient operation of the labor market over the longer term. These topics are discussed in the section on building markets.

• First, however, is the question of what resources are available to support such policies.

Maintaining Macroeconomic Balance

Even in times of severe fiscal constraint, public spending can be justified to the extent that it can pay for itself (by reducing the need to pay unemployment benefits) or is the best available means of dealing with acute social distress and potential political unrest arising from high or persistent unemployment. More generally, constraints on public expenditure focus attention on the role of private finance. These are questions of mobilizing resources. Then there are issues of cost-effectiveness, that is, of attempting to ensure that resources are used to best effect and outcomes achieved at minimum cost. These are issues of containing costs.

MOBILIZING RESOURCES. Resources to support active labor policy may come from both private and public sources, but during the early transition the scope for private finance was very limited.

Private financing is not uncommon in market economies. Japan depends extensively on enterprises to finance the redeployment of labor. In the U.S. automobile industry, members of the United Auto Workers who are made redundant are eligible under their labor contract to receive training funded by company contributions before being released. More recently, the collective bargaining agreement signed with all three major U.S. auto manufacturing companies established income support programs which guarantee members almost full wages for the life of the contract. Arguably, private companies will insure their workers against the risks of fluctuations in the demand for their labor only if the companies themselves can off-load these risks onto the capital market. In Central and Eastern Europe, such capital markets do not exist for much of the emerging private sector, and unions would be ill-advised under current conditions to insist on these types of provisions in employment contracts. In any event, most of the unemployment results from labor shedding by state-owned enterprises rather than by private firms.

State-owned enterprises in Central and Eastern Europe often continue to support workers who have been laid off by allowing them to continue using social benefits supplied by the enterprise (in particular, housing) and, in the Russian Federation, even by financing public works on which they can be employed. But such activities are not very systematic and, in any event, their costs are likely ultimately to fall on the state budget.

Public funding may be primarily from general tax revenues (Australia, Ireland, and Poland) or from employer/employee wage taxes (Germany, Hungary, and Romania). In the case of employer/employee wage taxes, the state administers what are, in essence, private resources, earmarked for the provision of employment services, but usually including the payment of unemployment insurance. As the economic situation improves, finance should increasingly be shared between the state budget and the contributions of employers and employees. (Hungary has moved in this direction, but with difficulty since hard budget constraints are being imposed on enterprises.) In some cases, and as full employment is reached, all financing may be shifted to the employer and employee, as in Japan.

In determining how different types of active labor policies are financed, particularly in countries with rapidly increasing unemployment, the finance of cash benefits should be separated from that of active labor policies which support displaced workers. This policy has been established in Hungary, but not in Poland. In the absence of such a policy, expenditure on cash benefits tends to crowd out labor redeployment and investment policies, as occurred in Poland in 1991. However, in countries such as Albania, which are facing a particularly severe budget constraint, massive unemployment, and collapse of the formal economy, government may be forced to place the majority of resources into income support and other poverty alleviation programs.

CONTAINING COSTS. Cost containment in active labor policy can be addressed in various ways. Careful targeting of programs improves cost-effectiveness, as does combining different measures and policies (OECD 1993). A common technique in many countries, including those in Central and Eastern Europe, is to limit the time individuals may participate in any active program. Romania, for example, imposes a limit on the time spent in retraining and on the proportion of resources which can be allocated to active programs.

Research in the West and initial evidence from Central and Eastern Europe suggest that the type of targeting for displaced worker programs strongly influences the degree of participation in the program, the overall cost, the timeliness of the assistance, and the overall effectiveness (OECD 1984a). Broadly targeted assistance (aimed, for example, at an entire sector) runs the risk of escalating costs. Excessively narrow targeting, however, may exclude workers who need assistance. Regardless of the approach to targeting, declaring a group of displaced workers eligible for special assistance raises problems of equity.

Most countries in Central and Eastern Europe have not developed comprehensive capabilities to monitor, target, and evaluate the cost-effectiveness of active labor policies and the allocation of resources to particular groups. Improved targeting will require clearer definition of the intended recipients of specific programs. Automation is critical to the success of any large-scale and comprehensive targeting and evaluation system. Such work is well advanced in Hungary and is under way in Poland.

Building Markets

It is necessary to strengthen the markets both for labor per se and for the various types of labor market programs. This involves policies to increase consumer choice, diversify supply, improve incentives, and raise the productivity of labor.

INCREASING CONSUMER CHOICE. The exercise of choice requires (a) that workers and employers have sufficient information on the labor market to make rational choices, (b) that labor exchange services exist to match job seekers with vacancies, and (c) that effective assessment and counseling are available, where necessary, to assist the exercise of choice.

In the past, most labor market information in Central and Eastern Europe, although of high quality, concentrated on providing data for controlling and directing the economy, rather than on monitoring changing economic conditions. The provision of information is crucial to building a market, and all the more so during times of change. To exercise choice, workers need to know what jobs are available and their characteristics. Firms need to know the skills and attributes of job seekers. Institutions implementing active labor policy need to know the number and attributes of people at risk of becoming unemployed. Labor market information, including information on households and enterprises, should show changes in employment and anticipate changes in demand. Such information must be easily understood by individuals or institutions attempting to develop labor investment programs.

Public labor exchanges in Central and Eastern Europe under central planning were not fully automated, were primarily local, and were based on a situation of labor shortage which no longer exists. Although of limited use in reducing unemployment when an entire economy is in recession, labor exchange services match vacancies with job seekers. They also contribute to labor mobility. Even during the worst of times, the job-seeking aspect of the labor exchange should be based in the public labor office, if that office is to be responsible for operating unemployment benefits. Public exchanges need to be continually refined and automated, making it possible to increase the number of vacancies listed at employment office labor exchanges. In addition, economies in both Central and Eastern Europe, for example Hungary, and the West are increasingly legalizing complementary private exchanges.

Assessment and counseling were virtually disbanded in most of Central and Eastern Europe before the transition, and the services which were retained were mainly for a small proportion of workers with severe physical or social problems. In a rapidly changing market economy, such services should be available to a much broader range of clients. The services are particularly important for redeploying labor and also for screening potential clients for expensive training programs. Improved employment counseling, including the assessment of aptitude and interest, and the provision of information about education and training opportunities are a central aspect of labor mobility (OECD 1984c). These services have proved effective in market economies (U.S. General Accounting Office 1991) but are not well developed in Central and Eastern Europe. Job search training and related job club activities for the long-term unemployed are also generally cost-effective in market economies because they help clients to define job objectives and discover employment opportunities (Commission of the European Communities 1990; Leigh 1990). Initial evidence in Central and Eastern Europe is also positive.

DIVERSIFYING SUPPLY. This part of the strategy deals mainly with diversifying the supply of labor market services—in particular services which assist the redeployment of labor—and training and retraining services.

Deploying labor more effectively. The liberalization of labor redeployment policy and, in particular, the regulation of private employment services, has been the subject of considerable debate, as exemplified by the International Labour Organisation convention no. 96, which provides for the regulation of private employment services. The major concerns relate to the licensing of employment agencies and training establishments, the accreditation of courses, and the monitoring of standards. However, the restrictions on private services addressed in the convention do not reflect current conditions, and the monopoly of the public employment service, where it exists, is being challenged (World Association of Public Employment Services 1992). There are two sets of reasons for this trend: (a) public employment services in Central and Eastern Europe are having increasing difficulty fulfilling all market needs, and (b) notwithstanding the extension of services offered by public institutions, private intermediaries are effectively filling the gap and are increasing in number and importance (EEC/ILO 1991).

Private agencies in the West tend to target their services to particular segments of the labor market, for instance, white-collar workers or workers who already have a job but are looking for a new one. Many countries, such as the United Kingdom and the United States, encourage private fee-paying placement services (although regulations of the International Labour Organisation permit fees to be collected only from the employer), and Hungary has also enacted legislation to allow private services. Some countries, such as Belgium, France, Germany, and Norway, apply considerable regulation to such services, although this regulation is being reduced in the light of the Social Charter of the European Union. A few countries, such as Denmark, Poland, and Spain, have laws which do not allow such services to operate at all. Countries in Central and Eastern Europe need to reevaluate the relationship between private and public service providers. Creation of an effective mix of private and public providers, together with an appropriate regulatory structure covering both sectors, is essential if the advantages of having both systems are to outweigh the disadvantages.

Increasing the skills of labor. Policies could also support private and public retraining programs. Experience indicates that the introduction of market mechanisms increases the responsiveness of retraining programs to labor market needs (Leigh 1992). Contracts are usually competitive and impose conditions which regulate quality: contracts can be for a selected number of training slots or for a number of days from a selected training agency; the training contractor may provide modular training with some degree of flexible entry and exit for clients; the training contractor may also provide additional counseling services; contractors may be required to accept responsibility for successfully placing a portion of the trainees (for example, 70 percent); and a specified minimum may be set below which the starting wage must not fall.

Several Western economies—Sweden and the United Kingdom—which used to maintain an internal training capacity for displaced workers have divested themselves of direct control and opened training to market forces. Poland has a highly developed private and semi-private training capacity (approximately 700 registered training associations in 1990), many of which provide training services for displaced workers on a contractual basis for local labor offices. This suggests a change in the role of employment services: instead of public provision of training, policy would rest on a dual foundation of public funding for displaced workers and an appropriate regulatory structure to ensure quality, with the training itself contracted out to public and private agencies.[5] Programs promoting occupational mobility, however, may not be a priority, particularly for older and married workers given the factors, such as housing, which restrict mobility among displaced workers. Moreover, evaluations in Western economies have not found these programs to be very effective (OECD 1984c).

IMPROVING INCENTIVES. The incentive system chosen to support the development of labor markets will depend on which labor adjustment model is selected and on which of the key players—the individual, the employer, or the government—takes the major role.

• Countries in Western Europe allow major government intervention in adjustment and in support of active labor programs. Many countries have enacted laws which make it difficult for enterprises to shed labor and require them to give lengthy advance notice and compensation to employees. Governments allocate considerable resources for active labor programs, the highest being Sweden, at just over 2 percent of gross domestic product (GDP).

• Japan's strategy relies heavily on private enterprise to bear the major responsibility for alleviating the adverse effects of restructuring. Concern with employment security has always been a guiding principle. For that portion of the work force covered by Japan's lifetime employment (roughly 30 percent), unnecessary workers are moved from job to job within a company and afforded the requisite training in the process. Unique to Japan is the concept of loaning redundant workers to other companies. The government absorbs some of the costs of these arrangements and allocates about 0.5 percent of GDP for labor market programs.

• In the United States and Canada, worker adjustment strategies give displaced workers a central role. The private employer, however, is playing an increasingly important part in assisting the readjustment process, as are state and federal governments, by enacting a variety of income support and employment and training schemes to ease the transition to new employment. The United States allocates an average of about 0.4 percent of GDP to active labor programs.

If individuals are to be the key players, they must be given appropriate incentives, such as substantial wage differentials and easy access to effective redeployment and retraining services. Disincentives must also be removed: excessively generous cash benefits; artificial blockages to employment, such as occupational certification or other forms of discrimination; and other barriers to mobility.

If employers are to take the initiative, again, the incentive structure needs to be right. So far as redeployment is concerned, they should have easy access to high-quality services and no unnecessary impediments to hiring and releasing labor or to transferring personnel within the organization. To encourage employers either to invest in labor directly or to support external labor investment policies (for example, through training levies financed by payroll deductions), government regulations and labor contracts should allow such investment to be integrated with related business decisions such as changing production processes. In addition, the quality of labor investment should be regulated, and training standards should reflect market needs.

Agencies providing services must also have incentives if they are to implement policies in a cost-effective way: some of the more important incentives are embodied in the training contracts just described, which prescribe performance criteria relating to the numbers of workers placed and their subsequent earnings.

RAISING LABOR PRODUCTIVITY. There are many routes to raising the productivity of labor, not least the improved management of existing enterprises. In this section, we examine two ways in which labor mobility can raise productivity: helping individuals make better use of their existing skills by moving from a low-productivity to a higher-productivity job and investing in human capital through training and retraining to improve the skills of the labor force.

Employment services are primarily related to the first route. In Central and Eastern Europe, as in most Western economies, they must increasingly go beyond their former simple matching function in order to (a) identify and react to imminent layoffs and (b) help individuals assess their capabilities, determine their need for new skills, and initiate self-directed job searches. The cost of providing these services is low, although cost-effectiveness varies for different categories of workers (Jacobson 1991; Johnson, Dickinson, and West 1985; for a fuller review, see Fretwell and Goldberg 1993).

Retraining policies are one aspect of skills improvement. The economic return to retraining is controversial, both because of its expense and because of its low return, particularly in excess supply situations. Because of the structural nature of a significant part of unemployment in Central and Eastern Europe, however, the option must be considered. The question, perhaps, is not if, but rather how and to what extent retraining can be accomplished. Too much training too soon is not warranted, but demand signals clearly show exactly where labor is in excess supply (mining and heavy industry) and where labor is scarce (the service sector, accounting, information technology).

Training for unemployed adults is made available to approximately 1 percent of the labor force each year in the highly industrialized countries (Scherer 1990), as opposed to about 0.1 percent in Poland and 0.5 percent in Romania. The differences reflect multiple factors: a lack of labor market information, a lack of requests for retraining, problems in determining what programs to initiate, difficulties in mounting flexible retraining programs, financial constraints, and a philosophy of waiting until demand emerges at the local labor exchange, which may never occur.

Research indicates that displaced workers may have particular needs which affect the design of retraining policies (OECD 1984a), and initial anecdotal evidence from Central and Eastern Europe suggests similarities with OECD trends. Public employment services in some countries (Romania and Russia, for example), like those in OECD countries, are involved in organizing, if not themselves providing, retraining to displaced workers. Public employment service offices which are also involved in administering retraining programs tend to achieve higher returns in terms of wages or placement rates than those which are not (U.S. General Accounting Office 1991).

Generalizations about the costs and benefits of government-sponsored training as a part of manpower policy are hazardous. Particular care is needed to distinguish between programs which provide little screening or job counseling prior to entering retraining and those which are more exacting. For example, initial retraining programs in Hungary encountered high dropout rates, but improved screening and delivery of training is now achieving up to 60 percent placement. A U.S. study (Manpower Demonstration Research Corporation 1993) examined the cost-effectiveness of training some 20,000 participants. The study included control groups and took account of such factors as salary forgone because of the time spent in retraining. The findings of the study show that providing general education skills to adults with less than a high school matriculation has a significant impact on earnings, that on-the-job training has a significant impact for both men and women, and that classroom training tends to have a more significant impact for women than for men.[6] If training is to make a difference, especially for higher-paid workers, it may have to be more intensive, or more long term, than that currently offered (Seitchik and Zornitsky 1989).

Many adult workers in Central and Eastern Europe may have limited general education (about eight years in Russia) complemented by narrow skill training. These workers will be faced with job changes, evolving management practices, and advanced and unfamiliar technology. Consideration should be given to retraining policies which include job counseling and assessment, as well as remedial general education, management training, and traditional vocational training.

In summary, training is expensive and should not be undertaken lightly or without first investigating other options, varying the type of training for different clients, and carefully screening clients before entry.

Regulating Market Forces

The major form of intervention in this context is policy to address unemployment, of which two sets of policies stand out: those which anticipate unemployment and those which seek to create jobs.

ANTICIPATING UNEMPLOYMENT. Policies to anticipate unemployment have two central elements: labor market information and industrial adjustment policies. Determining the extent of large-scale layoffs and developing policies to reduce their impact are among the most critical issues in labor market adjustment in Central and Eastern Europe. Evaluations of worker adjustment programs in Western economies indicate that advance notice of a plant shutdown or large layoff is a useful first step in promoting smooth adjustment (OECD 1984a, 1988c). Layoff notices are usually backed by the provision of other relevant guidance and employment services. Advance notice legislation, regulations, and services have been enacted in most Western countries undergoing economic restructuring and have been developed in Central and Eastern Europe in countries such as Hungary and Poland.

Industrial adjustment services are often cost-effective in Western countries (Employment and Immigration Canada 1984), and initial evaluations in

Poland indicate that similar approaches can be developed in Central and Eastern Europe. The services involve the transfer of authority to the local level, so as to involve the employer and the local community. One of the most compelling reasons for introducing such services has been the damage caused to a local community by the disappearance of its sole or dominant employer. Policies seek to establish a framework under which (a) public funds are concentrated on the most vulnerable areas and (b) local organizations, trade unions, employers, and local and central governments work together to solve practical labor problems. One of the objectives, both for incentive and for social solidarity reasons, is to demonstrate to the worst-affected areas that public funds, indicative of the broader community's involvement, will be used to buttress local efforts to resolve difficult adjustment problems (OECD 1985). Representatives of both the public and the private sectors, including the affected enterprises, should be involved; an example is the *Beschaftingungsgesellschaften* in Eastern Germany. However, efforts to have the private sector undertake the full responsibility for leading and financing industrial adjustment have not proved successful because of the resource constraints in communities where enterprises are closing their doors.

CREATING JOBS. The creation of jobs is economically difficult and politically highly sensitive. Policies center around encouraging self-employment, providing temporary jobs, and promoting local economic development.

Entrepreneurship. Throughout Central and Eastern Europe, there is considerable interest in creating jobs, particularly for the unemployed, by developing small businesses. The option of developing micro-enterprises is successfully used in the West by only a small portion of the unemployed (2 to 3 percent; OECD 1988b), and initial experience in Poland reflects the same percentage. This option might, however, be of greater significance in Central and Eastern Europe because of the poor development of small businesses, although evidently the scope for new entrepreneurial activity depends very much on deregulation policies and on the availability of finance.

Entrepreneurship programs usually reach only a small segment of the unemployed (individuals who are middle-aged, have a postsecondary education, are predominantly male, or who possess professional and technical backgrounds). In the context of Central and Eastern Europe, however, women tend to be better qualified relative to men than in the West, in both their education and work experience, particularly in the service sector. These programs may therefore have a role in encouraging entrepreneurship among women. In the West, however, the survival rates of such firms are often low (50 percent are still operating after thirty-six months in the United Kingdom), although success can be increased by providing technical assistance and credit (Mangum, Mangum, and Bowen 1992; Purdy 1987).

Entrepreneurship is supported in several countries by capitalization of unemployment benefits (Czechoslovakia, France, Hungary, Poland, United Kingdom, and United States) and by provision of technical assistance to the unemployed through the local labor offices (Hungary, Poland, and Romania).[7] Poland and Hungary have had difficulty administering these schemes through local labor offices because of a lack of staff and expertise. There appears to be a need to strengthen private and local nongovernmental organizations so that they can provide more effective assistance to the unemployed who wish to start firms; the role of the labor ministry would be limited to assessing the interests of the unemployed.

Transitional private employment. Schemes of this type offer wage subsidies to existing enterprises to hire the unemployed and give them on-the-job training, with some guarantee that a fraction of the persons hired will be retained at the end of the subsidy. This arrangement may be difficult to promote among large firms, which are shedding rather than adding workers. These policies have received mixed evaluations in the West (OECD 1982). The Socially Purposeful Jobs Scheme, which has been the main element of active labor policy in the Czech Republic, is of this form. Hungary and Poland have also designed limited programs which pay part of the cost of adding new employees for a limited time, if those employees are retained for a minimum period after the subsidy is removed. Similar job insertion programs have proved successful in Turkey (United Nations Development Programme 1992). Arguably, such subsidies may encourage enterprises to hire unemployed workers entitled to the subsidy for existing jobs rather than for newly created additional jobs. Nor are policies of this type effective in speeding up the adjustment process (OECD 1988b).

Temporary public sector jobs. Temporary job creation, often referred to as public works or public service employment, may be useful in Central and Eastern Europe because such programs can provide labor for environmental cleanup and infrastructure development, while temporarily lowering unemployment, reducing social stress, and keeping workers attached to the labor market. The difficulty is that such schemes may perpetuate public funding of what may be perceived as non-productive work. Poland and Hungary have enacted policies to encourage the creation of temporary employment, and eastern Germany and the Czech Republic have instituted extensive programs. Experience in eastern Germany suggests that these programs should be targeted to the disadvantaged and the long-term unemployed, that training elements should be included where possible, and that there should be concurrent job search requirements (Spitznagel 1992). These programs have also been used in many OECD countries, but their popularity has waned because they do relatively little to improve an individual's long-run job prospects. If alternative opportunities become available, however, public service employment may serve as a bridge between jobs. The risk is that dependence on public employment can develop, leading to longer tenure in the public employment service job than budgets warrant (OECD 1984a).

Cook (1985) conducted an extensive evaluation which concluded that public service employment programs in the United States are not appropriate for individuals who lack even the most basic skills necessary for employment. Based on the experience of such programs, public service employment should use carefully designed selection criteria, be prescribed for a limited time to minimize dependence, and impose a limit on the maximum wage, both to contain costs and to ensure that take-up is confined to those genuinely unable to find work elsewhere.

Economic development. It should not be forgotten, finally, that the primary ingredient for increasing the demand for labor is economic growth, and the major objective of active labor programs is to help ensure a supply response. Labor programs may also play a role in developing demand. Local labor offices in Central and Eastern Europe were historically often little more than passive repositories for information on unskilled or semi-skilled individuals looking for employment. As major restructuring occurs, however, local labor authorities could become more active partners in local economic development, as in some market economies where labor offices help to screen employees and organize custom-made training programs for firms agreeing to make new investments and to support community development. Examples include the Canadian Community Futures Programme and the Australian Regional Development Programme.

Implementing the Reforms

It should be clear from the multiplicity of programs discussed (table 8-5) that active labor policies are administratively demanding. They are also, and very obviously, highly political. Implementation of these policies requires significant development of both institutional and political capacity.

BUILDING INSTITUTIONAL CAPACITY. Public institutional capacity for implementing active labor policy was narrow and incomplete during the period of central planning: unemployment was not recognized, enterprises played the dominant role in providing social services, and private services were nonexistent. Institutional development is therefore needed in three broad areas: the supply and finance of private alternatives, the broadening of the legal and administrative structure to support both private and public services, and the development of personnel.

The restrictions discussed earlier on the operation of private employment services should be lifted and private training organizations allowed to develop. In addition, in countries like Poland the status of nongovernmental organizations remains unclear: in countries like Romania the laws regulating their activities are outdated and should be reviewed. These laws should be rewritten to ensure that nongovernmental organizations, which are major providers of active labor services in many countries, can deliver the services effectively and, given the fiscal constraints, can receive private as well as public funding.

Legal and administrative structures raise two sets of issues, one related to centralization and decentralization and one related to the links between the administration of cash benefits and that of active labor policies.

Active labor policies, particularly those related to unemployed and displaced workers, are generally administered and often financed by ministries of labor or public employment services. Various administrative structures can be used, depending, among other things, on the mix of policies. The trend is toward creating quasi-independent and autonomous government bodies, such as Germany has had and as Poland put into place in 1992, as opposed to maintaining the administration within a government ministry, as Romania has done. The advantage of keeping the administration within a ministry is that the ministry coordinates policy and administrative action; the advantage of adopting the more autonomous approach is that it strengthens nongovernmental input and cooperation (particularly if it is tripartite) and promotes greater operational flexibility. A totally autonomous or decentralized approach, however, can complicate the development of national labor policy.

The linkages between the administration of cash benefits and active labor policies are contentious and complex. Countries such as Australia, Germany, Romania, and the United Kingdom integrate the administration of active labor market policies and passive policies (cash benefits). Others such as Canada and the United States separate the two; Albania is considering such a policy. It is difficult to identify an overall trend. Indeed, Canada and the United Kingdom have moved back and forth between integrated and separated approaches.

Administering all programs together promotes administrative coordination and assures that clients receiving cash benefits have direct and immediate access to active programs. In addition, as more services are created, the existing network can provide a ready-made administrative vehicle to organize them, as has happened in Romania. The disadvantage of such an approach is that it tends to centralize program operation, sometimes reducing flexibility. In addition, during times of rising unemployment, most resources and staff are often shifted to the cash benefit programs at the expense of active programs.

In addressing these issues, countries in Central and Eastern Europe should bear in mind two striking features of the Western experience: the frequent changes in structure which occur and the rather different cultures of the two services. Labor redeployment and investment policies need to be dynamic and responsive to local economies and are mostly concerned with positive access; passive cash benefits policies are highly regulated, and part of the responsibility is to control fraud, or negative access.

The administration of active labor policy requires trained and experienced staff, a major issue in Central and Eastern Europe, where new policies are being enacted in the face of rapidly rising unemployment. Staffing ratios vary widely, as shown in table 8-6. The differences are explained partly by the types of services offered in different countries and partly by their quality. The data should be interpreted with care, however, because they do not always compare like with like. For example, a country like France, in which the public employment service operates adult retraining itself, will tend to have a higher staff/client ratio than a country like Poland, in which adult training services are contracted out. Nevertheless, in Central and Eastern Europe, limited funds, increasing workloads, and civil service hiring practices have all created serious problems with staff recruitment and training. The types of employees needed to deliver the new labor redeployment policies and administer labor investment policies are very different from those of the old system.

BUILDING POLITICAL CAPACITY. Given the high labor force participation rate under communism, unemployment can be expected to cause even more personal hardship in Central and Eastern Europe than in the West, particularly in those countries with no prewar history of being a market economy. As discussed in chapter 13, this makes the politics of unemployment a sensitive, indeed potentially explosive, area.

In part, the political capacity to deal with this situation must involve a general buildup of experience with democratic politics. This buildup will involve judgments about the balance between fiscal restraint, economic rationality, and political expedience, all three of which can be in conflict. Part of

Table 8-6. *Ratio of Labor Staff to Unemployed in Selected Countries, 1988*

Country	Ratio of staff to unemployed
Sweden	1:9
Austria	1:33
Germany, Fed. Rep. of	1:37
France	1:78
Italy	1:88
United Kingdom	1:98
Portugal	1:120
Poland	1:225
Turkey	1:375[a]
Mexico	1:400[a]

Note: Labor staff were those in public employment services only, including the administration of unemployment benefits.

a. Ratio of staff to applicants, no unemployment benefits, limited core services, and active programs.

Source: Scherer 1990.

the answer also involves building labor market institutions. Setting up a tripartite dialog between government and representatives of workers and employers is critical in this context.

Constraints

The major political constraint is the patchy understanding among the social partners about the need for active labor policies, the different types of policy, and their costs. This can make it difficult for one partner to implement policy or to defend it in the wider debate about expenditure on the social sectors versus other sectors.

The main economic constraints are acute fiscal limitations, the pressure to implement cash benefit programs, a shortage of private resources, and the difficulty of determining where and when to make labor investments in a time of major recession. These constraints frustrate attempts to implement active labor policies at the level found in the Western economies. Enterprises are reluctant to invest in labor because of financial pressures, uncertain needs, and layoffs. In many cases, enterprises are closing internal retraining facilities, or lowering enrollments, and turning the facilities over to labor and educational authorities, who also lack funds to operate them. Finally, because of inflation, slow wage decompression, and continuing problems of access to credit, few individuals have resources to invest in retraining or employment promotion initiatives.

Institutional capacity constraints are also severe. Central and Eastern Europe has insufficient private sector capacity, inadequate legal and regulatory structures to support implementation of private or public programs, and inappropriate public infrastructure. Policymakers tend to believe in simple and complete solutions, an urge to find the silver bullet, rather than to create a range of policies which can be targeted to different clients.

Conclusions: Priorities and Sequencing

The early years of the transition were characterized by a reluctance of enterprises in most of Central and Eastern Europe to shed labor on a large scale. When output fell, the main adjustment was on hours worked or on wages rather than on employment. As a result, unemployment in the first instance concentrated on new entrants to the labor market (youths), on women with young children, and on older workers. This phase provided an opportunity for governments to adjust policies before major hard layoffs were made.

What should be done? As noted at the start of this chapter, it is unrealistic to aim to eliminate all unemployment. In most countries, significant rates of unemployment are likely to continue for many years and will play an important role in restraining wage demands, in encouraging work effort, and in facilitating labor mobility. Nevertheless, the economic devastation of industries and communities threatens to impose unacceptable economic, social, and political costs. In the short term, the policy strategy entails financial assistance to the locality to alleviate these costs; in the long term, it entails measures to improve labor market flexibility in response to changing demand.

Short-term Policies

The overriding short-term priorities are to monitor emerging unemployment, to anticipate possible enterprise closures and mass layoffs, and to develop policy actions which shorten and ameliorate the impact of unemployment through job creation programs. A start should also be made on developing institutions to facilitate labor mobility in a market economy. This includes parallel development of appropriate wage and employment policies (chapter 7) and effective cash benefit programs (chapters 9 and 10).

Early action is needed to anticipate unemployment:

- Advance notice by employers of closures or mass layoffs is essential to enable policymakers to design appropriate responses.
- Close liaison is needed with other policies addressing unemployment. In particular, the design and administration of active policies should be carried out in liaison with the delivery of unemployment benefits.

Policy options should be identified at an early stage. Policies to assist industrially devastated areas include:

- Entrepreneurship and small business assistance
- Employment subsidies
- Public works or public service employment and
- Identification of opportunities for local economic development.

The merits and costs of each will need to be evaluated. For example, if the prospects for economic revival in the long term are good, a strong economic case can be made in the short run for temporary work on environmental and social infrastructure improvements; if the prospects are poor, employment subsidies may be more cost-effective. The choice of instrument will also be influenced by the politics of unemployment.

Social dialog is critical and entails the following:

- Trade unions and employers should work with government to identify policy options and implement policy.
- The dialog should go beyond reaction to short-term crisis. Unions and employers can and should play a key role in the design and delivery of retraining schemes, including in-service retraining, and in the provision of labor market information.

Medium-term Policies

Policies with a medium-term dimension are of two sorts. They serve, first, to develop institutions, from improved employment exchanges to retraining schemes, which increase labor mobility and thus enhance the ability of workers to respond to changes in the demand for labor. Second, they deal with large-scale layoffs which in many cases are an inevitable outgrowth of privatization.

Retraining schemes should be established. In the short run, retraining suffers from constraints both on the supply side (lack of skilled trainers) and on the demand side (lack of jobs for trainees). There is no doubt, however, that in the long run a major problem in restructuring will be a skills gap, and that substantial resources will have to be devoted to retraining. Various practices have been employed by Western countries with respect to the organization and finance of training. Countries in Central and Eastern Europe will, for budgetary reasons, be interested in schemes in which some or all of the cost can be shared with employers, trainees, or trade unions.

The administrative framework needs to be refined.

- Initially, most active labor policy emanates from labor ministries. Once short-run policies have been enacted, however, different administrative structures for ministries of labor and public employment services should be reviewed. The flexibility which decentralization makes possible should be balanced with the need to foster labor mobility through national labor market information and labor exchange services.
- Plans should be made to automate services and to develop a management information system to facilitate the planning and evaluation of different policies.

Policies should be created to deal with large-scale layoffs. These policies should build on experience during the early transition, with three objectives: to identify staff affected by restructuring, to develop tailor-made labor adjustment plans for displaced workers, and to provide services for individual workers:

- All the social partners (representatives of workers, employers, and government) should be involved.
- Measures should be linked to the privatization program and tailored to diverse industrial sectors and geographic circumstances.
- Resources should be allocated to local organizations which would have the major responsibility for administering the measures.
- Operations should be coordinated by an appropriate body at the national level, perhaps a subgroup of an overall privatization steering committee.

The initial shock of emerging unemployment has been considerable. Individual citizens and policymakers are rethinking the speed of reform, particularly since the institutions for social protection and labor redeployment are having difficulty coping with the influx of displaced workers. Experience is showing that it will take time and resources to reorient labor institutions toward providing the human capital necessary to support the transformation. The absence of the necessary change in labor policy and institutions will slow down the economic reform. The question is how to balance the speed of the overall economic reform with the reform of labor and social programs.

The problem is made more difficult because the returns to active labor policy are hard to quantify: the costs are clear enough, but the benefits of reduced unemployment include many intangibles. The real costs of unemployment are not only the lost output and the cost of income support. Unemployment is also a personal matter affecting human dignity, imposing stress on family life, and creating many of the problems discussed in chapter 4. When it affects the young, the unskilled, ethnic minorities, and older workers, it falls on those least able to absorb the cost. The challenge is both economic and political. It must be met if restructuring is to proceed.

Notes

1. In 1991, Poland spent 1.7 percent of GDP on labor market programs (0.3 percent for active policies and 1.4 for income support); comparable figures were 3.3 percent in Sweden (2.0 percent active), 3.2 percent in Spain (0.8 percent active), and 1.5 percent in the United Kingdom (0.5 percent active) (see OECD 1991b).

2. For example, the staff-to-client ratio in public employment services in Romania was 1:470 in November 1992, compared with an average of 1:64 in selected OECD countries.

3. In Poland, for instance, Lehmann (1993) found that participation in training programs appeared to increase the re-employment of men, but not of women.

4. Conspicuously, in the Czech Republic unemployment started to fall quite sharply when benefit entitlement was cut in January 1992.

5. The distinction between the state (a) as provider of publicly produced services and (b) as the source of finance, and possibly also regulation, for privately produced services is crucial. See chapter 2 for a fuller discussion.

6. A possible explanation of the last finding is that women tend to be more highly educated than men (see, for instance, Fong 1993, table 7).

7. Some countries, for example, allow individuals to receive their entire entitlement to unemployment benefits (twelve months of benefits) in a single lump sum.

Nine

INCOME TRANSFERS: SOCIAL INSURANCE

NICHOLAS BARR

A KEY MESSAGE OF THE PREVIOUS TWO CHAPTERS is that wages and employment should assist the efficient use of labor. Some other instrument is therefore needed for distributional purposes. Cash benefits are thus central to the economic aims of the reforms.[1] They are also, and obviously, highly political. If unemployment benefits are too low or paid too late, or if, more generally, too many people remain too poor for too long, the reforms could founder for lack of political support. The backlash against the reforms in some countries is already clear. This chapter and the next seek a meeting ground between the economic pressures to contain costs and the political pressures to pay higher benefits. They contain four main messages:

- The state in Central and Eastern Europe has a continuing and substantial role to play in cash benefits, as it does in the highly industrialized countries (see table 1-3).
- The major purpose of the system during the early transition was to address emerging poverty. In particular, during the current fiscal crisis, poverty relief should be given priority over the other aims of cash benefits, such as insurance and income smoothing (see the glossary). This emphasis on poverty relief is intended not as a value judgment about the proper role of cash benefits, but as a response to the fiscal crisis. As soon as the fiscal situation allows, other objectives will come into play, the balance of which is a matter for each country to decide.
- Both to achieve these short-run aims and to ensure that cash benefits assist the operation of labor markets, the inherited arrangements throughout Central and Eastern Europe need to be reshaped. Targeting can be improved by increasing expenditure on unemployment benefits and poverty relief, while cutting spending on some other benefits.
- The savings from the cuts in other benefits could finance much of the additional cost of the increase in unemployment benefits and poverty relief.

Throughout the two chapters, following conventional usage, cash benefits are divided into three types:

- *Social insurance* is awarded without an income test on the basis of (a) a person's contributions record and (b) the occurrence of a specified event, such as becoming unemployed or reaching a particular age. The essence of social insurance is that it offers protection against what, at least in principle, are insurable risks.

- *Universal benefits* are awarded on the basis of a specified event, without regard to a person's income or contributions record.[2] Examples include family allowance and free health care.
- *Social assistance* is awarded on the basis of (a) an income test (see the glossary) and (b) the occurrence of a specified event, without a contributions test.[3]

This chapter discusses social insurance. Chapter 10 looks at family benefits (a substantial part of which are universal benefits) and social assistance. Income transfers should be regarded as an integrated system, so the two chapters should be read as a whole. The system should be seen in its broader context. Historically, cash benefits made up a much smaller share of gross domestic product (GDP) in Central and Eastern Europe than in the West, with consumer and producer subsidies and public investment making up a much larger share.[4] The impact on welfare of the cancellation of most subsidies and the dramatic fall in public investment is an important part of the backdrop.

Poverty relief is a recurring theme. Its definition and measurement pose a series of questions for policymakers. First, how is the poverty line defined? Under an absolute definition, poverty is defined relative to the cost of buying a given quantity of goods and services; with a relative definition, it is compared with the income and consumption of others; with a subjective definition, it is compared with a person's expectations and perceptions. It is well known that a "scientific" definition of poverty is a mirage; all definitions of poverty, ultimately, are political (see Atkinson 1987, 1989, chap. 1, 1991a; Barr 1993b, chap. 6; World Bank 1990b, chap. 2, 1992c, chap. 1). Nevertheless, the state cannot avoid the issue merely because there is a conceptual impasse. Policymakers need to consider what is necessary for survival, the fact that people require different amounts, and the need to work out the cost of the minimum consumption basket. There is no easy answer: the poverty line has to be based on the best available data on people's incomes, coupled with a view of political realities.

The yardstick by which poverty is assessed raises a separate set of problems. First, what income should be included? Should home-produced goods, such as vegetables, be included? In principle they should, but the resulting administrative burden is acute. Second, whose income should be included? A narrow definition includes only individuals and married couples; a household definition includes the income of, say, grandparents sharing the family home; an extended family definition includes the income of close relatives like grown-up children living elsewhere. Narrower definitions reduce the risk of excluding the genuinely needy but are more expensive. A separate problem is how to treat families of different sizes. Should the poverty line for a family of four be four times that for a single person, or should it be lower, for instance, because small children consume less than adults?

Measuring the extent of poverty is difficult even where a definition of the poverty line has been agreed upon. Policymakers are interested in *how many* people are poor (the headcount measure), in *how much* their income falls below the poverty line (the poverty gap measure), and in *how long* they are poor (that is, whether their poverty is transitory or long run). All three measures require large amounts of up-to-date information on individuals and families.

A final problem is that the meaning of poverty relief is ambiguous. Poverty relief has two possible objectives: (1) eliminating poverty by bringing everybody above a poverty line or (2) reducing the poverty of those below a given poverty line by increasing their income, without necessarily eliminating the poverty gap. The purpose of the first policy is, in principle, to reduce both the headcount measure and the poverty gap to zero. The second strategy seeks to reduce the depth of poverty of poor people, without necessarily implying any reduction in their number.

Targeting is another recurring theme. As discussed in chapter 6, it has two aspects:

- To ensure effective poverty relief, benefits should go to all who need them (that is, there should be no significant gaps in coverage). This aspect of targeting is known as horizontal efficiency.
- To contain costs, benefits should go only, or mainly, to those who need them (that is, there should not be excessive leakage of benefits). This aspect is known as vertical efficiency.

Most political discussion focuses on vertical efficiency, in that it relates to the cost of benefits, but horizontal efficiency is equally important, since it has a direct bearing on the effectiveness of poverty relief.

There is a mistaken view that accurate targeting is possible only by awarding benefits on the basis of an income test, where the amount of benefit is directly related to individual or family income. Income tests have major costs, however: they are administratively demanding, they can be intrusive and hence stigmatizing, and they can create major disincentives to work effort and saving. Moreover, accurate targeting does not always require an income test. The poor can be identified in two ways: through an income test or via indicators of poverty, that is, possession by the individual or family of one or more easily observable characteristics which are highly correlated with poverty, for example, poor health, old age, or the presence of children in the family. Social insurance benefits are based on factors such as unemployment, poor health, or old age and thus use indicator targeting. Similarly, family allowance, being based on family size, is also well targeted, notwithstanding the absence of an income test.

The difficulties should not be underestimated: defining a poverty line is problematic, tight targeting can create a tension between the economic and political sustainability of the reforms, and indicators of poverty may be imperfect. For all these reasons, targeting is a matter of considerable subtlety (see Atkinson 1993).

The Inheritance

The system of social insurance which countries inherited at the start of the reforms was well established. This was one of its great strengths. But the system was also wasteful and poorly adapted to the needs of a market economy.

Strengths

As discussed in chapter 3, the old arrangements comprised a wide-ranging and mature system of social insurance, including sick pay, pensions, and generous family allowances. The system covered the great majority of workers and their families. These benefits, combined with more or less guaranteed employment, gave workers and their families a considerable degree of security.

Weaknesses

The major adverse political inheritance was the view that the state and the enterprise were jointly responsible for the well-being of the individual. The major adverse economic inheritances were lack of targeting and weak administration (for studies of Hungary, Poland, Romania, and the Russian Federation, see World Bank 1992b, 1992d, 1993b, 1993c, respectively; see also Kopits 1992). The lack of targeting manifested itself particularly in the form of ineffective cost containment, holes in the social safety net, and adverse incentives.

INEFFECTIVE COST CONTAINMENT. Expenditure can be excessive because the level of benefits is high or because benefits are easy to obtain; it may also be possible to combine work with receipt of benefits on generous terms.

High benefits arise, first, because many benefits are fully related to earnings. It is not self-evident that this should be so. Flat-rate unemployment benefits are becoming more common, as in Bulgaria and Poland, and Estonia and Latvia have both flat-rate unemployment benefits and flat-rate pensions. Another possibility is to have a basic, flat-rate pension plus a second-tier earnings-related pension which rises less than proportionately with earnings. Furthermore, most benefits are increased in line with changes in earnings. When real earnings fall, costs decline, but so does the minimum benefit. This reduces the effectiveness of poverty relief. When real earnings rise, however, the use of an earnings index adds to the cost of benefits. Thus, tying benefits to earnings hinders poverty relief if real earnings fall and makes cost containment less effective if real earnings rise.

A second class of problem with cost containment is the ease of access to some benefits. Early legislation tended to define unemployment broadly. The 1989 Polish law, for instance, awarded benefits to all persons without a job, even if they had no recent attachment to the labor market. Access to pensions is also easy. Normal pensionable age in Central and Eastern Europe (sixty years for men, fifty-five for women) is low by Western standards, and some groups, such as miners, teachers, and ballerinas, are eligible for pensions even earlier.[5] These groups can be numerous. As will be discussed shortly, they can generally receive pensions while continuing to work, often in their old job. In addition, years of work while receiving a pension count as years of service and add to a person's pension entitlement. These arrangements, again, are understandable in their historical context, given pervasive labor shortages, but their cost is high. In Poland in 1990, one-third of expenditure on old-age pensions benefited individuals below the normal pensionable age (World Bank 1993b, chap. 4).

Invalidity pensions raise parallel problems. Recipients are numerous for two separate reasons. First, for the reasons discussed in chapter 5, little emphasis was put on safety at work, so that many individuals now have genuine and serious health problems. This situation requires separate remedial action. Second, benefits are relatively easy to obtain, not least because the authorities typically have little power to control access. Western European experience suggests that rising unemployment will put increasing pressure on invalidity pensions.

Cost containment can be ineffective also because of the ease with which receipt of benefits can be combined with work. Old age pensioners can generally work full time, or nearly full time, with no loss of benefits. Similarly, people can qualify for invalidity pensions even if they have suffered no loss of earning capacity. An individual who loses a limb, but whose long-run health and capacity to work are not affected, will generally receive a continuing, partial invalidity pension while performing his or her old job and receiving a full wage.

HOLES IN THE SOCIAL SAFETY NET. Despite high levels of spending, poverty relief can be incomplete for four reasons: data may be lacking on the extent of the problem, some benefits may be too low, there may be gaps in coverage, and some people may not receive the benefits to which they are entitled.

The absence of data on people's incomes exacerbates the problem of defining and measuring poverty. Some countries have better data than others, but information can be seriously deficient.

The adequacy of the minimum benefit is questionable. The problem arises because the minimum level of the major benefits is often tied not to a poverty line but to the average wage or the minimum wage. If the minimum benefit is tied to the average wage, the minimum benefit falls if average wages fall, and real wages fell significantly throughout Central and Eastern Europe during the early transition. Indeed, the whole purpose of the incomes policies which normally accompanied the early transition was to ensure that wages were *not* fully protected against inflation. If the minimum benefit is tied instead to the minimum wage, the issue becomes highly politicized. The minimum wage is highly political in the best of times, and even more so when subsidies on basic commodities are being reduced. If there are additional knock-on effects for the minimum level of benefits, it becomes even more political.

Gaps in coverage arise, first, through inadequate linkage between benefits. Under the old system, virtually all cash benefits were part of social insurance. The reforms typically produced three sets of benefits: unemployment benefits, social insurance benefits, and social assistance, organized by three separate authorities.[6] Gaps arise where the three sets of benefits are incompletely linked with one another, which is a particular problem for recipients of unemployment benefits. A broad-ranging system of social assistance is needed to cope with these imperfect links, to cope with poverty in families of different sizes, and to cope with the narrowing scope of social insurance. The weakness of social assistance, the second important gap, is one of the central topics of chapter 10. A third gap is incomplete coverage of workers in the private sector. The problem is no longer one of eligibility (workers in the private sector typically face the same regime as workers in the state sector), but one of enforcement. In an unknown number of cases, private sector employment remains undeclared, resulting in incomplete or missing contributions records and hence loss of benefit entitlement.

Non-receipt of entitlement is potentially a further problem. Not everyone who is entitled to benefits may receive them. Take-up can be incomplete either because people do not apply for benefits, or because they apply and are wrongly refused (see Atkinson 1989, chap. 1). No quantitative evidence exists on take-up in any of the countries in Central and Eastern Europe. Problems have been observed, however, particularly with social assistance, but sometimes also with pensions, when funding sources have dried up.[7]

ADVERSE INCENTIVES. Expenditure depends on two sets of factors: the average level of benefits and the number of recipients. Both are significantly influenced by the underlying incentive structure.

Benefits can create adverse incentives, particularly in respect of work effort. In many countries in Central and Eastern Europe, unemployment benefits pay up to 75 percent of the worker's previous wage; in Ukraine and Belarus early in the transition it was 100 percent for the first three months of unemployment. Similarly, sick pay is often 100 percent of the worker's wage. The resulting tendency to take extended sick leave is aggravated by weak monitoring. The incentive effects of unemployment benefits have been studied extensively in the West (for surveys, see Atkinson and Micklewright 1991; Layard, Nickell, and Jackman 1991). The prevailing view is that the ratio of benefits to previous earnings (known as the replacement rate) should not be given undue emphasis as an influence on the level or duration of unemployment. Other factors—for instance, the fact that unemployment benefits are cut off after, say, twelve months—are regarded as more important. That research, however, is based on the range of replacement rates typical in Western Europe and the United States, which are between 35 and 70 percent, and therefore cannot be applied uncritically to Central and Eastern Europe.[8] Other adverse incentives arise where the definition of unemployment is broad, where the duration of entitlement is unlimited, and where the condition that recipients must be actively seeking work receives little attention.

The contributions regime also creates adverse incentives. The social insurance contribution is generally paid entirely by the employer, the absence of a worker contribution having been regarded as one of the victories of socialism. In the West, contributions are normally shared between worker and employer. If the entire contribution is paid by the employer (and even worse where employers face a soft budget constraint), everyone thinks that benefits are paid by someone else. This is the third-party payment problem, in which no one has an incentive to restrain benefits. The situation is exactly parallel with fee-for-service medical care financed by insurance.

Contributions are typically about 40–45 percent of the gross wage, although they can be considerably higher (81 percent in Ukraine in April 1992), not only interfering with the competitiveness of enterprises but also depressing the demand for labor and, in particular, given the high cost of benefits for mothers with young children, the demand for female labor. A further problem is that contributions are fragmented. Throughout Central and Eastern Europe there was considerable earmarking through multiple contributions: some pay slips could have more than twenty separate contributions for different benefits. Such contributions used to be transferred to the general budget, which allowed for sizable redistribution between the insured and the general population. Governments routinely used social insurance resources for other purposes. One ill effect is fiscal inefficiency. Depending on how contribution rates are specified by legislation, for example, some benefits may be overfinanced, while others are underfunded. In addition, separate collection procedures for each of the funds are administratively costly.

WEAK ADMINISTRATION. In the past, benefits, such as family allowance and pensions, were mostly long term. Individuals receiving short-term benefits from the public authorities were few and could therefore receive Rolls Royce treatment. The main Employment Office in Warsaw at the start of the reforms was paying unemployment benefits to five people. In addition, little emphasis was put on administrative efficiency. For both reasons, benefit structures tended to be highly complex. This is not sustainable with large-scale unemployment.

The second key administrative characteristic was the major role played by the enterprise. Enterprises provided information to employees about entitlement, assisted with pensions claims, and organized the payment of family allowance and short-term benefits as part of an individual's pay packet. The first problem with this arrangement is that administration of benefits by enterprises is becoming more and more unsustainable as job mobility increases, as enterprises increasingly face hard budget constraints, and as large numbers of people work for small-scale employers or become self-employed. These changes necessitate a major transfer of administration from the enterprise to the public benefit authorities. A second problem is the extent to which the social insurance authorities are ignorant of individual records. All but the smallest enterprises pay social insurance contributions as a single lump sum covering all their workers. The social insurance authorities *have no knowledge whatever about an individual's contributions record*. Individuals who are approaching retirement must put together, with the aid of their employer, a dossier of their entire work history to establish their contributions record. For the same reason, the social insurance authorities generally *have no knowledge of individuals who receive short-term benefits and family allowance*. Both problems impose major limitations on the range of reforms which are administratively feasible in the short run.

Administrative capacity was, and remains, limited. Records are generally of the pencil-and-paper variety. The main method of getting short-term benefits to recipients was through the employer. For benefits such as pensions, the main vehicle of payment was, and is, cash sent through the postal system.

Perennial Problems

Viewed from a Western perspective, the old system paid insufficient attention to resource constraints and to the types of tradeoffs discussed in chapter 1. A key aspect of the reforms, therefore, is to improve the efficiency of resource use. As chapter 3 made clear, however, most of the countries in Central and Eastern Europe will need a long time to catch up with living standards in Western Europe. Additionally, and in sharp contrast with Marx's thinking, higher standards of living will never wholly solve the problem of scarcity. Resource constraints will persist, and institutional capacity constraints will be eliminated only slowly. For both reasons, the social safety net will not be as effective as soon as people would wish. Perennial problems will remain. As experience in the West makes abundantly clear (Atkinson 1992), the twin problems of unemployment and poverty will never completely be solved.

The Forces Driving Change

The economic and political driving forces discussed in chapter 4 apply in a powerful and very simple way to cash benefits. On the economic side, price and wage liberalization create two sets of effects:

- The widening distribution of income is associated with increased unemployment and rising and more visible poverty. In addition, the administration of benefits becomes more complex: under the former, fairly flat distribution, benefits could be targeted at specific *groups*, such as families. With the current, widening distribution, benefits must be targeted also by level of *income*. The resulting system has greater coverage and more complex targeting.
- Declining output is creating a fiscal crisis which exposes many of the weaknesses of the old system. It emphasizes the need to improve targeting, both to relieve poverty more effectively and to contain costs. Difficulties with tax collection and social insurance contributions expose the administrative weakness of the old system.

An important implication of the widening distribution of income is the need for a more sophisticated structure of benefits. The social insurance system, which in the past also delivered most social assistance (such as pensions to invalids with little or no contributions record), will, as a result of the reforms, no longer be able to do so. Insurance benefits (with benefits bearing an explicit relationship to contributions) must be separated conceptually from noncontributory benefits such as family allowance and social assistance.

The political driving forces are equally clear. The imperative of achieving higher living standards creates pressures for increased benefits; and higher benefits are in direct conflict with the need to contain current expenditure in the interests of longer-term growth.

Policy

Different cash benefits contribute to the objectives of policy in different ways, depending in part on their construction. Unemployment benefits and sickness benefits contribute to the insurance objective by reducing the extent of sharp, unexpected falls in income. If the formula for determining benefits is weighted toward lower-paid individuals, benefits also help to relieve poverty, particularly short-term poverty, and redistribute income from rich to poor. One of the major purposes of explicit social insurance contributions is to give recipients an entitlement to benefits, thereby fostering social integration. If properly constructed, the benefits minimize adverse incentives to labor supply.

Pensions have two aspects. The primary purpose of invalidity and survivors' pensions is insurance. Retirement pensions offer insurance, since people do not know how long they will live, and also enable people to redistribute income to themselves over their life cycle, thus contributing to income smoothing. Both types of pension help to relieve poverty, particularly long-term poverty, and may also redistribute income from rich to poor. Like other social insurance benefits, they also contribute to social integration.

Family allowances have a major objective of income smoothing and also contribute to poverty relief. To the extent that the benefit is financed out of progressive taxation, it also contributes to vertical redistribution. If the benefit is universal, it assists social integration; if it is paid to the mother, it redistributes toward women. Social assistance primarily seeks to relieve poverty and also contributes to vertical redistribution.

The Strategy

The effects of the driving forces include rising unemployment, increased poverty, severe fiscal constraint, and exposure of administrative limitations. So far as cash benefits are concerned, the resulting policy strategy follows naturally and is entirely consistent with that presented in chapter 6. The primary policies concentrate on:

- Strengthening poverty relief
- Containing costs
- Raising administrative capacity.

Additional policies are concerned with:

- Improving incentives
- Diversifying supply, particularly in connection with private pensions.

The first two policies are the twin aspects of targeting discussed at the start of the chapter: benefits should go to everyone who needs them and, at least up to a point, only to those who need them. There is a clear distinction between short- and medium-term policies. While the fiscal crisis lasts, poverty relief should be given priority over the insurance and income-smoothing objectives. Thus fiscal constraints blur the distinction between insurance and non-insurance benefits. As soon as resources permit, a clear relation between individual contributions and benefits should be restored; the extent to which the relationship is strictly actuarial is a policy choice.

Before turning to a detailed discussion of how this strategy might be translated into policy, it is worth asking whether the social safety net, even if reshaped along the lines suggested below, is affordable. In trying to answer the question, it is important to distinguish public costs from social (that is, economy-wide) costs. Put crudely, if the state ceases to pay a wage to a worker who is producing little or nothing, and instead pays him or her a lower unemployment benefit, there is a net saving.

Two cases should be considered: where value added is smaller, but not much smaller, than wages, and where value added is effectively zero. As discussed in chapter 7, if value added is significant, the cost-minimizing policy *might* be temporarily to subsidize employment. If redundant workers produced zero value added in their former job and previously earned the average wage, gross savings are then the number of redundant workers times the average wage times the average duration of unemployment. Since unemployment benefits are generally less than the worker's previous wage, it follows (unless the costs of training and other active labor market policies are high) that the total savings to the enterprise sector will be greater than the total costs of unemployment benefits. From the viewpoint of the economy as a whole, there is a net saving. The problem is that most of the costs fall on government and most of the savings accrue to the enterprise sector, so that more rapid adjustment has budgetary implications, although these are less acute if wages had previously been subsidized out of tax revenues. The general point, however, is valid: public expenditure on the social safety net should not be considered in isolation but viewed alongside the resulting direct and indirect savings to the enterprise sector. This argument confirms the point that income support, quite apart from its distributional objectives, has a key role to play alongside other labor market policies in assisting the overall economic objectives of the reforms.

A different aspect of affordability is that replacing untargeted general price subsidies with well-targeted income subsidies also creates some leeway for financing the social safety net.

Maintaining Macroeconomic Balance

Maintaining macroeconomic balance is pursued both through efforts to mobilize private resources and through policies to contain costs.

MOBILIZING RESOURCES. As discussed in chapter 1, most income transfers in the highly industrialized countries are organized by the state. The only significant exception, private pensions, is discussed in some detail later in the chapter.

CONTAINING COSTS. The objective here is to restrict benefit spending without undermining poverty relief. Policies include sharing the costs of sick pay, tightening access to pensions, and improving the targeting of uprating of benefits (that is, increases in benefits in response to increases in earnings or prices).

Sharing costs. Sick pay would be more cost-effective if workers were made responsible for the initial period (say, one to three days) of absence from work for health-related reasons, with employers responsible thereafter for a specified period (perhaps two to four weeks). Only then would people with health problems become a charge on the social insurance fund. The social insurance authorities would establish minimum standards for such arrangements.

Tightening pension eligibility. Reducing access to pensions would yield large savings while still offering effective poverty relief to the old and disabled.

First, awarding an invalidity pension to anyone whose earning capacity has not been reduced could be discontinued, with compensation for permanent loss of a faculty which does not reduce earning capacity taking the form of a lump-sum payment. Although the policy is clear, implementation is problematic because earning capacity is difficult to assess. However, it is administratively less difficult to stop paying invalidity pensions to people who continue in their old job at their old rate of pay, which is a common occurrence in Central and Eastern Europe. The proposal is not to withdraw invalidity pensions generally, but only where earning capacity has not been adversely affected.

A second aspect is pensionable age. The ratio of pensioners to workers is high both because the official pensionable age is low by Western standards and because for significant numbers of people the effective age at which they first receive a pension is even younger. Early retirement in Poland absorbs one-third of all expenditure on old age pensions. Similar problems exist in all the countries of Central and Eastern Europe. The facts are clear. Policy, however, must achieve a difficult balance between different objectives. In the medium term, pensionable age should be increased, with early retirement only on the basis of an actuarially reduced pension or actuarially higher contributions (see the glossary). This would remove artificial incentives to early retirement, improve the equity of the system as between different groups of workers, and ease administration. Short-term policy is more difficult. The argument for raising the effective pensionable age is the resulting substantial savings in public expenditure. There are three arguments for caution. First, there might not be any savings; this could happen if people whose retirement is delayed are paid a higher pension when they finally do retire. Second, increasing the pensionable age might not be politically feasible in the short run. Third, raising the age of retirement, even if it does yield fiscal savings, may have other costs. Western experience shows the discouraging and debilitating effects of long-run unemployment, especially for the young. A case can therefore be made for encouraging older workers to retire because this would assist restructuring and open up jobs for younger people.

Combining work with receipt of a pension is a third problem area. Again, there is a conflict of objectives. On the one hand, the ability to continue working while receiving a more or less full pension is costly. This suggests the need for a retirement test, at least for people below normal retirement age. For example, the pension could be withdrawn from—or at least lowered for—anyone who earns more than a fairly small amount. An effective retirement test of this sort would create savings in two ways. If a person does not retire, the pension payment is saved; if he or she does retire, a job is released which can be filled by an unemployed person, thereby reducing expenditure on unemployment benefits. The argument against this approach is that early retirement allows firms to make older workers redundant and thus assists restructuring. This suggests that rather than impose a *retirement test*, it is worth considering a *change-of-job test*, whereby individuals who retire but find another job are allowed to keep at least part of their pension. Once more, though the policy is clear, implementation is problematic; Soviet enterprises were adept at organizing fictitious job changes. This sort of problem can be sidestepped for people who become self-employed, where a strong case can be made for allowing them to keep some or all of their pension.

It is important to clarify what is being criticized and what is not. The problem is not the outcome—early retirement—but the process leading to that outcome, in particular the absence of any clear connection between the contributions paid by an individual and the benefits he or she receives. The problem can be resolved in the medium term by strengthening the relationship between contributions and benefits. In the short term, however, there is an acute conflict between restructuring and employment for the young, on the one hand, and cost containment, on the other.

Better targeted uprating of benefits (see the glossary). As discussed earlier, benefits tied to earnings are problematic. If real earnings fall, they fail to relieve poverty; if real earnings rise, they conflict with cost containment. A policy which combines cost containment with effective poverty relief has two legs:

- The minimum level of all the major benefits and of social assistance is fully protected against price increases.
- While the need for fiscal stringency persists, benefits above the minimum are increased only to the extent that the budgetary situation permits.

The first part ensures effective poverty relief, the second ensures that it is achieved at minimum cost. As the economic situation improves, real benefits above the minimum could be partly or wholly restored. Although the strategy is clear, the administrative and political difficulties of controlling increases in benefits during times of high inflation should not be minimized.

Building Markets

The reform of cash benefits can support markets (a) by diversifying the supply of some social insurance benefits and (b) by improving incentives.

DIVERSIFYING SUPPLY. The main benefit for which markets are relevant is private pensions. These are a medium-term issue which is currently the subject of heated debate in Central and Eastern Europe. The subject is taken up in detail below.

IMPROVING INCENTIVES. Three areas stand out: improving the incentive structure of benefits, reducing the burden of payroll contributions, and sharing the social insurance contribution between worker and employer.

The structure of benefits can be improved, first, by reducing the high replacement rate (see the glossary) for unemployment benefits and the 100 percent replacement rate for sickness benefits. The right to combine full benefits with work for individuals below retirement age should be reexamined for reasons of incentives as well as costs. The ideal qualitative relativity of the different benefits as earnings decompress and the fiscal situation improves is that the poverty line should be below the minimum unemployment benefit, which should be below the minimum wage, which should be below the average wage. In the short run, the relativities should be kept under review to create the best balance between the need to relieve poverty and the need to minimize adverse incentives.

The level of payroll contributions needs to be reduced and their structure simplified. It might be argued that with real wages so low, high contribution rates are not a major problem. That view is mistaken. High employer contributions give incentives against new employment, encourage worker and employer to collude in fraud (by declaring a lower wage than the worker actually receives), create incentives to unofficial employment, distort the relative price of labor and capital, and take from enterprises the resources they need to invest in new plant and equipment. The employer's contribution can be reduced in three ways: by financing through general taxation benefits like family allowance which do not relate to any insurable risk, by reducing benefits, and by sharing the contribution between worker and employer.

Shared contributions would follow the typical pattern in the highly industrialized countries. The social insurance contribution should be divided between the employer and the individual worker, with the worker's contribution appearing as a deduction on his or her pay slip. Although the theoretical importance of the division can be debated, worker contributions give important political and economic signals.[9] The political signal concerns responsibility and is particularly important during the transition: that the state is responsible for establishing the general framework of social protection, but that the individual, through earnings and contributions, is substantially responsible for himself or herself. Where contributions are paid wholly by the enterprise, and even more if the enterprise faces a soft budget constraint, neither workers nor employers face any incentive to moderate their claims. This is another instance of the third-party payment problem. With a worker's contribution, any increase in benefits instantly increases the contributions deducted from individuals' pay packets. The resulting economic signal is that it is the *worker* who pays, at least in part, for social insurance benefits.

Regulating Market Forces

Market forces need to be regulated through policies to relieve poverty, through the development of an enabling environment for insurance and income smoothing, and through the introduction of a regulatory structure for private providers of benefits.

RELIEVING POVERTY. The policies discussed below are based on three propositions:

- In the short run, for fiscal reasons, poverty relief should be given priority over the objectives of insurance and income smoothing.
- The major components of cash benefits which should be expanded are unemployment benefits and poverty relief.
- Wherever possible, poverty relief should be based on indicators such as age or family size to minimize the need for administratively demanding income testing. This suggests that the minimum level of the major social insurance benefits should not be below the poverty line.

Strengthening unemployment benefits. Early legislation tended to exhibit problems requiring substantial amendment of employment laws to tighten eligibility, to strengthen the incentives for individuals to seek work (for instance, by limiting the length of time for which benefits were paid), to strengthen the checks to ensure that individuals were seeking work, and to simplify administration. A second set of activities concerned building administrative capacity to ensure that the system could deliver benefits to increasing numbers of recipients. In Poland, unemployment rose from negligible levels to more than 1 million during 1990; and there was a proportionately similar increase in Bulgaria during 1991 (see table 1-1). The system, by and large, coped well with delivering benefits but less well with organizing active labor market policies.

Flat-rate unemployment benefits, such as those in Poland and Bulgaria, have short-run advantages: reduced cost, improved incentives to seek work, and ease of administration; in the short run, sick pay could be paid at the same flat rate and certainly at less than 100 percent of the worker's wage. Incentive arguments suggest that a person should not normally be eligible for unemployment benefits for more than a specific number of months, although such a provision could not be put into effect until social assistance is able to provide poverty relief for those whose entitlement to the insurance benefit has expired. As the fiscal situation improves, government may choose to retain a system of low, flat-rate unemployment benefits or to increase the flat-rate benefit or to return to benefits related to individual earnings.

Improving poverty relief. Poverty existed under the old system, and there is increasing evidence that the transition aggravates it, particularly for the unemployed, the elderly, and families. More and better information is needed, both to make poverty relief more effective and to help contain costs. The data should be sufficiently detailed to show the changing position of different groups such as children, women, the unemployed, and the elderly.

Key policy reforms include the protection of minimum benefits, the avoidance of gaps in coverage, and the creation of a system of social assistance. The starting point is to establish and maintain appropriate levels of minimum benefits. Given the doubts about the level of the minimum benefit discussed above, the minimum level for all major benefits requires continuing review. The aim should be to set unemployment benefits and the major social insurance benefits at or above an *individual* poverty line; social assistance should be defined relative to a *family* poverty line. To the extent possible, the minimum benefit should be fully protected against inflation.

Coverage should also be as complete as possible. Unemployment benefits and most social insurance benefits, if the minimum benefit is adequate, address individual poverty but not necessarily family poverty. Recipients of such benefits should therefore be eligible for social assistance. The administration of social assistance must thus be able to cope with large numbers of applicants, necessitating administrative reform and additional administrative resources. Another aspect of coverage is the protection of employees in the private sector through action to reduce large-scale evasion of contributions.

Clarifying the distinction between insurance and non-insurance benefits. Action of this sort has already occurred in several countries. Although policy in the short term may have to concentrate scarce resources on protecting the minimum benefit, the relation between contributions and benefits could usefully be strengthened as soon as economic conditions permit. The move to strengthen the relation between contributions and benefits increasingly rules out the use of social insurance contributions for non-contributors or to finance in-kind benefits. It thereby removes from social insurance much of the social assistance function. The narrower scope of social insurance has already excluded many people who were previously covered, such as disabled people who have no contributions record, or part of the agricultural population who, after the dissolution of cooperatives, are unable or unwilling to pay contributions. Alongside, and connected with, the stronger relation between contributions and benefits is the increased self-governance of social insurance funds, whose independent management is responsible for maintaining actuarial balance and representing the interests of the insured.[10]

Though there is general agreement that the relationship between contributions and benefits should be strengthened, the extent to which it is *strictly* actuarial is a policy option. Social insurance has two distinct roles: (a) as a device to address failures in private insurance markets and (b) as a device which allows redistribution from rich to poor within a generation or between generations. The extent to which social insurance embodies redistribution is fundamentally and importantly a value judgment: the lower the weight given to redistribution, the stricter the relationship between contributions and benefits. Where redistribution is not a goal, the relationship will be strictly actuarial. Where social insurance incorporates redistribution (as it typically does in the West), benefits should still be related *at the margin* to individual contributions, and contributors and beneficiaries should perceive this to be so. The argument is important. Policymakers may have a pension formula which is redistributive in the sense that worker A, who has twice the earnings of worker B over his working life, gets a pension which is higher than B's, but less than twice as high. If either A or B retires early, however, his pension should be actuarially reduced relative to the pension he would have received at age sixty-five.

The relationship should be strengthened for at least three reasons: to achieve equity (for example, the early retirement provisions for miners discriminates against other groups in the work force), to minimize distortions to individual retirement decisions, and to counteract the strong incentive for workers to evade contributions when they do not see a clear relationship between contributions and benefits. It is important that some of the impetus to enforce contributions comes from incentives rather than administrative activity, not least because enforcing contributions, particularly in the growing private sector, will be a continuing problem. In the medium term, therefore, a closer relationship between contributions and benefits is a key component of the social insurance strategy. Poverty relief should increasingly be based on social assistance and other tax-funded benefits such as family allowance.

PROVIDING AN ENABLING ENVIRONMENT FOR INSURANCE AND INCOME SMOOTHING. A central role of the state for this purpose, as more generally, is to keep prices sufficiently stable to enable financial assets to act as effective stores of value, thereby encouraging voluntary savings and insurance.

INTRODUCING A REGULATORY STRUCTURE. Private insurance in general, and private pensions in particular, depend crucially on well-regulated financial markets generally and a regulatory structure of the pensions industry specifically. A key purpose of such regulation is consumer protection.

Diversifying Supply: The Role of Private Pensions

Discussion so far has concentrated mainly on the short run and mainly on addressing problems with the existing system. This section turns to an area with a more medium-term horizon. State pensions, it is said, pay too little to too many at too high a cost. One suggestion has been partially to replace the state system with private pensions, which would accumulate a fund out of which future pensions would be paid.

WHY PRIVATE PENSIONS? This is an area of continuing controversy among economists (see Aaron 1982; Auerbach and Kotlikoff 1990; World Bank 1994). Notwithstanding arguments to the contrary, this is another area with no silver bullet, that is, no single, simple, completely effective solution.

A number of definitions are important. With *pay-as-you-go* (PAYG) schemes, this year's pensions are by and large paid out of this year's contributions. Most state schemes are organized in this way. With *funded* schemes, in contrast, the individual accumulates a fund over his or her working life; upon the worker's retirement, this is generally converted into an annuity, that is, a flow of income for life. Funded schemes thus combine saving during a person's working life with insurance (the annuity). Most private schemes operate in this way; a few state schemes also have funded or partially funded components. Two types of funded schemes should be distinguished. Under a *defined contribution* scheme, a person's pension depends only on the size of his or her pension accumulation. The risk of varying rates of return to pension assets and the risk of inflation after retirement are therefore borne entirely by the individual. With a *defined benefit* scheme, usually run at the level of the firm or industry, the individual's pension is normally based on the number of years of service and his or her final salary. In this case, the employer bears at least part of the risk of different rates of return to pension assets.

Several advantages are claimed for private, funded pensions:

- To the extent that they take over from state pensions, they reduce public spending.
- They lead to increased savings and hence to higher economic growth, thus making it possible to pay higher pensions or to pay pensions in the face of an aging population.
- They help to develop capital markets and, in the context of Central and Eastern Europe, may assist privatization.
- They protect the government from political pressure since, in a funded scheme, future increases in pensions are possible only if contributions increase in the present.

Such claims engender powerful lobbies in favor of private pensions in a number of countries.

BACKGROUND THEORY AND EVIDENCE. Several conclusions provide helpful background to subsequent discussion of experience in the highly industrialized countries and elsewhere. These relate to the lack of consumer information, the effects of inflation, the budgetary impact of private pensions, and their effect on savings and economic growth.

Consumer ignorance is a major problem. Financial markets are highly complex, and consumers are ill-informed of how they work in principle and even less informed of how they operate in practice. Regulation of financial markets generally and of the pensions market in particular is therefore required to protect consumers in an area where they are usually not sufficiently well-informed to protect their own interests.

Inflation is not an insurable risk for the reasons set out in chapter 2. Nor is the private sector generally able to offer other indexed assets on which to base pensions. Although there is controversy as to why no private sector financial instruments offer a risk-free real return (that is, a return which is consistently greater than the rate of inflation), the evidence is clear. Bodie's (1990, p. 36) survey points out that "virtually no private pension plans in the U.S. offer automatic inflation protection after retirement." Gordon's (1988, p. 169) cross-national conclusion is that "indexing of pension benefits after retirement . . . presents serious difficulties in funded employer plans." Private pension schemes in the West have experienced problems with inflationary shocks even in the absence of structural change. It is true that many countries have also experienced problems with their PAYG schemes in the face of inflation; such problems, however, are not the result of inflation per se but rather of unrealistic promises in the past and fiscal deficits in the present.[11]

Sometimes these problems can be sidestepped. Private schemes can cope with inflation during a worker's contribution years, and internationally diversified funds can resist purely domestic inflation. The latter, however, is not necessarily useful in Central and Eastern Europe, given the risk of capital flight if private pension funds are allowed to hold foreign assets. More generally, the greater the extent of a common inflationary shock such as an oil crisis, the less well-placed are funded pensions to resist it. Without intervention by the state or some other benevolent entity, individual pensioners face the entire inflation risk.

Effects on the state budget in the short run are fairly clear-cut; any significant move to funding generally *increases* public spending. The reason is straightforward: under a PAYG scheme, contributions by workers are used to pay the pensions of the elderly and so cannot be put into the workers' pension fund; in a switch to funding, an additional contribution is therefore needed to begin building the fund. Because it is generally not politically possible to impose a double contribution on the current generation of workers, substantial budgetary transfers are necessary in the early years of a move to funding. Even in the long run, the budgetary gain is reduced to the extent that the state guarantees a minimum pension or offers any guarantees against inflation or other risks, for example, by issuing indexed government bonds.

The effect of funding on savings and growth is arguably the most controversial area. The large literature on the experience of countries in the West is inconclusive both theoretically and empirically (see Aaron 1982; Auerbach and Kotlikoff 1990; World Bank 1994).

EXPERIENCE IN HIGHLY INDUSTRIALIZED COUNTRIES. As table 1-3 makes clear, state spending on old age security in all the highly industrialized countries is substantial, mainly in the form of PAYG schemes, and financed by earnings-related social insurance contributions (the Western experience is well documented; see OECD 1988a, 1992b on highly industrialized countries; for the OECD experience and possible implications for Central and Eastern Europe, see Holzmann 1991). State-funded schemes, which exist in very few of the OECD countries, have not fared well.[12] An international survey concluded that they "offer powerful evidence that this option may only invite squandering capital funds in wasteful, low-yield investments [which] should give pause to anyone proposing similar accumulations elsewhere" (Rosa 1982, p. 212).

Private funded schemes exist in all the highly industrialized countries as a complement to state schemes, often in the form of defined benefit schemes run on occupational lines. Their performance has been sensitive to overall economic developments. In particular, they are sensitive to the effects of unanticipated inflation. Overall, however, they have performed reasonably well.

Given the problems of consumer ignorance and the inflation risk, private pensions in Western Europe can be broadly described as having four characteristics. They may be optional, in that the individual can choose to remain entirely within the state scheme. They are supplemental, in that they replace only part of the state pension. They are constrained in two ways: individual choice is limited, and the conduct of pension companies is regulated to protect consumers.[13] Finally, virtually all private pension schemes are subsidized through major tax advantages. In addition, because of the uninsurable inflation risk, the state may give private pensions at least a partial guarantee or may issue indexed government bonds.

Public social spending in the West proved robust during the 1980s despite determined attempts, at least in the United Kingdom and United States, to roll back the frontiers of the state. One explanation is that these areas do not accord well with the conditions necessary for private markets to work efficiently (the major topic of chapter 2). The demographic prospects in most of the highly industrialized countries, coupled with continuing recession, have prompted searching reviews of public spending. The result so far has been limited reduction in the scale of activity, but no significant change in its structure (see U.K. Department of Social Security 1993).

EXPERIENCE IN LATIN AMERICA. Most Latin American countries have well-established PAYG pension systems, and state pension spending is significant, as table 1-2 shows (for a detailed discussion, see McGreevey 1990 and Mesa-Lago 1990a, 1990b). Most of the schemes have had endemic financing problems of precisely the sort faced by countries in Central and Eastern Europe, mainly as a result of promises which have proved fiscally unrealistic. These problems, as in Central and Eastern Europe, have led countries to seek better ways of financing pensions and, in particular, to advocate funded pensions. State-funded schemes in Latin America, however, have not performed well.

> In general, pension reserves have not been invested . . . in instruments with the highest economic returns; social insurance institutions have not been designed to play the role of financial intermediaries, their personnel lack experience in this field, and no investment plans have been formulated. Capital markets are poorly developed in the region, and inflation has had a pervasive effect on the value of the reserves (Mesa-Lago 1990b, p. i).

Of eight representative countries in the region, only Chile has had a significant real return. Most of the rest "had negative yields as low as −21 percent annually, hence decapitalizing the reserves" (Mesa-Lago 1990a, pp. i–ii).

Experience in Chile is worth discussing separately because its pension reforms of the early 1980s are a recurring topic in Central and Eastern Europe (for detailed discussion, see Diamond and Valdés-Prieto 1994; Gillion and Bonilla 1992; Myers 1992). Employees are required to join an approved scheme; all such pension plans are private, funded and defined contribution (see the glossary); workers pay 10 percent of their earnings, plus a commission; and there is no employer or government contribution. Workers can choose which scheme to join and can change schemes. Each pension company manages a fund, the return to which is credited to the accounts of individual contributors, implying no redistribution from rich to poor. Upon retirement, the worker either can make a series of phased withdrawals from his or her accumulation (with maximum withdrawal limited by formula) or can buy an annuity from an insurance company. The pension is indexed to price inflation, largely though not wholly on the basis of government-indexed bonds. A sizable minimum pension, paid by the state, is guaranteed for any worker with twenty years of contributions, and generous, government-funded transitional arrangements are made for workers transferring from the old PAYG system to the new scheme. In September 1991 the new scheme had 4 million members, but relatively few pensions (100,000) had been paid out. These arrangements are usefully assessed under four heads: their effect on the incomes of pensioners, their effect on the budget, their impact on savings and growth, and the form of their administration.

Effects on the incomes of pensioners. In a defined contribution scheme, pensions depend on the worker's accumulation at retirement, which in turn depends on (a) the rate of return over his or her contribution years and (b) his or her contribution density, that is, the number of contributions paid over the years. The average real rate of return to pension savings in Chile during the 1980s was 12.6 percent a year. This is very high, and a key question is whether it is sustainable. Chile implemented an intensive privatization program during the 1980s, generating large capital gains; that program is now largely complete. On contribution density, matters are more problematic. In 1990, the proportion of members who paid contributions regularly was 53 percent, down from 76 percent in 1983. The compliance rate varied by income level, being 45–55 percent for funds which cater to lower-paid workers, 80–90 percent for those with better-paid members.

Gillion and Bonilla's (1992) average case assumes a 3 percent long-run real rate of return, leading to a replacement rate (see the glossary) of 44 percent. This is low and implies that only about half of all workers will receive pensions above the state minimum. A long-run real return of 4.5 percent is needed to reach a 70 percent replacement rate. Both sets of figures assume a complete contributions record. With a 3 percent real return plus a 60 percent contributions density, 65 percent of workers would receive only the state-guaranteed minimum pension. Thus a contribution rate of 10 percent may be too small.

Several features of the system are noteworthy. First, because the system is based on defined contributions, the entire risk above the minimum pension is borne by the individual worker. Second, the scheme is individualistic. There is redistribution neither within a generation (there is no redistribution from rich to poor except through the guaranteed minimum pension) nor between generations (pensions are indexed to prices, not wages, so that pensioners do not share in economic growth occurring after their retirement).[14] Third, there are significant gaps in coverage, both because of non-compliance and because formal employment embraces only about 65 percent of the work force, most of the remainder being outside the scope of the new scheme. Finally, outcomes are highly sensitive to real rates of return and compliance rates, and the scheme's twelve-year history is only a small fraction of its total lifetime.[15]

Effects on the government budget. Diamond and Valdés-Prieto (1994, p. 257) point to the fiscal costs of the old pension system and the promises associated with the transition. Such a fiscal sacrifice, they point out, is inevitable if the flow of payroll tax revenue into privatized mandatory savings accounts is to be genuine new savings. Pension contributions now go into individual accounts rather than into social insurance funds. Government guarantees include the cost of pensions for older people who never transferred to the new system, the cost of the transitional contribution for workers who switched to the new system, the cost of indexed bonds, and the guarantee of a minimum pension, which helps to protect contributors against poor performance of their chosen fund and pensioners against bankruptcy of the insurance company paying their annuity. The first two costs will eventually decline; the last two will continue in steady state. It is significant that the reform was introduced at a time when the government budget was running a surplus of more than 5 percent of GDP, giving room for the up-front public expenditure costs of the new system. Public pension spending in the early 1990s was variously estimated at between 4 and 6 percent of GDP and is likely to remain at that level for at least another ten years.

Effects on savings and economic growth. The accumulation of new savings, together with high real rates of return, has led to a large accumulation of funds. In June 1992 the total holdings of pension funds amounted to 35 percent of GDP, about 40 percent of the total being public debt. These facts prompt two questions: What has been the effect on the total volume of savings? What has been the effect on economic growth? The answer to the first question is by no means clear. The findings of theoretical and empirical research have been inconclusive. The savings rate in Chile was 21 percent in 1980 and 20 percent in 1989, with a major dip in the early 1980s (Santamaria 1991). The answer to the second is that economic growth depends not only on the quantity of savings but also on the effectiveness with which they are used. It is widely agreed that the pension reform significantly improved the effectiveness of capital markets in Chile. The pension funds have become a source of equity capital and long-term finance for private enterprises; and they have stimulated the growth of other institutional investors. Although the matter cannot be quantified, there is agreement that such deepening of capital markets has contributed to economic growth. To some commentators, this is the only substantial benefit of the pension reform which could not have been achieved by redesigning the old PAYG system.

Governance and administration. The pensions industry in Chile is oligopolistic: three of thirteen pension funds account for 65 percent of all insured persons; four of the largest companies are controlled by foreign corporations. Both features imply a heavy regulatory task for the supervisory body. Administrative costs are "much larger than costs in uniform government-managed conventional systems" (Diamond and Valdés-Prieto 1994, p. 288). Administrative costs could be as high as almost 3 percent of taxable earnings; thus, if the replacement rate is 50 percent, administrative costs are nearly 6 percent of pensions paid. The comparable figure for the state scheme is 1 percent of pensions paid in the United States and 1.25 percent in the United Kingdom. At least part of the difference is the added administrative costs of multiple funds (the Chilean figures are not out of line with the administrative costs of private insurance in the West) and the significant expenditure on marketing and sales.

EXPERIENCE IN ASIAN ECONOMIES. Experience in the high-performing Asian economies varies considerably (for a convenient summary, see Phillips 1992; MacPherson 1992). Taiwan (China) has a system of social insurance transplanted from the China of the 1930s. The only benefit it offers for old age is a single lump sum, and the system is supplemented by social assistance, mainly for emergency relief, on the basis of an extended family means test (that is, a means test based not on the income of the individual or nuclear family, but on the income of the extended family). Hong Kong has no contributory scheme: there is a tax-funded universal benefit, at a level well below subsistence, for the elderly and disabled, supplemented by a very limited scheme of social assistance based on the Poor Law tradition. "Protection from hardship and poverty relies on self-help and on the family. The role of government social security is to meet only the most basic needs of the most deserving, in the direst need of financial assistance" (MacPherson 1992, p. 57). In 1988, the Republic of Korea introduced a contributory pension scheme for all employees and the self-employed, but no benefits have yet been paid. A system of social assistance pays benefits only to people over sixty-five who meet a stringent income and assets test and for whom no one is legally responsible.

Singapore has adopted a totally different approach, whereby all workers and employers make compulsory contributions (22.5 and 17.5 percent of wages for workers and employers, respectively, in 1991) into the Central Provident Fund. Members can use their accumulations to buy a house, to pay hospital bills, or to meet living costs in old age. The underlying principle is similar to that of Chile to the extent that it is based on compulsory individual savings; it differs in that (a) members can use the savings for purposes other than income support in old age, and (b) there is only a single, state-organized fund. This is a state, not a private, scheme.

Effects on the incomes of pensioners. Poverty in old age is a continuing problem in all the high-performing Asian economies. In Taiwan (China), the pension is only a modest lump sum. In Hong Kong, "most poor people are old, their numbers are growing, and their situation is worsening" (MacPherson 1992, p. 57). In Singapore, "the Central Provident Fund has . . . only a marginal impact on poverty among the old" (MacPherson 1992, p. 58); and there are still significant gaps in coverage, for instance, for the self-employed and for casual workers.

Redistribution is minimal. The arrangements in Taiwan (China) are very individualistic: the main social insurance scheme covers only the workers themselves, not their dependents, which poses a particular problem for health care. In Hong Kong, except for very limited social assistance, there is no redistribution from rich to poor. In Singapore, there is virtually no public expenditure on income transfers, and "the provident fund approach is one which allows of no redistribution, neither vertical nor horizontal. This . . . accords with the dominant features of development in Singapore. The price of rapid economic growth is high" (MacPherson 1992, p. 55).

Effects on the government budget. There is little to be said. By definition, pension schemes which involve little or no public spending have little impact on the government budget. If the gaps in coverage are significant, however, there may be knock-on effects for alternative benefits such as social assistance.

The central role of the family. Kopits (1993, p. 23) stresses the importance of the cultural and social context in policy design. "At one end of the spectrum there are societies where the extended family . . . still operates rather actively as an informal social security scheme, obviating the urgent introduction of large-scale public pensions and assistance schemes." In the South Asian context, the Chinese emphasis on family responsibility is a continuing influence. As a result, "powerful family networks of obligation and reciprocity have continued to be of major importance" (MacPherson 1992, p. 51). In Taiwan (China), "about three-quarters of elderly people were living with their children or close kin" (Chan 1992, p. 140). In Hong Kong, "the family . . . is still providing the most important 'caring community' for elderly people" (Chow 1992, p. 74). In Singapore, "the family has always been the primary care giver for its aged members. No evidence has been found in Singapore to suggest that the family, as an institution, is shirking its responsibilities" (Cheung and Vasoo 1992, p. 96). Moreover, "the government is prepared, if necessary, to institute legal protection for the welfare of elderly people if families appear, in the future, to be abrogating their responsibilities" (Cheung and Vasoo 1992, p. 98). In the Republic of Korea, "about half of older Koreans . . . are dependent on their children for their living expenses" (Choi 1992, p. 151). In Thailand, "the extended family household [is] an important internal mechanism for raising the consumption opportunities of low-earnings individuals" (Warnes 1992, p. 196).

The fabric of family support, however, is coming under strain as family size declines and more and more women join the labor force. There is concern in Taiwan (China) about the sustainability of the system as "traditional Chinese assumptions about the role of the extended family are challenged" (Hill and Lee 1992, p. 69). Lack of support of the elderly "is one of the emerging aspects of the ageing problem in Korea" (Choi 1992, p. 151). One interpretation of these facts is (a) that the Southeast Asian system has been able to work without substantial state intervention only because of the major role played by the extended family, but (b) that such arrangements are on the decline even in Southeast Asia and are therefore not necessarily a model which can easily be copied.

IMPLICATIONS FOR CENTRAL AND EASTERN EUROPE. A few conclusions can be asserted fairly strongly; others are more in the nature of a list of decisions which will need to be made.

Reform of the state pension is essential. The problem of pension finance in Central and Eastern Europe is the cost of the existing pension regime. However, it is important to avoid the argument that "PAYG schemes have problems; therefore the solution is funding." The word "therefore" does not follow in logic.[16] In reality, PAYG and funding both address the same problem, namely, how to divide national output between workers and pensioners. It is therefore not surprising that neither approach offers a complete solution. With PAYG, the problem manifests itself as political pressures to pay higher pensions; with funding, political pressures arise when inflation reduces the capital of pension funds, eroding the funded basis of pensions, as shown by the Latin American experience. Whatever policy is adopted, private pensions are no short-run substitute for dealing directly with the excessive expenditure of the state scheme.

Reform should be as simple as possible. The management of private pension funds is complex. So, more generally, is the operation of financial markets. Both sets of activities require (a) the necessary operational expertise and experience and (b) the expertise to construct an appropriate regulatory regime. These activities require planning and operational skills which are not yet common in Central and Eastern Europe, and they require such skills on a very substantial scale. The severe institutional capacity constraints, and the many demands made on institutional capacity by other aspects of the reform program, make simplicity essential.

Private pensions require a regulatory regime. The management of pension funds raises complex technical issues which are poorly understood by consumers even in the West. Regulation of pension funds, like other financial institutions, is therefore taken for granted, although the effectiveness of such regulation is under continuing review, for instance, in the United Kingdom in the wake of the Maxwell imbroglio and in the United States after the savings and loan scandal.[17]

Private pensions require some protection against inflation. The inflation risk, which is problem enough in the West, is even greater during fundamental structural change in an economy with no experience of financial markets. For both reasons, uncertainty about the prospects of pension funds during the transition will be high, raising potential problems of capital flight. Thus a strong case can be made for extending at least a partial guarantee from the state, correspondingly reducing the potential budgetary savings. Either this could be a direct guarantee, or the state could issue indexed bonds in which funds could hold a significant fraction of pension savings. The argument for offering some protection is twofold: horizontal equity suggests that pensioners should not face substantially more inflation risk than wage earners, and the collapse of private pension schemes during the infancy of a market economy puts at risk the political consensus underpinning the reforms. As a practical matter, the political pressures for some sort of guarantee will be strong; and Western experience suggests that the pressures will be even stronger for government to help any pension fund which runs into difficulties.

Reform requires key policy decisions. How important are social solidarity goals? Should the state scheme be residual or substantial? Should it embody minimal or substantial redistribution? Should the risk of private pensions be faced by the individual alone (implying a defined contribution scheme) or shared with employers (implying a defined benefit scheme) or shared more broadly (implying at least some state guarantee)?

How restrictive should the regulatory regime be? First, what restrictions should be put on individual choice: in ascending order of freedom, should citizens be obliged to belong to their employer's scheme? Should they be allowed to choose between authorized pension funds? Or should they be allowed to manage their own pension portfolio? Second, what restrictions should be put on the conduct of pension funds, in particular their choice of assets? Should a pension fund be allowed to hold some (or all) of its portfolio in foreign assets? Should it be allowed to hold risky domestic assets? In all these cases, how will compliance be monitored and enforced?

How extensive will state guarantees be? Will the minimum pension be guaranteed? What type and volume of indexed assets will the state issue? And what help, if any, will be given to pension funds which do poorly? All three questions face a major tension between political pressures and cost containment; in addition, the third question may raise a significant issue of moral hazard. The more generous is the treatment of funds which face difficulties, the weaker are the incentives for managers to act prudently.

Private pensions, in conclusion, should not be thought of as a solution to the short-run budgetary crisis. Their real advantages are twofold. First, the long-term improvement in the operation of capital markets is significant. Second, the introduction of private schemes might make it politically easier to allow privileged groups to keep their higher pensions, but only on the basis of higher contributions. Other advantages include increased individual choice and an eventual budgetary saving.

Implementing the Reforms

To implement the reforms, both institutional and political capacity require development.

BUILDING INSTITUTIONAL CAPACITY. Strengthened institutional capacity will be needed, both because of the increased number of recipients and because a wider distribution of income necessitates targeting benefits not just by group, but also by the income level of individuals and families. Legislative changes should take account of administrative constraints. As chapter 13 makes very clear, individuals with implementation experience should be involved from the start of the design stage in all policy areas.

The first, and overriding, priority in the short run is to ensure that unemployment benefits and social assistance can cope with potentially large numbers of unemployed people. The unemployment benefit should be simplified as much as possible for administrative reasons, and sufficient personnel and adequate equipment, such as calculators and computers, are needed to administer it. All the countries in Central and Eastern Europe have started, or intend to start, computerizing the calculation and delivery of unemployment benefits and employment services. The effective implementation of such developments, which requires substantial investment, is rightly accorded high priority. Staff should be adequately trained and detailed regulations available to guide staff on how the law should be implemented.

A further administrative priority is action to minimize defaults and delays on contributions. This is obviously problematic in the growing private sector, but such problems in the state sector are becoming endemic and even critical in some countries, particularly the Baltics and some other former Soviet republics.

In the medium term, a single, coherent administrative structure is needed to replace the present fragmented system of cash benefits. Four aspects stand out: moving the administration of benefits from enterprises, developing a system which can track the social insurance contributions of individuals, improving relations with the public, and devising more cost-effective methods of delivering benefits. Again, these areas all require heavy investment.

The reform process will necessitate a move from enterprise-based to individual-based benefits and contributions. At present, enterprises administer most short-term benefits, and the social insurance/pension authorities generally know nothing about individual recipients. This creates confusion about the respective roles of wages and benefits; workers need to understand that wages are determined mainly by productivity, not by distributional goals. In addition, administration imposes a substantial compliance cost on enterprises, which will harm their competitiveness, especially in an international context. In Western Europe, the social insurance/pension authorities organize the great bulk of benefit payments.

On the contributions side, if the social insurance/pension authorities are to administer benefits, they need to know about individual contributions. This is not at present generally the case. Automating contributions and benefits would lead to two sorts of savings: administration per se would be cheaper, and policy options, particularly for containing costs through better targeting, would be possible which are not possible with manual systems. One example is the capacity to raise the basic pension fully in line with inflation and pensions above the minimum by a smaller amount.

Relations with the public become more important as contact with individual beneficiaries (as opposed to enterprises) increases. Offices should examine how they provide information to current and potential beneficiaries. The information should be easily understandable by people who are not particularly well educated or used to being responsible for their own claims. Forms should be easy to understand to reduce the time staff need to spend helping people with their claims. The authorities should gear up to give such additional support and, where possible, should simplify procedures.

BUILDING POLITICAL CAPACITY. The political issues raised by social insurance, family benefits, and social assistance are very similar. Discussion of the need to build political capacity is therefore deferred until chapter 10.

Constraints

The widening distribution of income increases the need for income transfers. The major economic constraints are the fiscal crisis, which reduces the ability of the economy to provide income transfers, and gaps in the legislation and regulation necessary to underpin a market system, particularly in connection with financial institutions and private pensions.

A political constraint of particular relevance to cash benefits is that the time scale of reform is long term, but the pressures of democratic politics are short term. People want a perceptible increase in their standard of living, and they want it soon. Thus the demand for increased benefits is strong, as is the resistance to policies which reduce them. The pressures toward decentralization are also relevant, in that fragmented systems of cash benefits can lead to a fragmented labor market (if some benefits cannot be transferred across different regions) and may also create perverse incentives.

Major institutional constraints arise, first, because the private sector attracts the best people, partly because the backlash against government has reduced morale in the public sector and partly because the fiscal crisis has reduced public sector pay. The result is a declining number of skilled civil servants. The second institutional constraint is a lack of implementation skills. In part this is connected with the first constraint, but it is also a result of the tendency to regard policy and administration as separate activities, carried out by separate groups of people. People with administrative experience are rarely involved when policy is formed and legislation drafted, leading to legislation which is unnecessarily complicated to administer.

Conclusions: Priorities and Sequencing

The enormous macroeconomic shock and consequent fiscal crisis mean that conventional social insurance arrangements, with an explicit relationship between contributions and benefits, are likely to be unsustainable in some countries during the early transition, blurring the distinction between insurance and non-insurance benefits. Thus short-term policy, which concentrates on surviving the early transition by focusing on poverty relief, is clearly distinct from medium-term developments, in which income transfers are more obviously a part of the structural reforms.

Short-term Policies

Three sets of policies follow directly from the strategy discussed earlier in this chapter. Increasing unemployment and poverty require urgent action on poverty relief, the fiscal crisis makes cost containment imperative, and action is needed in the short run, as well as in the medium term, to ensure that the necessary benefits are paid.

Poverty relief should take precedence for fiscal reasons over other objectives of income support such as insurance and income smoothing:

- Urgent priorities are (a) to strengthen the system of unemployment benefits and, as discussed in chapter 10, (b) to develop a system of social assistance capable of delivering poverty relief to potentially large numbers of individuals and families.
- Expenditure should be concentrated on protecting the minimum level of benefits. Depending on the severity of fiscal constraints, a case can be made for paying unemployment benefits and pensions at a flat rate.

- Unemployment benefits and the major social insurance and pension benefits should be at least equal to the poverty line for an individual, social assistance should be defined relative to a family poverty line, and the minimum benefit should, if possible, be fully protected against inflation. In choosing a poverty line, political pressures and fiscal realities are in direct conflict.

Measures to contain costs are vital, notwithstanding their high political visibility:

- In the short run, benefits above the minimum should be protected only to the extent that resources permit.
- Invalidity pensions should no longer be paid to individuals whose earning capacity has not been affected.
- The right to receive a pension while continuing in one's old job should be withdrawn as rapidly as political realities allow. To assist restructuring and to protect young people from long-term unemployment, it might be appropriate to allow people who find a new job or who become self-employed to keep part of their pension.
- Private pensions should not be introduced for short-run reasons. They create no short-run fiscal savings and are no short-run substitute for dealing directly with controlling expenditure on state pensions.

Administrative capacity must be adequate in the short run to ensure that unemployment benefits can cope with potentially large numbers of applicants:

- Unemployment benefits should be simplified as much as possible, staff should be adequately trained, and detailed regulations should be available to guide staff on how the law should be implemented.
- Large-scale income testing may exceed short-run administrative capacity. Where possible, therefore, benefits should not be based on income but rather on indicators of poverty such as family size, unemployment, or age.

Medium-term Policies

Policies with a medium-term dimension relate to actions which should start in the short run but will take effect in the medium term.

Administrative capacity requires continuing action. The shortage of implementation skills implies that policymakers should not be over optimistic about the speed and scale of major reform. Nevertheless, administrative reform is vital, and not just because of the administrative savings which would result, important though they are. A modernized administration strengthens the capacity to collect contributions, speeds up the delivery of benefits by reducing the lag between applying for a benefit and its first payment, opens up options for containing costs which are not possible with manual methods, and increases the flexibility of policy. Thus:

- Work already under way on decoupling the delivery of benefits from enterprises should continue.

- An early start should be made on the planning necessary to increase administrative capacity through a coordinated process of policy development and upgraded delivery, including computerization.
- Such increased administrative capacity is needed not least to make it possible to introduce individual social insurance records.
- Methods of delivering benefits more cost-effectively than sending cash through the postal system should be introduced.

The structure of benefits should be refined. In particular, as fiscal and administrative constraints start to relax, the system should evolve toward one with a more explicit relation between contributions and benefits:

- Pensionable age should be raised over time. One way of phasing in such a change is to raise the retirement age by one year every second year. Fiscal pressures suggest an early starting date for such action; political pressures and a desire to minimize unemployment among younger workers suggest a more cautious approach.
- Social insurance contributions should be shared between worker and employer, with the worker's contribution appearing on his or her pay slip.
- To avoid incentives against employing women, the cost of maternity and family benefits should, so far as possible, not fall on the employer.
- As soon as economically and administratively feasible, the relationship between social insurance benefits and individual contributions should be strengthened. As part of this process, growing administrative capacity will allow greater reliance on income-tested poverty relief. The combined result would be to restore the distinction between insurance and non-insurance benefits.

Private pensions require continuing work on design and implementation. There are two firm recommendations:

- Private pensions should not be introduced until the necessary regulatory structure has been put in place. The design and implementation of such a regulatory regime is the first step toward developing private pensions.
- Private pensions should offer future pensioners at least some protection against inflation. This is particularly the case in Central and Eastern Europe, where restructuring poses a serious inflation risk. One mechanism for achieving this objective is through indexed government bonds.

Beyond that, policymakers have a choice about the mix of public and private pensions:

- A mainstream Western European system would have three tiers and would be very much a partnership between the public and private sectors. The foundation of the system would continue to be a PAYG social insurance pension, with a wider role than mere subsistence, although limited by appropriate ceilings for contributions and benefits. The state system would be complemented by a mandatory system of appropriately regulated private pensions, leaving open the question of whether they would be defined contribution or defined benefit schemes. Tax incentives could also be used to encourage voluntary private, defined contribution schemes, subject to the same regulatory regime as all financial institutions.

• A more individualistic system would have two tiers. These would be a minimal state PAYG pension, whose major purpose would be to provide a minimum guarantee for private pensions. Pensions for most people would be provided by one or more funded, regulated, defined contribution schemes, probably organized in the private sector.

In the three-tier system the separate components address different purposes of pensions: the state scheme is concerned with poverty relief and vertical equity; the second tier, by allowing people to redistribute income to themselves across the life cycle, facilitates income smoothing; and the third allows for the expression of individual preferences. The scheme accords a significant role to social solidarity, and risks are shared fairly broadly. If pensioners are to share in economic growth occurring after their retirement, the state pension could be indexed to earnings. The two-tier approach gives less weight to social solidarity. Apart from the minimum pension, there is no redistribution from rich to poor, nor any redistribution between generations; indexation, if provided at all, would be to price inflation.

The choice between the two approaches depends primarily on (a) objectives, (b) political arrangements, (c) the economic environment, and (d) the social context. On the first, Chile and a number of the high-performing Asian economies had a single objective, economic growth, which was given priority over all other objectives including social solidarity. On the second, government in the high-performing Asian economies is authoritarian; and the same was true in Chile at the time its reforms were introduced. On the third, those countries face fewer constraints than the countries of Central and Eastern Europe: they have long-established market systems, a sophisticated banking system, highly developed capital markets, and relatively stable prices; they need no substantial restructuring. In addition, the Chilean reform was introduced at a time of substantial budgetary surplus. Finally, the importance of the extended family in the high-performing Asian economies has already been noted. Such family structures exist in the Central Asian republics of the former U.S.S.R., but not in the other countries of Central and Eastern Europe.

Notes

1. The term social security is avoided because of its ambiguity. In the United States, it refers to retirement pensions; in the United Kingdom, it refers to the entire system of cash benefits; and in mainland Europe (in accordance with the usage of the

International Labour Organisation), it refers to all cash benefits plus health care. The term cash benefits is therefore used throughout.

2. There is no convenient shorthand for this type of benefit: they are often referred to as universal (Gordon 1988, p. 37), although in reality they are not, since they depend on the occurrence of the specified contingency.

3. These are the pure cases. In reality the categories can become blurred: both social insurance and universal benefits may be partly income tested by being included in taxable income.

4. Kopits (1991) found that in Central and Eastern Europe, subsidies comprised as much as 22 percent of GDP, while cash benefits comprised only 10 percent. In the European Union, the comparable figures were 6 and 16 percent; see also Kopits 1992.

5. The exception is Poland, where pensionable age is sixty-five for men and sixty for women.

6. Some countries have avoided this problem. In Albania and Latvia, unemployment benefits were incorporated from the start in the social insurance arrangements.

7. During the summer of 1992, with rapid inflation and a consequent shortage of cash, President Yeltsin, visiting a number of Siberian towns, was reputed to have had a planeload of cash, specifically for paying wages and pensions, as part of his entourage.

8. The figures in Layard, Nickell, and Jackman (1991) range from a notional 2 percent in Italy to 90 percent in Denmark. After Italy, the lowest is the United Kingdom (36 percent). A number of countries are above 60 percent, but only Denmark, Finland, Spain, and Sweden are above 70 percent.

9. In economic theoretical terms, the incidence of the contribution depends on the relative factor and product demand and supply elasticities and is independent of where the contribution is legally imposed. In theory, therefore, it does not matter how the contribution is shared between worker and employer. This result does not, however, necessarily hold in the short run, nor in situations where markets are not competitive. For further discussion, see Stiglitz 1988, chap. 17.

10. In Hungary in May 1993, national elections were held for self-governing bodies for social insurance. Seven trade union groups put forward lists of candidates, and it was difficult to distinguish between general elections and the social insurance elections. Those elected feel a strong mandate to respond only to their constituency.

11. Other things—particularly national output and the number of pensioners—remaining constant, PAYG schemes are not affected by inflation. With earnings of $1,000 the yield of a 10 percent pension contribution is $100. If prices double, so, generally, will nominal earnings. The yield of the 10 percent contribution is now $200; nominal pensions can therefore be doubled without changing the rate of contribution. In real terms nothing has changed.

12. The earnings-related component, but not the flat-rate pension, in Japan and Sweden is run on funded lines.

13. A U.K. reform in 1988 allowed individuals to manage their own pensions portfolio; very few people have taken up the option. Individual Retirement Accounts in the United States, though only a small part of private pension provision, similarly allow individual control.

14. In principle, the converse is also true, namely, pensioners are insulated from negative growth. This may not be the case, however, if negative growth is associated with inflationary pressures, as it currently is in Central and Eastern Europe.

15. Based on life expectancy, a first-generation worker who joined in 1981 will contribute to 2026, after which he and his surviving spouse will receive pensions for a further twenty years. The scheme's twelve-year history accounts for less than 20 percent of this period.

16. "Once upon a time there was a king who wished to hire a troubadour. The first performed so badly that he hired the second without even bothering to hear him sing" (Holzmann 1991, p. 75).

17. The results of an investigation by the U.K. Securities and Investments Board of a 1988 reform under which individuals were allowed to make their own pension arrangements were disquieting; see, for instance, "Nine in Ten Pension Deals Suspect," *Independent (London)*, December 17, 1993, p. 27. A British government enquiry (U.K. Pension Law Review Committee 1993, p. 10) found that "the present law . . . has a number of shortcomings. These include considerable complexity and a lack of structure and organisation. The law allows such wide powers and discretions to be left in the hands of the employer and the trustees that the interests of scheme members are not always sufficiently protected. . . . There is no form of compensation to cover loss through misappropriation of assets . . . [leaving] members at risk of hardship. Finally, there is no regulatory body." See also U.K. House of Commons Select Committee on Social Security 1992 and Nobles 1994. For disturbing claims about the vulnerability of private pensions in the United States, see Bartlett and Steele 1992, chap. 9.

Ten

INCOME TRANSFERS: FAMILY SUPPORT AND POVERTY RELIEF

SÁNDOR SIPOS

CHAPTER 9 DISTINGUISHED THREE TYPES OF INCOME TRANSFERS: social insurance, universal benefits, and social assistance. Social insurance awards benefits for particular contingencies on the basis of past contributions, universal benefits operate without a contributions test or an income test, and social assistance awards benefits on the basis of an income test. Social insurance addresses income losses which, at least in principle, are insurable. This chapter looks at the companion income transfers: family benefits, which are aimed at mothers and children and are generally awarded without an income test, and social assistance, which is aimed at poor people generally. Neither set of benefits relates to risks which are even remotely insurable. The discussion of poverty relief through income transfers should be seen alongside other policies to prevent poverty, notably, developing the human capital of the poor and providing them with earning opportunities, issues which are discussed in chapters 7, 8, and 11. An important omission is any detailed discussion of housing.

Family benefits in Central and Eastern Europe are generally high as a percentage of average wages in comparison with the highly industrialized countries. Two lines of argument exist for reducing their level and scope: to assist the move to a market economy in which wage differentials play an increasing role and to help address the fiscal crisis. Family allowances are controversial. The institution is important in Western Europe but has a much smaller role, if any, in some other highly industrialized countries, such as the United States, and in Latin America. This chapter takes as its starting point the Western European conviction that children are not merely a private utility but also a public concern. Thus family policy addresses not only poverty relief but, at least to some extent, also horizontal equity, for example, between families of different sizes. This concern leads to the recommendation that family allowances should be reduced as a percentage of wages, but not abolished.

Poverty relief, in contrast, should be expanded. As discussed in chapter 4, one of the effects of the transition has been to increase the number of poor people, at least in the short run; and, as discussed in chapter 9, social insurance benefits will increasingly be related fairly directly to individual contributions, thus reducing their scope as a device for poverty relief. The need for poverty relief to expand during a time of fiscal stringency means that benefits will have to be carefully targeted. Poverty relief has two possible objectives: to eliminate poverty by bringing the incomes of all poor people up to a poverty line (policy type A) or to ameliorate the poverty of those below a given poverty line by increasing their income, without necessarily completely eliminating the poverty gap (policy type B). Most social assistance schemes in the United States were designed with the second objective in mind; those in Western European countries tend more to type A policies (see Lødemel 1992; Northrop 1991).

The Inheritance

The major characteristics of family benefits and poverty relief bear a direct relationship to the inheritance. When income differentials are small, benefits can be targeted at groups rather than individuals; it is thus natural that there should have been a system of fairly large family benefits to meet the income needs of families of different sizes. Since poverty did not officially exist, it is not surprising that the system of poverty relief was rudimentary and aimed mainly at groups with permanent major health problems and at the frail elderly.

Strengths

The purposes of family benefits are significantly different from those of poverty relief. It is therefore helpful to discuss them separately.

FAMILY BENEFITS. Family support in Central and Eastern Europe was wide-ranging before the transition. In-kind benefits included the provision (usually free) of crèche facilities, kindergartens, day care centers, school meals, and the like. Cash benefits included:

- Family allowances (a monthly payment, tax free and normally not income tested; see the glossary), paid until the child reached a certain age
- A birth grant (a single payment at the time of birth)
- Maternity leave, often on full pay and usually for three to six months
- Parental or child care benefits (a monthly payment for which the mother was eligible after her entitlement to maternity leave expired, until the child reached a given age, varying between eighteen months and three years)
- Paid leave for the care of a sick child
- Various tax allowances and credits
- A death grant.

Under the old system, the distribution of earnings was fairly flat, with differences in family size accommodated by generous family benefits. This arrangement had significant advantages. It prevented many families from falling into poverty. It empowered households as consumers as opposed to passive recipients of in-kind provisions.[1] It allowed women to participate in the labor force or to stay at home with some income and job security or to opt for part-time work (see Cornia and Sipos 1991). It promoted gender equality, which later became an explicit policy goal (see Ferge 1993; Fong 1993; Fong and Paul 1992; Neményi 1990).

As an increasing body of empirical literature shows, a final advantage is that these benefits were generally well targeted, in that they reached most of the people who were poor or at risk of being poor. Sipos (1992) shows that in Central and Eastern Europe young people comprised 30 percent of the population but 40 percent of the poor. Zám (1991, p. 183) finds that social expenditure in Hungary in 1989 provided 82 percent of the income of large families with the poorest 5 percent of incomes but only 22 percent for families in the top 5 percent. Ravallion, van de Walle, and Gautam (1993) conclude that family allowances reduced poverty in Hungary in the late 1980s and were well targeted, but that if they been focused on younger and larger families, they could have been even better targeted. Jarvis and Micklewright (1993) argue that making family allowances universal in Hungary in 1990 improved targeting. Similar conclusions have been drawn for Czechoslovakia (Dlouhy 1991), Poland (Topinska 1991; World Bank 1993b), Yugoslavia (Vukotič-Cotić 1991), and for the whole region, except for the Russian Federation (Zimakova 1992).

An extensive research project found that

family allowances are strongly pro-poor in absolute terms in all [Central and Eastern European] countries except Russia. This means that poor households receive more of them not only in relative terms . . . but also in absolute amounts. Family allowances are the only income source in Eastern European countries that is both important and strongly focused on the poor. They achieve a significant reduction in inequality, lowering the overall Gini coefficient by approximately 3 percentage points in Hungary and CSFR [Czechoslovakia] and about half a point in Bulgaria, Poland, and Russia. For comparison they reduce income inequality by 1.3 percentage Gini points in France (where they are comparable in size to Central European countries) and by 0.8 points in the U.K. [Milanović 1992, p. 15]

POVERTY RELIEF. The main strength of poverty relief is that it was delivered to *groups* of people (families, the elderly) through more or less guaranteed employment and a comprehensive set of cash and in-kind benefits, rather than to *individuals* through income-tested social assistance. Poverty relief thus took place *before* large sections of the population fell into poverty. Although poverty, contrary to official orthodoxy, was not fully eliminated, it was not a mass phenomenon (Milanović 1991; Sipos 1992). Notwithstanding these strengths, this preventive approach became unsustainable in the face of the slowdown in growth which occurred in the 1980s.

Weaknesses

The significant weaknesses of both family benefits and the system of poverty relief are in many ways the other side of the coin of their strengths.

FAMILY BENEFITS. The serious weaknesses of family benefits parallel those of the social insurance system.

Ineffective cost containment. Table 10-1 shows the high cost of family benefits which, in the mid-1980s, cost over twice as much as the OECD (Organization for Economic Cooperation and Development) average of 1.2 percent of gross domestic product (GDP).[2] Table 10-2 shows the extent to which family benefits in 1980 were a high fraction of earnings in Central and Eastern Europe in comparison with those in Western Europe. A decade later, early in the transition, the family allowance for each child was 8 percent of the average monthly wage in Poland and 11 percent for each child in a two-child family in Czechoslovakia. Across the OECD, the average level of support from all family benefits was 7.5 percent of average earnings *per family*. In some countries, benefits were also paid for a dependent spouse. As a result, family allowances played part of the role which in industrial countries is played by wages. The pattern in Western Europe is to pay such benefits at a lower rate, complemented by income-tested family support (see Commission of the European Communities 1991).

Maternity benefits were generally at or close to 100 percent of the mother's previous wage (in Czechoslovakia, the figure was 90 percent; in Hungary, 65–100 percent; and in Romania, 50–94 percent). Again, this is high compared with Western countries (usually 50–85 percent of the mother's wage) but similar to the figure in many Latin American countries (Argentina, Brazil, the old Chilean system, Costa Rica, Mexico, and Uruguay; see Mesa-Lago 1991).

Table 10-1. *Public Expenditure on Family Support as a Percentage of GDP in OECD Countries and Central and Eastern Europe, 1980-88*

Group	1980–84 average	1985–88 average
OECD	1.2	—
Poorer countries[a]	0.6	—
Social welfare countries	1.5	—
Central and Eastern Europe	2.5	2.5
Czechoslovakia	3.3	3.2
Hungary	2.3	2.7
Poland	2.2	1.8

— Not available.
Note: Family support includes family allowances and parental or child care benefits.
a. Poorer OECD countries include Greece, Ireland, Portugal, Spain, and Turkey.
Source: Rutkowska 1991, pp. 10, 21.

Table 10-2.   *Family Allowances for Two Children as a Percentage of Average Earnings, by Country in Eastern and Western Europe, 1980*

Region and country	Percent
Eastern Europe	
Hungary	22.2
Bulgaria	20.0
Czechoslovakia	19.6
Poland	17.0
German Democratic Republic	3.9
Western Europe	
Austria	16.9
Belgium	10.7
Netherlands	9.0
Sweden	8.7
United Kingdom	8.2
Switzerland	6.9
Germany, Fed. Rep. of	6.6
France	6.5
Norway	6.4
Italy	5.4
Denmark	3.0

Note: The figures for Denmark, Germany, Norway, and Sweden refer to 1981; the figure for Poland, to 1984.

Source: International Labour Organisation 1989, p. 55.

The high level of family benefits was understandable in the past, when the distribution of earnings was flat and differences in family size were accommodated by generous family benefits. This policy becomes increasingly inappropriate as the economic reforms take effect, and wages become the main source of family support.

Holes in the social safety net. The effectiveness of family benefits in relieving poverty had significant deficiencies: inadequate coverage of marginal groups, particularly those working outside the state sector; lack of refined targeting; inadequacies in respect of gender considerations; and, on occasion, ethnic discrimination.

Gaps in coverage were a particular problem for maternity benefits and family allowances. Maternity benefits depend on a contributions record in almost all the highly industrialized countries and throughout Latin America (where twenty-one of the twenty-two countries had some sort of maternity benefits; see Mesa-Lago 1991). In Central and Eastern Europe, however, large agricultural populations and the self-employed were excluded or received significantly lower benefits. Declining agricultural populations and gradual extension of coverage improved matters, but gaps remained in some countries, for example, Poland and Romania.

Family allowances had even greater gaps. Whereas social insurance is intended to offer insurance against risk, an important purpose of family allowances is to help a family redistribute income across the life cycle. Precisely because having children is not in any sense an insurable risk, family allowances in most Western European countries are universal (that is, paid without a contributions record or an income test; see Commission of the European Communities 1991). In Latin America only the richer countries (Argentina, Brazil, Chile, and Uruguay) provide some family allowance, usually for those covered by social insurance (Mesa-Lago 1990, table 19). In Central and Eastern Europe, family allowances were normally part of social insurance and thus excluded families outside formal employment. In some countries, for instance Poland, family allowances were related to salary and hence were regressive (see Wiktorow and Mierzewski 1991); in others, such as the former Soviet republics, means-tested benefits supplemented or replaced benefits based on social insurance. During the 1980s, coverage of family allowances gradually broadened, but the first country to introduce a universal allowance, Hungary, did so only in 1990.

Lack of refined targeting was pervasive. First, family allowances were not differentiated according to the age of the child, and older children in secondary and higher education were eligible for a range of uncoordinated benefits. Such arrangements simplified administration but had major disadvantages for relieving poverty and containing costs.

Second, given aging populations and acute labor shortages, benefits were often disproportionately higher for larger families. This type of pro-natalist policy can be criticized on several grounds: it is expensive, it has only a small and temporary effect on birth rates (Andorka 1991; Chernozemski 1991), and, given the high correlation between poverty and family size (Milanović 1992; Sipos 1992), high benefits become an indispensable element in the budgets of many large families. In the short run, pro-natalist family benefits may be the least bad way of keeping large families above the poverty line. As the distribution of earnings widens, however, the approach will need to be reviewed.

The idea of extended child care benefits, first introduced in Hungary in 1968 as an extension to maternity leave, is well founded, but its design could be improved. The introduction of the scheme was prompted in part by fears that contemporary moves to a more market-oriented economy (the New Economic Mechanism) would cause substantial female unemployment. Most countries in Central and Eastern Europe and many in Western Europe quickly followed Hungary's lead. The scheme has undoubted appeal because it encourages more flexible participation of women in the labor force and enhances early child development. In most countries of Central and Eastern Europe, however, it was introduced as a social insurance benefit, based on an appropriate work history and differentiated according to previous earnings.

These features were not consistent with social justice and the needs of children. It seems more appropriate to pay women seeking to withdraw from the labor market for an extended period a flat-rate benefit for raising a child.

Gender considerations were not always adequately taken into account. Family allowances were usually paid to the head of the family, that is, the father. Given women's high participation in the labor force, this was unnecessary and made the scheme vulnerable to abuse (for example, by alcoholic fathers). At the same time, the eligibility of fathers for maternity or child care allowances was generally more restricted in Central and Eastern Europe than in the West.

Ethnic discrimination arose in some countries. Pro-natalist policies paid a high benefit for the third child, but benefits sometimes declined sharply thereafter because very large families usually belonged to minorities: gypsies in many countries or Turks in Bulgaria (Chernozemski 1991; Kroupová and Huslar 1991).

Adverse incentives. A vast literature treats the effects of family benefits and social assistance on the supply of labor (see Burtless 1986; Moffitt 1992). Its application to Central and Eastern Europe is unclear, however, not least given very high rates of female participation in the labor force. The only benefit for which there is evidence of a reduced supply of labor is parental or child care allowances, which support extended maternity leave: unskilled women used the benefit fully; those with higher earnings resumed work earlier than required (Zimakova 1992). Since the objective of extended maternity leave is to allow very young children and their parents to spend more time together, this outcome was not necessarily a disadvantage.

Weak administration. Most family benefits were administered by enterprises, raising the same set of problems as with social insurance. The problems were compounded in some countries by the multiple forms of family benefits (for example, in 1992, Ukraine had about sixty forms of largely overlapping family benefits).

POVERTY RELIEF. The approach to poverty relief during the communist era assumed that social assistance was not needed. Nevertheless, poverty persisted. Most of the poverty lines in table 10-3 are officially calculated subsistence or social minima. In contrast with countries in the European Union, where a guaranteed minimum income reflects some sort of subsistence level, in Central and Eastern Europe these indicators seldom served as a threshold for policy intervention (see Commission of the European Communities 1991; the United States is unusual among the countries of the OECD in that its main income-tested benefit—Aid to Families with Dependent Children—is not paid to families at the poverty line).

Despite persistent poverty, social assistance was highly residual. The system was fragmented and given low priority. Most benefits designed for temporary income support were discretionary, resulting in arbitrary decisions, a high

Table 10-3. *Poverty Indicators and Rates in Central and Eastern Europe, Beginning and End of the Period, 1958–91*

Country and year	Expenditure on food as a percentage of poverty line	Poverty line as a percentage of average wage	Percentage of population in poverty
Bulgaria			
1978–89, subsistence minimum	—	65–53	—
1978–89, social minimum	34–42	82–67	—
Czechoslovakia			
1958–88, subsistence minimum	60	42[a]	6–0
1958–88, social minimum	50	55–56[a]	11–7
1976–90, social minimum	60	42–56[a]	6
Hungary			
1967–82, subsistence minimum	35–37	45–55[b]	10–7
1967–82, social minimum	29–31	54–66[b]	14
1983–90, subsistence minimum	30–38	40–58[b]	7–10
1983–90, social minimum	25–32	48–70[b]	15
Poland			
1981–89, subsistence minimum	50–60	—	14–17
1981–90, subsistence minimum	60	36[c]	14–33
Romania			
1991, subsistence minimum	—	—	13
1991, social minimum	—	—	42
U.S.S.R.			
1967, subsistence minimum	56	58	16 (urban) 39 (rural)
1967, social minimum	36	76	
1980–90, subsistence minimum	49[d]	72–59 (urban) 90–67 (rural)	24–2
Yugoslavia			
1978–89, subsistence minimum	54[e]	28–32 (urban)[f]	17–24

— Not available.

Note: The number before the dash is the value at the beginning of the period, and the number after the dash is the value at its end.

a. Percentage of average household income.
b. Minimum and maximum figures for the period.
c. 1987.
d. 1985.
e. 1989.
f. 1982–86.
Source: Sipos 1992, p. 56.

level of stigma, and low take-up rates (see the glossary). Thus a high percentage of individuals in need failed to receive benefits. Poverty was regarded as a pathology, with high personal responsibility, rather than as a lack of income. Social assistance was therefore targeted at social problems, often without an adequate means test or proper assessment by a social worker.

Organizations outside government were invited to provide neither complementary sources of care nor financial assistance to state channels. The few charitable nongovernmental organizations (see the glossary) which were allowed were mainly church organizations. They operated in a hostile environment because of ideological constraints on charities and political suspicion of civilian initiatives.

The administration of social assistance was poorly staffed and usually local. Most countries had few social workers, and those in place were not adequately trained. Ignorance was reflected in a restricted range of policy instruments: gaps included overnight shelters and temporary work opportunities.

The Forces Driving Change

Rising poverty, one of the main short-run outcomes of the process of transformation, has become one of the central political forces. Brief discussion of the overall forces driving change is therefore followed by a more detailed discussion of available evidence about the extent of poverty under the old system and during the early years of the reforms.

Economic and Political Forces

Among the major driving forces discussed in chapter 4 are the widening income distribution in general and increasing poverty in particular. The transition involves large and unusual dislocations, such as mass layoffs, abolition of cooperatives, and the sudden decline of single-industry towns, and has not always been peaceful, requiring additional assistance. Social insurance will not suffice, not least because its scope is being narrowed, so that many people previously covered are now excluded. Thus social assistance has a central and growing role as the benefit of last resort.

At the same time, the fiscal crisis narrows the scope for income transfers. Social policy has to be redefined. Better targeting is needed both to avoid gaps in coverage (horizontal efficiency) and to contain costs (vertical efficiency). Social assistance, being income tested, is at least in principle highly effective in this context, although policy development is necessary to prevent administrative costs from absorbing much of the savings resulting from tighter targeting.

Empirical Evidence on the Extent of Poverty

Table 10-3 shows the extent of poverty prior to the transition relative to an absolute poverty line. The figures measure the income of households or individuals compared with a minimum basket, often called the subsistence or social minimum, which reflects an official perception of poverty (on the merits and shortcomings of these calculations, see Atkinson and Micklewright 1992; Milanović 1991; Sipos 1992). In most countries of Central and Eastern

Europe, poverty declined significantly between 1960 and 1980. In the 1980s, as growth started to slow, poverty rates stagnated or started to rise again, with the exception of the U.S.S.R., where poverty rates continued to fall until the end of the decade.[3] Poverty rates varied considerably both between and within countries. Although not a mass phenomenon, poverty was too large for the prevailing residual poverty relief regimes and was aggravated by the wage and price liberalization which occurred in the early transition.

In Poland, according to official statistics, real income per head fell more than 41 percent between 1989 and 1991, and the poverty rate (the number of poor people) doubled from 17 to 34 percent (see table 10-4). These figures are almost certainly too high both because of the statistical problems discussed in chapter 4 and because of the specifics of the Polish situation.[4] Nevertheless, the direction of change is clear, and the effects on child poverty were sharp:

> Because the increase in poverty was greater among large-size families, the change in the percentage of children who are poor is even greater than the change in the percentage of the poor households or the population. In 1989, about 17.5 percent of children under six years of age lived in poor households; this increased to more than 50 percent in 1991. Poverty incidence among children of workers and mixed households tripled. [Milanović 1993, p. 15]

Milanović also found that the poverty gap (the total cost of bringing the income of all poor households up to the poverty line) almost tripled between 1989 and 1991, from 1.4 percent of GDP to 4.6 percent. Thus the poor are more numerous than before and, subject to caveats about the statistical data, are becoming poorer.

In Hungary, the experience was somewhat different: the increase in poverty rates took place without an increase in the poverty gap. According to household surveys, the incidence of poverty increased from 10 percent in 1990 to

Table 10-4. *Poverty and Real Income in Poland, 1989–91*
(percentage change)

Categories of households	Decrease in real per capita income	Poverty rate	
		1989	1991
Workers	33.3	15.8	38.1
Mixed households	41.0	7.9	21.2
Farmers	51.6	17.2	39.4
Pensioners	11.6	36.2	33.0
Total	41.5	17.3	34.4

Source: Milanović 1993.

15 percent in 1991, and to 27 percent in 1992 (Andorka, Kolosi, and Vukovich 1992; Ferge 1993; KSH 1992). Using a narrower definition, poverty increased from 8 percent to nearly 16 percent between 1989 and 1992 (KSH 1993a). Even if the impact of hidden incomes and activities in the informal sector is considered, the poverty rate was estimated to be in the range of 20–25 percent, more than double the level before the transition. Nevertheless, mainly because of offsetting social policy measures, the poverty gap did not widen substantially (see Sik and Tóth 1992). The changes, nonetheless, are profound. The average income of the richest 10 percent rose from 4.9 times that of the poorest 10 percent in 1987 to 6.0 in 1992 (Andorka, Kolosi, and Vukovich 1992). Again, the number of children and the incidence of poverty are highly correlated. In 1992, only 9 percent of people without children lived below the poverty line, rising to 21 percent of two-child households, and 49 percent of families with four or more children. More than 15 percent of children in poor families lived on less than 70 percent of the narrowly defined poverty line (KSH 1993a). The incidence of poverty was also high among the unemployed (nearly 50 percent) and gypsies. In 1992, about two-thirds of gypsies aged sixteen or older were poor (and even more if children are counted; Ferge 1993).

In Russia, about 12 percent of the population was poor in 1990; by September 1992, the figure had risen to 37 percent (although the data are not strictly comparable due to differences in methodology). Once more, the great majority of poor households were families with children (Popkin 1992; Popkin, Mozhina, and Baturin 1992; Sipos 1992). In the first quarter of 1993, 42 percent of all families with children under sixteen years of age were below the poverty line, rising to 72 percent of families with three or more children (Radio Free Europe/Radio Liberty, *Daily Report* 99, May 26, 1993). Women were traditionally overrepresented among poor and vulnerable households in the former U.S.S.R. This trend apparently intensified during periods of rapid inflation for households dependent on cash benefits and with low earnings (see Fong 1993). In contrast, in late 1992, only 19 percent of elderly households were below the poverty line. In the case of Russia, not only the poverty rate but also the poverty gap was large. In September 1992, the average shortfall below the poverty line was about 34 percent (World Bank 1993c).

Less is known about poverty trends elsewhere. In Bulgaria in 1991, the poverty rate was about 34 percent; in Romania, it was between 13 and 42 percent, depending on the choice of poverty line (Barbu, Gheorghe, and Puwak 1992; Sipos 1992).

Although evidence is still fragmented, the main patterns confirm the analysis presented in chapter 4: the distribution of income is widening, and poverty is spreading, especially among children, young adults, the unemployed, and disadvantaged minorities (see also Burrows 1994). Problems which were marginal prior to the transition are becoming a common experience.

Policy

This section starts with a brief discussion of the overall strategy. After discussing poverty relief elsewhere, it turns to the reasons why family benefits and social assistance should be continuing parts of the economic landscape.

The overall strategy set out in chapter 9 applies equally here. It is necessary at this stage, however, to remind ourselves of the distinction made at the start of this chapter between type A policies, which aim to prevent poverty, and type B policies, which aim to ameliorate the most extreme poverty by reducing but not eliminating the poverty gap. The choice between the two strategies is critical in addressing the tension between fiscal constraints and political imperatives. Not many countries in Central and Eastern Europe can afford true type A policies, and some definitely cannot afford them. A likely outcome in many countries is a mixed regime. The correct floor for policy A during the early transition was a parsimoniously defined subsistence minimum. Policy B requires several decisions: How much of the poverty gap should be reduced? Should all groups be helped equally, or should policy focus on particular groups, such as the elderly, mothers, and young children? Should policy focus on particular needs, such as nutrition, health, or education?

The traditional methods of family support and poverty relief are by no means above criticism. The arguments, which are not new, assert that benefits create adverse incentives, both because taxation is necessary to pay for them and because their existence reduces people's incentives to be self-sufficient; benefits, particularly benefits in kind, have high administrative costs; and benefit administrations have a vested interest in expanding the schemes. Many commentators, therefore, advocate poverty relief in cash: they empower consumer choice; they facilitate the market mechanism; and they reduce administrative costs.

Some commentators argue further that there should be no specific benefit administration; benefits, instead, should be integrated with personal income taxation. Such negative income tax schemes (sometimes known as a guaranteed minimum income) work in the following way: each individual or family receives a given income from the state (the guaranteed income) and then pays tax on all other income.[5] The size of the guarantee can vary with family size and can be higher for specific groups such as pensioners or the disabled. The system is administered by the tax authorities (for the classic statements, see Friedman 1962; Friedman and Friedman 1979). Depending on the size of the guaranteed income, both type A and type B interventions could be delivered via negative income taxes, which could replace all other benefits. The idea has support from all parts of the political spectrum: from the right, who see it as a way of breaking what they regard as the stranglehold of in-kind benefits, and from the left, who see the guaranteed income as a right of citizenship, awarded without the stigma of an income test. It is therefore important to ask whether family support and social assistance are necessary or whether the new market economies should eliminate them altogether. After a brief description of systems elsewhere, this chapter turns to this basic question.

The Strategy

Most Western European countries pursue type A policies (this section draws on Esping-Andersen 1990; Esping-Andersen and Micklewright 1991; Lødemel 1992; Lødemel and Schulte 1992). However, their approach to poverty relief varies substantially.

In the Nordic countries, an employment-oriented policy has kept the number of poor small. Social assistance is residual, largely the responsibility of local governments, and discretionary inasmuch as national legislation needs to be interpreted by local governments and social workers. Poverty is perceived not only as a lack of income but also as a social pathology. Poverty relief is therefore delivered via cash benefits coupled with the active, often discretionary, involvement of social workers.

In continental Europe (Austria, Belgium, Germany, and the Netherlands), social insurance, based on a contributions record, is distinguished sharply from social assistance, financed from general taxation. In these countries, poverty relief is seen largely as income maintenance. It is usually a legal entitlement but is administered and financed locally with little caseworker discretion. Benefit administrators outnumber social workers.

The British system of social assistance is nationally organized as part of the administration of social insurance (though it is separately financed). The focus is on income maintenance. Benefit administrators pay social assistance to a large clientele on the basis of well-defined rights, with little discretion. Social workers are not involved in income maintenance; rather, they focus on special problem cases at the local level.

The Latin or Southern European countries (France, Italy, and Spain) are latecomers and, with the exception of France, have no general schemes for the poor. Greece and Portugal, despite persistent poverty, have no sizable social assistance schemes. In consequence, local welfare organizations such as the church and other charities play an important role. Existing schemes are characterized by a categorical approach which defines groups of people, such as the elderly who receive little or no pension and the disabled, as beneficiaries of social assistance (see Tavazza and others 1990). In 1988, France introduced a national system of assistance (the *revenue minimum d'insertion*) which limits caseworker discretion and emphasizes reintegrating the poor into society by making income maintenance conditional on participation in reintegration programs.

In the United States, social assistance is fragmented, with programs geared to help different categories of poor people, such as individuals without health insurance and children in poverty, or to meet particular needs, such as food and medical attention. Critics argue that the system is too expensive, is open to abuse, and does not reduce poverty because it fosters an attitude of dependency (see Friedman 1962; Friedman and Friedman 1979; Murray 1984). Others show that U.S. poverty relief programs are all type B. Benefits are set at levels which only ameliorate poverty after the fact but do not bring recipients above the poverty line (Northrop 1991).

Most developing countries pursue type B policies focused on prioritized needs. In some countries, the priority is food; in others, it is primary health care or education (see Ahmad, Drèze, and Hills 1991). In some Asian economies—mainly those where Confucian traditions play a prominent role, such as Korea, Taiwan (China), and to a certain extent mainland China—social assistance tends to focus on health, education, and some limited nutritional intervention (see Leipziger and others 1992; Yeun 1986). In Latin America, pervasive poverty is hardly mitigated by social assistance. The poor are rarely covered by any form of social insurance and receive very limited support from the state and charitable organizations (often the Catholic Church; McGreevey 1990; Mesa-Lago 1991; Psacharopoulos 1990; Psacharopoulos and others 1992).

WHY FAMILY BENEFITS? Although most of the highly industrialized and higher-income developing economies have a broad range of family benefits, some benefits are absent in some countries: Australia, New Zealand, and the United States, for example, have no system of universal family allowance. It is therefore necessary to ask why these benefits should continue notwithstanding fiscal constraints.

Historical forces are not an overriding argument, but neither should they be ignored. Family benefits exist throughout Central and Eastern Europe as adjuncts to wage setting and constitute an unusually large part of family income. Unlike price subsidies and many in-kind benefits, they do not cause major distortions apart from the impact of their cost on tax rates. For the reasons set out in chapter 2, an overwhelming case can be made for undertaking substantial privatization and for phasing out most price subsidies to gain the efficiency advantages of market allocation. The same arguments do not apply to family benefits, at least not with the same sense of urgency. Indeed, when radical cancellation of producer and consumer subsidies and some state-provided services is on the agenda, and when falling output makes it impossible to raise wages or to compensate fully for the cancellation of subsidies and services, family benefits cannot be eliminated without exacerbating the impact of the transition on family incomes. In addition, the abolition of family benefits in such circumstances, even if government thinks it desirable, is likely to meet strong political resistance or to invite disillusionment in democratic institutions.

Income smoothing introduces a second set of arguments: family benefits stabilize the consumption of beneficiaries during difficult times. It is true that hunger is an efficient incentive to seek employment (if jobs are available). However, hungry employees are rarely productive, let alone innovative; both qualities are badly needed in Central and Eastern Europe. Children cannot wait until better times to grow physically and develop their abilities. Poor nutrition during their early years can permanently stunt their intellectual development. If they drop out of school for financial reasons, they lose the most productive years of cognitive development. The lack of income smoothing can easily become irreversible at a personal level (for a fuller discussion, see Cornia and Sipos 1991). Income smoothing thus has major implications for human capital formation, quite apart from its obvious moral and political dimensions.

Poverty relief raises a third set of issues. Since poverty is highly correlated with family size, family benefits, being awarded on an easily observable indicator of poverty, are well targeted. This proposition is supported by convincing evidence from the early transition. It remains to be seen over what time horizon the correlation between poverty and family size might loosen, or when cost-effective and easy to administer means-testing techniques might become feasible in Central and Eastern Europe. If and when either of these developments occurs, the case for using family benefits to relieve poverty will correspondingly weaken (although the other arguments will remain).

Public choice, the fourth element in the discussion, is the link between the historical context, on the one hand, and the tax burden of paying for family benefits, on the other. Family benefits are a significant tax burden and so, like any item financed from taxation, are subject to public control in a democratic society. Thus family benefits must reflect public preferences. Values vary across nations and cultures, depending on prevailing attitudes toward children, families, the role of women, and so forth. Societies which perceive children more as a private than as a public benefit are less sensitive to horizontal equity, and the public is less willing to share the costs of child rearing. Countries where women's rights are a major issue are likely to emphasize benefits for single mothers. A drive for gender equality is usually reflected in equal access to family benefits—maternity and parental or child care benefits for fathers and family allowances for mothers. A quest for higher participation of women in the labor force usually leads to more child care facilities. Finally, pro-natalist incentives, regardless of their motive and efficiency, also fall into the category of public choice.

The connection of family benefits to negative income tax is often overlooked. The United Kingdom's switch in the late 1970s from child tax allowances to universal, tax-free child benefits was, in effect, a partial move toward an income guarantee scheme.[6]

The conclusion to which these arguments lead is that there is no universal recipe for the optimal level or form of family benefits. The appropriate choice will depend on a compromise between public choice and economic sustainability. Two statements can be risked, however:

- During the transition, a relatively high level of family benefits should remain for reasons related to history, poverty relief, income smoothing, and public choice.
- Although the public expects higher wages to pay for an increasing share of child rearing when economic growth resumes, the pendulum probably will not swing toward total abolition of family benefits. A sense of horizontal equity seems to be present in most countries of Central and Eastern Europe.

Against this backdrop, it is probably more useful to concentrate on increasing the efficiency of family benefits through better targeting and streamlining than to focus on persuading the public that such benefits should be eliminated.

WHY SOCIAL ASSISTANCE? The case for social assistance is even less ambiguous. The case is not that social assistance is better targeted than family benefits, but rather that it has important additional aims.

Labor market policy and social insurance rest in important ways on social assistance. Entitlement to unemployment benefits generally expires after a fixed period, varying by country between six and eighteen months. Given the shortage of jobs during the early transition, many people remained unemployed longer than the maximum duration (in the case of older workers, perhaps permanently). An important function of social assistance, therefore, while maximizing incentives to find a new job, is to provide poverty relief for the truly needy. This is partly to enhance equity, partly to sustain political support for the reforms, and partly to avoid the social costs—increased crime and erosion of human capital—which would otherwise occur. The level of and access to benefits are much more controversial issues and, as discussed earlier, vary widely in different countries. The choice among particular models and their mixes is again affected by public preferences, but the need for social assistance is not questioned.

An additional and forceful argument in favor of social assistance is its ability to take account of family size. In reformed social insurance schemes, benefit is generally awarded on an individual basis rather than being fully related to family size. This is a rational development, in that a closer relationship between contributions and benefits contributes both to efficiency and to horizontal equity, even if some redistribution occurs in most schemes. However, households differ greatly in size and composition. Social assistance is the instrument with which to take account of any of these differences which other benefits fail to accommodate.

A third purpose of social assistance is as the social safety net of last resort, providing highly targeted poverty relief.

WHY NOT A NEGATIVE INCOME TAX? It is necessary, finally, to consider the case for and against a negative income tax. One way of thinking about such systems is that they pay everyone a guaranteed income equal to the poverty line, recouped through the tax system from those who do not need it. The problem with the idea is not its logic, whose simple clarity is part of its great appeal, but the empirical fact that in almost all countries the distribution of income is heavily skewed toward lower incomes (there are many poor people and few rich ones). The heavy lower tail of the pre-transfer income distribution necessitates high tax rates to pay for the benefit; and tax rates that high are unsustainable because they discourage work effort. The inexorable logic of the point is important. Atkinson (1983, p. 275) illustrates it in terms of a typical Western European income distribution. If the guaranteed income for an average family is x percent of average income, and if income tax currently raises y percent of average income for purposes other than income support, the average income tax rate must be $x + y$. With plausible values for x and y (say, 35 and 15 percent), the *average* rate of income tax, ignoring all indirect taxes, is 50 percent. Although it is, up to a point, possible to finesse the issue by introducing more complex schemes (see Parker 1989), the underlying point remains valid.

For these reasons, negative income taxation has been introduced in the West only on a small scale. In the context of Central and Eastern Europe, in addition, (a) the average tax rate would have to be significantly higher,[7] and (b) large parts of the population would remain outside the income tax net. Moves from child tax allowances toward a universal family allowance would be a partial step along the road.

Maintaining Macroeconomic Balance

There is an obvious need for policies to contain costs. But possibilities for mobilizing additional resources should not be overlooked.

MOBILIZING RESOURCES. There are pressing needs and new opportunities for mobilizing additional resources. The pressing need is to address the increase in poverty. The new opportunities are of three sorts: the phasing out of consumer and producer subsidies creates substantial savings; in addition, new democracies offer increasing opportunities to raise funds from private sources through nongovernmental organizations such as charities, self-help groups, and the like; and the reforms open up more flexible arrangements between central and local government.

The abolition of subsidies hits vulnerable groups the hardest. Their abolition should therefore be coordinated with an increase in social assistance budgets and a buildup of the administrative capacity necessary to deliver services. The attractive part of this logic is that since a general, untargeted subsidy is being replaced by a targeted benefit, only part of the resources released by the abolition of subsidies is needed to reduce poverty; the rest could be used to reduce budget deficits.

In extraordinary situations, social solidarity usually strengthens. Central and Eastern Europe suffers a handicap in this respect. The church and nongovernmental organizations, and charities generally, require resources to restore their capacities, an area which often requires considerable political tact.[8] Policy needs both to encourage and to regulate nongovernmental activity. Traditions are very important in this field. It is not by chance that countries with charitable traditions (Croatia, the Czech Republic, Hungary, Poland, and the Slovak Republic) have dozens of soup kitchens, while poorer countries like Albania have none. In addition to providing a conducive environment, governments can actively encourage international nongovernmental organizations and other charities to help. However, quality control is important in the light of potential abuse, for example, scandals connected with the foreign adoption of children in Romania or the contracting out of orphanages by militant religious groups in Albania under conditions which threaten the children's interests. Thus both the traditional charities and new nongovernmental organizations require immediate legislative and regulatory changes. Regulation of nonprofit activities should be high on the legislative agenda.

The central-local mix in financing and providing social assistance is a subject of heated debate throughout Central and Eastern Europe. Local initiatives, such as public works and feeding programs for the elderly or for children, could be crucial in expanding the resources available for social assistance. Local governments, charities, and nongovernmental organizations are better suited than the state to provide some aspects of social assistance: they are closer to the problems, and they can devise more flexible responses than the central government. The existence of a variety of providers could empower individuals to use the most appropriate forms of social assistance. Social assistance benefits should, to the extent possible, be self-targeting, for instance, by offering low-quality goods such as used clothes or low remuneration for public works. The stigma associated with relying on social assistance is also a self-targeting device, although more controversial than the others since it might reduce take-up to an unacceptable extent.

CONTAINING COSTS. So far as family benefits are concerned, cost containment involves (a) reducing the real value of universal family allowances, (b) eliminating these benefits for a dependent wife, and (c) rationalizing the duration of certain family benefits.

Reducing the real value of family allowances per child does *not* imply that the benefit is an unwanted luxury at present levels of development in Central and Eastern Europe. Rather, it relates to the changing role of wages and other income and to retargeting public transfers. The emerging structure of market-determined wages will cover a larger part of the costs of child rearing than was the case under the old regime. In an ideal transition this would be a painless process, with family allowances falling as real wages rise. The problem is that during the actual transition, and especially during the stabilization phase, real wages fall, which precludes the painless route. Income smoothing arguments suggest that family benefits need to be *increased*. That, however, would be the wrong route, given fiscal constraints. The policy which makes sense would reduce the real value of family allowances (but not necessarily of other family benefits), reflecting the changing role of wages at the higher end of the wage structure. At the lower end of the wage spectrum, universal family benefits would be supplemented, but not replaced, by income-tested family support. The process should be gradual. Its phasing should take account of changes in the wage structure and of the development of administrative and other target-ing capabilities.

Family allowances to dependent wives should be eliminated altogether. The practice makes little sense and is paternalistic. The resulting savings would be better used for labor market policies designed to help women find a job or for child care services designed to assist the participation of women in the labor force. If the purpose of the transfer is to acknowledge child rearing as an alternative to gainful employment, then the transfer should be made condi-tional on the presence of children and the costs of the scheme should be monitored closely.[9]

Reducing the duration of some benefits offers additional savings. Family allowances in some countries can be paid for a child up to twenty-six years of age who is in continuing, full-time education. The problem of young adults in education is better addressed, however, by an explicit system of student support involving grants and loans designed to encourage participation in higher education, increase self-reliance, and improve incentives. Student loans (discussed further in chapter 11) are another example of a state-organized device to facilitate income smoothing. Sequencing is important. During the early transition, when wages were low and insufficiently differenti-ated by educational level, loan repayments were likely to be problematic. Student support which does not have to be repaid should remain, at least in part. The duration of other family benefits (for example, maternity and child care allowances) could also be reviewed.

Containing expenditure is also relevant to social assistance. The topic arises in discussion of the effects of different financing mechanisms (in the context of incentives) and of the level at which the benefit is pitched (in the context of poverty relief).

Building Markets

Family support and poverty relief can assist the working of a market system by improving incentives and by increasing labor productivity.

IMPROVING INCENTIVES. Incentives can be improved through careful design of both the structure and financing of benefits.

Benefits. Smooth links between wages and different types of benefits are essential not only for poverty relief, but also for reasons related to incentives. Social assistance should be harmonized with social insurance benefits and wages. Thus unemployed recipients of social assistance should be subject to the same requirements to seek work as are recipients of unemployment benefits. Also for reasons related to incentives, the working poor should be potentially eligible for social assistance, at least during the early transition. This is especially relevant for the increasing numbers of self-employed. Self-employment is risky, particularly during a time of rapid change; yet it is an important ingredient in the success of the transition and should be encouraged. Allowing the self-employed to be eligible for social assistance would reduce the risks they face. The administrative problems should not be minimized, however: the income of self-employed people is difficult to assess, making means testing very difficult (see Atkinson 1991b).

Universal family benefits improve labor supply incentives, particularly with a compressed wage structure. If benefits are income tested, families lose up to $1 of benefit for every extra $1 they earn (this is the so-called poverty trap); the poverty trap creates major labor supply disincentives. If, in addition, the distribution of wages is compressed, many families will have incomes just above or just below the poverty line; a family might therefore be little better off working than if it were receiving a means-tested benefit. If the difference between the means-tested benefit and the prospective wage is small, the family might prefer activity in the household or the informal sector to formal employment (the unemployment trap). Making family benefits universal, albeit at a lower level, helps to avoid problems related to both the poverty trap and the unemployment trap.

Sharing costs. The division of costs between different levels of government has incentive effects which are crucial both to running services efficiently and flexibly and to controlling expenditure. Although discussed here in the context of social assistance, the matter has much wider relevance. The issue is that local involvement in social assistance can increase available resources or improve the efficiency of their use, but it can also create difficulties. The problem is at its most acute where local government administers a benefit paid out of an open-ended transfer from the central government. In such cases, local government, which is often politically opposed to the central government, has little or no incentive to restrain expenditure and faces strong local electoral pressure to pay generous benefits. The result is a recipe for exploding costs (this is another example of the third-party payment problem discussed in chapter 2). The problem is somewhat less acute where social assistance is centrally funded *and* centrally administered, for example, through local offices of the relevant central ministry. This approach, however, increases administrative costs, and administrative capacity is frequently inadequate for such an arrangement; and without additional measures, it would not wholly guard against exploding costs. Two standard methods—matching grants and block grants—are used to build in caps to avoid uncontrolled spending. The first can be applied only to local governments; the second, both to local government and to local offices of a central institution.

Matching grants work best where local government controls a significant tax base and has the capacity to collect taxes. The central government could then offer to cofinance local social assistance schemes, on condition that appropriate quality control was exercised locally. The flaw in this solution is that it is usually the better-off local governments which can afford to pay their share of social assistance costs, whereas the needy cannot. For this reason, matching grants could be applied only selectively and only in some of the better-off countries. A departure from the standard matching grant solution, but still within its domain, would recognize nonfinancial inputs from local communities, such as labor and logistical support for public works. This type of solution might be more useful to lower-income countries in the region.

Block grants can be applied both to local governments and to local offices of the central government. Their advantage is that they avoid the adverse incentives of matching grants, since they leave local government facing the full cost of any significant increase in spending on poverty relief. Their disadvantage is that the central government needs considerable knowledge of the distribution of poverty across the country in order to determine the size of the block grant for each area. Such information is difficult to obtain, especially in countries with little or no experience of household surveys and where administrative statistics are not reliable. Nevertheless, block grants are the preferred option, with special emphasis on improving the data on which their distribution is based and on devising and implementing systems of quality control.

An added attraction of block grants is that they divide responsibility between central and local government: local governments or local offices can use these resources relatively flexibly within the limits of legislation and available resources, and the central government does not have to worry about third-party payment incentives which can cause costs to explode. The role of the central government is to consider the appropriate level of spending on poverty relief. In countries where the local tax base is weak, which for the time being is the case throughout Central and Eastern Europe, the extent of poverty relief depends largely on the size of block grants.

RAISING LABOR PRODUCTIVITY. Reform policies should bear in mind that family benefits can be useful in maintaining the productivity of labor, increasing the participation of women in the labor force (see Fong and Paul 1992), and increasing individual choice. Family benefits affect labor productivity in different ways, both through their effects on productivity per se and through their effects on labor force participation.

Beneficial effects on labor productivity arise if paying a family allowance encourages children to pursue their education beyond the minimum age for leaving school. Although loans and grants were proposed for higher education, such solutions are not feasible for secondary education (students fourteen to eighteen years old). Family allowances can help to reduce the extent of dropping out. In rural Albania, for example, which has no family allowances, the drastic decline in living standards, together with the privatization of land, has led to a sharp drop in secondary school enrollment and increased truancy from elementary schools because of the newly rediscovered practice of using children in the fields. Under these circumstances, conditioning family allowances on school attendance until the age of sixteen or eighteen might be a useful device, especially in highly agricultural countries.

Effects on labor force participation arise because the structure and finance of family support form part of labor-market policy (see Fong 1993 and the discussion in chapters 6 and 7). In particular, support like child care facilities could help to maintain the participation of women in the labor force. This might not sound like an important policy at a time of rising unemployment. However, the demographic prospects, particularly an aging population, suggest that high participation of women in the labor force will soon be back on the agenda, as it clearly is already in most of the highly industrialized countries. Undoing some child care services would be counterproductive from this perspective. Additionally, and separately, family benefits help workers, especially women, to withdraw temporarily from the labor market and subsequently to return to full- or part-time work. This increases individual and family choice.

Regulating Market Forces: Poverty Relief

In the short run, poverty relief is a central focus of the reforms. The issues raised by family benefits and social assistance are somewhat different, and so are discussed separately.

FAMILY BENEFITS. Family benefits are not designed for poverty relief per se. They do, however, substantially relieve poverty (a) because they accrue disproportionately to poor people and (b) because their share in family budgets is high. Family allowance should therefore be retained as a universal benefit.

Scope exists, however, for improved targeting. As discussed earlier, this does not necessitate income testing. Universal family allowances should be retained because they are well targeted at fairly low administrative cost. That does not, however, mean that no improvement is possible.

Child tax allowances. Tax allowances and tax credits require discussion as background:

• A tax allowance (for example, allowing $250 of tax-free income for each child) is a deduction from taxable income. Such an allowance is worth nothing to a family below the income tax threshold, is worth $50 to a family whose marginal tax rate is 20 percent, and is worth $125 to a family whose marginal tax rate is 50 percent. Since income tax rates generally rise with income, child tax allowances are thus worth more to richer families and are worth nothing to the poorest families. The more progressive the tax system, the more the rich benefit from any tax allowance.

• A tax credit, in contrast, is a deduction not from taxable income, but from a person's tax bill of, say, $250 for each child in the family. A tax credit is generally worth the same to everyone. Tax credits can be arranged so that even the poor benefit, so long as the system allows for negative tax payments (the payment of $250 per child to someone whose income is so low that he or she pays no income tax). Tax credits, however, face practical problems. Taxes are paid annually; thus credits would normally also be paid annually, which would cause problems for the poor. The payment of tax credits monthly is possible, but administratively cumbersome.

• In contrast, a tax-free family allowance, which is not income tested, is worth the same to all income groups and raises no significant administrative problems.

Thus child tax allowances are an expensive and poorly targeted form of family support.

Taxing family benefits. Making family benefits taxable is a promising approach to improved targeting. If family allowance is taxable, a poor family would receive the benefit (say, $250) in its entirety; a family facing a marginal tax rate of 20 percent would receive a net benefit of $200; and a family with a 50 percent tax rate would receive a net benefit of $125. Thus a larger share of total expenditure goes to lower-income groups. Attractive though the idea sounds, it needs to be adopted with care:

• It has the desired result only if the personal income tax is progressive, that is, if marginal tax rates are higher at higher incomes.

• If the assessment unit for income tax is the individual rather than the family, as is often the case, subjecting family benefits to income tax can be problematic (see Jarvis and Micklewright 1993).

• There are significant administrative problems in taxing family benefits. If the benefit is taxed monthly, the agent paying the benefit needs to know the recipient's marginal tax rate each time a payment is made; with diverse sources of income, this is administratively complex. Alternatively, families could receive the gross benefit each month and pay tax at the end of the year. This is easier administratively, and it would work well for the better off by reducing their net benefit. For lower-income families, however, the need in some cases to make additional tax payments at the end of the year could cause serious problems.

• Administrative difficulties are compounded by the fact that personal income taxation is a new institution in most of Central and Eastern Europe, so that administrative experience is still building up.

For these and other reasons, family benefits in many of the highly industrialized countries are not subject to income tax (although Hungary is considering the possibility of taxing family allowances beginning in 1994).

Relating family allowances to family size or the age of the children. This type of policy could improve poverty relief but raises issues of both design and implementation. If the costs of child care are high, family allowances should probably fall with age; where child care costs are not generally borne by the family, family allowances should rise with age because older children consume more food, clothing, and so forth than younger children. In both cases, it would be administratively helpful to have relatively large age brackets (such as under six, six to fourteen, and fifteen to eighteen) to avoid the need for frequent reclassification.

SOCIAL ASSISTANCE. The design of a workable system of social assistance is complex. Policymakers have to decide on the level of benefits, on the division between benefits in cash and in kind, on the specifics of design and implementation, and on the sequencing of reform.

The level of benefits. Setting the level of social assistance benefits confronts head on the tension between resource constraints and rising poverty. Ideally, benefits should be set at or above the poverty line, that is, a type A policy (as discussed at the start of this chapter a type A policy seeks to eliminate poverty; a type B policy seeks to ameliorate it). This should be less than the minimum unemployment benefit or minimum pension or minimum wage to maximize incentives to seek work. With low minimum wages and a compressed wage distribution, however, a meaningful margin between social assistance and the rest is difficult to determine. Although the calculation of poverty lines could be improved, in many countries minimum wages are set at or below the poverty line. The core of the problem is the remaining vestiges of socialist wage determination, under which an average wage for an individual earner could not pay the subsistence minimum of even the smallest family. This is still true in most of Central and Eastern Europe. Under these conditions, it is doubtful whether a parsimoniously redefined poverty line could create the meaningful margin which is missing from the equation. Either the level of social assistance is at the same level as the minimum wage, with adverse effects on incentives, or it has to be below the poverty line—a type B policy—which does not eliminate poverty but only ameliorates it.

Having said that, it is worthwhile to reiterate earlier arguments. Hungry people are not efficient workers, and hungry job seekers are not efficient job seekers either. Competitive economies require a competitive labor force, and social assistance recipients should have a chance to maintain or improve, by retraining, their productivity. An inappropriate social assistance system generates a downward spiral from which few, if any, can return to the world of work. The poor can easily become marginalized during a long spell on inadequate social assistance, risking not only a lack of income but also exclusion and the related dangers of developing behavioral patterns which perpetuate their exclusion. The symmetry with the discussion of long-term unemployment in chapters 7 and 8 is clear.

To avoid this outcome, poverty relief policies should be strengthened in at least three ways: through a wider choice of instruments, through a well-designed system of social assistance capable of dealing with large numbers of recipients, and through smooth linkages with social insurance benefits and assistance for the working poor.

Cash versus kind. The choice of instruments is crucial (this section draws heavily on Hughes 1993 and Jimenez 1993). Many options (see table 10-5), an analytical framework (chapter 2), and a wide array of international experience are available.

Cash benefits are generally preferable to benefits in kind if the objective is poverty relief. They are more cost-effective (in-kind programs are more difficult to administer), and they provide recipients with resources which they can use freely. Cash benefits help to create and maintain markets and consumer discretion. In-kind programs, in contrast, can inhibit private production of the commodities concerned. Since new democracies aim to increase liberty and reduce paternalistic solutions and stigma, a strong case can be made for letting the poor choose by granting them cash benefits.

Cash is generally more effective (a) the greater the fraction of the population it is intended to help, (b) the easier it is to trade the commodity being redistributed, (c) the more fungible is the commodity in family income (that

Table 10-5. *Instruments of Social Assistance and Their Relation to the Market*

Instrument	Provider	Price	Examples
Cash transfers	Generally private traders	Market prices	Family allowance, social pensions, cash assistance
In-kind transfers			
Direct public provision	Government, state-owned enterprise	Free or highly subsidized (below government cost)	Public education, public health, public utilities
Vouchers	Private traders	Market prices	Food stamps, housing, health
Producer subsidies	Private traders	Market prices	Social funds
Administrative price controls	Private traders, state-owned enterprises	Subsidized prices	Food, housing, utilities

Source: Jimenez 1993.

is, the easier it is to redirect spending), (d) the better the information consumers have about the commodity, (e) the greater the extent to which the commodity is available at market prices, and (f) the lower is institutional capacity to organize in-kind redistribution.

Benefits in kind also have a role, particularly if the aim is to stimulate the consumption of certain types of goods or services among previously identified groups. As discussed at length in chapter 2, the issue of cash versus in-kind benefits requires careful analysis. Cases where distribution in kind might be appropriate include commodities which the private market supplies poorly, if at all: an example is primary education because of imperfect consumer information and the lack of complementary markets, such as for student loans; another is vaccination against communicable diseases (because of externalities). In addition, some in-kind benefits can be self-selective. If the poor consume identifiable bundles different from the rich, self-selection can be efficient. Carefully targeted noncash benefits can combine good coverage with cost-effectiveness.

Considerable care is needed, however, when designing in-kind programs. Such programs generally have two aims: to alleviate absolute shortages, by giving out food directly, or to constrain the consumption of the poor to types of consumption which policymakers regard as appropriate. On the former aim, the consensus on famine relief (Ravallion 1987, 1991) tends toward income transfers rather than direct transfers of food, although a case may be made for limited direct transfers of food during the transition if shortages become acute. Food stamp programs and the like, however, may not be successful if their aim is to constrain the consumption of the poor. The group aimed at is large; the commodity is easily traded, even if the law specifies otherwise; family income may be diverted to other uses from the food on which it would otherwise be spent; and significant resources are needed to administer the program.

Two possible arguments remain. One is that, in political economy terms, it may be easier to redistribute benefits in kind than in cash. The argument has to be recognized, but its strength should not be overestimated. The other argument is a strong one: in very specific circumstances, direct in-kind transfers are both well targeted and nontransferable, particularly where there is a captive target group. Examples include pregnant women and infants (nutritional programs such as free orange juice at maternity clinics, medical checkups) and schoolchildren (free milk, meals, and health checks). Such programs are aimed at a precise and largely captive group, and they are not readily tradable. More generally, targeted family support, particularly for nutritional and medical purposes, can be a useful complement to cash benefits in addressing emerging poverty (for a further discussion, see Grosh 1994).

The design and implementation of a system of social assistance. Putting into place a workable system of social assistance is the major priority alongside unemployment benefits. The system should have the following broad characteristics: it should be universal (eligibility should not depend on whether a person is in the work force); it should provide an acceptable level of income (the exact level of which will depend on whether a type A or a type B policy is being pursued); it should be flexible; it should be simple to understand and administer; and, to the maximum extent consistent with political realities, it should have a national framework.[10] On the last point, the legislative framework (eligibility criteria and the introduction of block grants to localities) should be national; implementation could be local.

The links to other benefits are important. Recipients of social insurance benefits should also be eligible for social assistance. For example, a person who receives a widow's pension and a family allowance paid for a dependent child should receive social assistance if the combined household income is less than the level of social assistance. The gap could be met in cash, or specific benefits such as financial support for the education of the child, school meals, and rent subsidy could be provided to ensure that certain priority needs are met. Especially with a type B poverty line, a cash payment designed to bring family income up to the poverty line would not be sufficient to cover the cost of these important needs. Here, again, policymakers have to make difficult and politically contentious choices.

The design of such a system requires a set of sequenced decisions:

• What categorical tests, if any, will there be (will benefits depend on criteria other than income, such as age or employment status)?

• How will the poverty line be set? This requires decisions about the basket of goods to be covered and about how families of different sizes are treated.

• What range of benefits will be offered (the mix of cash and in-kind benefits)?

• How will means testing be organized? This involves a series of decisions: What income is relevant to determining benefits (for example, how will home-grown food be assessed)? Whose income is relevant (will the income of an elderly parent sharing the family home be included)? How much income, if any, will be ignored for the purpose of determining the benefit? And how much benefit will be lost for each dollar of income above any disregarded income?

• Who will operate the benefit? Will it be a new operation (which will increase administrative costs and complicate the system for the claimant), or will it be an addition to the current administration (which reduces administrative costs but adds to the complexity for benefit offices)?

• What are the legislative requirements, procedural instructions, and staff recruitment and training needs?

The design should lead to implementation as soon as possible.

Sequencing. In this area, as elsewhere, sequencing is crucial. Phasing out general subsidies produces savings, part of which could finance social assistance. But the phasing out of subsidies and the introduction of social assistance should be harmonized. The standard mistakes are cutting subsidies without increasing the allocation of social assistance or the institutional capacity to deliver services. For example, the abolition of hostels for commuting workers subsidized by the enterprise or from taxation is a major reason for the increase in homelessness in Central and Eastern Europe. Measures which take this burden off enterprise and state budgets should be coordinated with measures which relocate the workers affected or establish night shelters for the destitute. An across-the-board abolition of price subsidies without adequate mechanisms for targeting and institutions for social assistance (means testing for cash benefits, food stamp schemes, soup kitchens, and so forth) would cause much avoidable hardship.

Removing social assistance functions from social insurance also requires effective sequencing. As discussed in chapter 9, the reform of social insurance is likely over time to reduce redistribution within the system, thereby shifting to social assistance the burden of caring for vulnerable groups. Moving the finance and delivery of social assistance benefits out of social insurance programs would greatly improve the transparency of both schemes. Any such change requires adequate funding for social assistance and the development of an effective alternative delivery mechanism. The latter usually requires the transfer of institutions or the development of new administrative capabilities.

Implementing the Reforms

Efforts to reform the system of family support and poverty relief will fail unless the serious constraints imposed by a lack of institutional and political capacity are addressed.

BUILDING INSTITUTIONAL CAPACITY. The need to build institutional capacity poses a major challenge. As with social insurance, the emergence of large numbers of small-scale employers and widespread self-employment means that the administration of benefits should be moved from the enterprise to the benefit authorities.

A poorly developed system of social assistance should be upgraded quickly to respond effectively to the rapidly expanding number of claims. The critical bottlenecks relate to (a) a lack of capacity to administer a means test to large numbers of people and (b) a lack of personal counseling. During the communist era, poverty relief was a residual area, with poverty regarded as a pathology afflicting the individual concerned. A network of social workers is largely nonexistent, with a few exceptions such as Croatia, Slovenia, and, up to a point, Hungary. In most of Central and Eastern Europe, local governments or local offices of labor ministries were staffed with people of general clerical education. Specialized training to enable benefit officers and social workers to deliver services more efficiently is a matter of some urgency.

To fill these gaps, and to run an extended system, reform is needed in at least three areas. First, the regulatory framework generally requires strengthening to guide the determination of benefits. A second change is the need carefully to distinguish the role of social worker from that of benefit officer. Social work (helping families with problems) requires high skills. The assessment of benefits can, for the most part, be carried out by officers with relatively little training. It is by no means clear that the two functions should be carried out by the same people; the balance between the two professions should up to a point depend on the traditions of each country. Third, the assessment procedure should be streamlined, given the large numbers of people who are likely to claim social assistance once their entitlement to unemployment benefits has expired.

Appropriate training could greatly improve the ability of all levels of government—from ministries down to the local level—to control the quality of the welfare apparatus. This is crucial for the running of orphanages, nursing homes, night shelters, and other welfare institutions, where the administration—whether or not it provides the service itself—should set, publish, and enforce standards.

BUILDING POLITICAL CAPACITY. The objective is to maintain support for the overall reforms in the face of the overarching tension between maintaining current standards of living, on the one hand, and releasing resources for investment, on the other. In addressing poverty, it should be remembered that the poor are citizens and are part of the electorate. Support for the most needy, beyond solidarity, is a tool to avoid a breakdown in the social fabric both in its political aspects (riots) and in its civil manifestations (crime, delinquency, and drug abuse). Although reintegrating the marginalized poor should be an aim of social assistance, for which certain incentives could be applied, depriving the "undeserving" poor of access to social assistance could inflict a variety of costs on society.

Maintaining a certain level of social protection is essential to secure support from the most vocal groups. That can hardly ever be done without sacrifices in the form of leakages of benefits to persons who are not truly needy but who might otherwise lose interest in lending support to poverty relief. This especially applies to family benefits. In general, this is a twilight zone where policymakers should strike an appropriate balance between targeting benefits very tightly on the poor (for fiscal reasons) and ensuring broader coverage (for political reasons; see Cornia and Stewart 1992).

At a more practical level, building political support has two main aspects: adopting policies which respond to public opinion and taking action to change public opinion. On the first, the state earnings-related pension is regarded as very important in most of the former U.S.S.R.; it should therefore be made clear that policy which, in effect, gives everyone something close to the minimum pension is a temporary expedient. Similarly, public opinion in the former socialist countries tends to favor family benefits, and any attempt to reduce their level should take account of these public preferences. To achieve that, and to avoid inflicting undue hardship upon the poorest segments of society, at least some of the savings from reducing family benefits must be diverted to increased expenditure on social assistance or to new means-tested family support. Policy should also seek to address people's strongest fears. As an example, in some countries inflation has largely wiped out people's savings, and many of the elderly fear that they will not be able to afford a funeral. A death grant sufficient to pay for a simple funeral would yield significant political dividends among both the elderly and their middle-aged children. The benefit would not be costly, both because it is a single payment and because the claim for a death grant automatically gives the authorities the information they need to discontinue the dead person's pension or to reduce it to that of a surviving spouse. The grant therefore has administrative, cost saving, and political advantages.

The second aspect of building political support is action to shape public opinion. Political acceptance from this perspective requires public understanding of the issues, in particular the tradeoffs discussed in chapter 1. A major misconception (common enough also in the West) is that the standard of living of pensioners, or that of families with young children, can be increased by giving them more money, that the government has lots of money, and that the government can always print more money. It is important to help the electorate understand that what matters is not money but national output, and that if pensioners receive a larger share of national output, then less is left to pay workers' wages or to invest in ways which lead to larger national output in the future. Such public education is an important guard against populist political pressures.

Conclusions: Priorities and Sequencing

Declining real wages and dislocation of large numbers of people have led to a sharp rise in poverty. Poverty alleviation is a clear imperative, both for political reasons and for maintaining human capital; it is particularly important in countries which inherited a very low standard of living or where the stabilization or supply shocks have been particularly severe.

Short-term Policies

Short-term policies should seek to strengthen or develop a system of social assistance capable of delivering poverty relief to potentially large numbers of claimants, to ensure the necessary administrative capacity, and to promote practical measures to target benefits tightly in the interests of cost containment.

Poverty relief, for both political and economic reasons, should seek to cushion the fall in real incomes through relatively high family allowances and various income-tested benefits:

- Family allowance should continue to be paid without an income test, at least during the early transition, since family size is a good indicator of poverty when wage differentials are relatively small. Family allowances should be supplemented by targeted family support, with particular emphasis on families with infants and very young children. Financial responsibility for all such benefits should be transferred from the social insurance fund to the general budget.
- Expenditure on social assistance should be increased in line with the rising numbers of claimants and the average level of benefits. A choice must be made between type A policies, which aim to prevent poverty, and type B policies, which seek to relieve the most acute poverty, but do not attempt to bring everyone up to the poverty line. Additional expenditure on social assistance could be financed by part of the savings released once general, untargeted price subsidies have been abolished.
- Given large regional differences in income, social assistance should be financed mainly by block grants from the central government to localities, based on indicators of poverty. It is a matter of choice whether social assistance is controlled by the central government or, subject to appropriate regulation, by local administrations.
- Legislative barriers to the involvement of domestic and international non-governmental organizations should be removed immediately. The involvement of these organizations in social assistance, subject to an appropriate regulatory regime, should be encouraged.

Administrative capacity must be adequate in the short run to handle a large flow of new claimants for social assistance and, thereafter, should seek to refine its targeting capacity:

- Where possible, administration should build on existing institutions such as local governments or social work centers rather than create new institutions.
- During the early transition, targeting of social assistance should as far as possible be on the basis of easily verifiable indicators of poverty (family size, age, social group, and crisis region) to minimize the administrative demands of income testing. The latter approach, however, should become the norm once the administration of social assistance has been modernized.
- Deficiencies in data gathering should be rectified to give government information about the number of poor people and the groups which disproportionately experience poverty. These data are important for a number of reasons, not the least of them being the need to project costs. In most countries, the short-term priority is to introduce the appropriate (sometimes

even makeshift) capacity to conduct poverty surveys. In countries with no social assistance administration prior to the transition, the development of administrative statistics is equally important.

Measures to contain costs, though vital, require a more than usually fine balance of economic and political considerations, given the size, visibility, and political sensitivity of cash benefit programs:

- In addition to its administrative advantages, using indicators of poverty to target benefits is likely to be more cost-effective and efficient, at least initially, than introducing means testing prematurely. In countries where incomes in the informal sector and remittances from abroad are important, this method could be supplemented by allowing authorities or individual caseworkers a measure of discretion in the award of benefits, subject to suitable regulation.
- As wage differentials widen and administrative capacity increases, universal family allowances could be scaled down and supplemented by means-tested schemes.

Medium-term Policies

Policies with a medium-term dimension should focus on consolidating social assistance as a complement to a reformed system of social insurance and other forms of income transfers.

Key choices must be made:

- Decisions are needed about the desired mix of type A and type B policies and, in the case of family allowances, about horizontal equity objectives.
- These issues and their cost implications should be made explicit in the political process and should form part of public debate.

Administrative capacity requires continuing action:

- Administrative costs should be kept down by simple, standardized procedures.
- When the appropriate administrative and financial institutions are in place, integrating social assistance and family allowances with the personal income tax system could be considered but should not be regarded as the only option. In any case, a backup scheme is needed to respond to individual circumstances. Such a scheme could be based on a series of rules or could operate within a framework of caseworker discretion.
- Poverty monitoring should be improved by integrating the conduct of statistical surveys with systems designed to collect information from caseworkers and data necessary for administration.

The structure of benefits should seek to coordinate poverty relief with other cash benefits:

- Where income levels permit type A policies, the emphasis should be placed on coordinating social insurance, social assistance, and other cash benefits relative to a well-defined poverty line and in a way which minimizes adverse incentives.
- In countries with type B policies, family allowances should be differentiated by income level and could possibly be withdrawn for higher-income groups.

The finance of poverty relief could be diversified:

- In the longer term, more local and private agencies such as nongovernmental organizations, charities, and self-help groups could be involved both in financing but especially in delivering social assistance.
- Regional disparities should continue to be addressed by differentiated central allocations in ways which do not undermine the motivation for increased local responsibility.

In the longer term, the poverty relief function of cash and in-kind benefits should be replaced to the extent possible by policies aimed at developing the human capital of the poor and providing them with opportunities to earn income.

The sorts of reforms outlined in this and the previous chapter have a series of advantages. Through better targeting, they improve poverty relief and simultaneously contain costs. The system is flexible and can be adapted to objectives besides poverty relief, such as income smoothing, in the light of the economic and political conditions emerging in each country. By freeing the labor market to perform its proper function of helping workers move to jobs in which they are more efficient, the reform of cash benefits constitutes an integral part of the overall reform process.

Notes

1. As discussed in chapter 3, the early communist principles favored collective over individual consumption. The distribution of benefits in kind derived from this view. Thus a move toward cash benefits which empowered individual consumer choice was a major departure from the early principles (see Rupp 1992; Zimakova 1992). It is therefore no accident that the scope of cash benefits was greater in countries where some market-oriented reforms took root or where the civil society was stronger, such as Czechoslovakia, Hungary, and Poland, and developed only partially and belatedly in countries which adhered to a more orthodox vision of the socialist society, such as Albania and the U.S.S.R.

2. These data face difficulties additional to the general problems of scarce and unreliable data discussed in chapter 4 and should therefore be interpreted with care. First, the difference between the OECD countries and Central and Eastern Europe is

partly the result of demographic factors: only Australia, Portugal, and Spain among OECD members have more than 30 percent of the population under nineteen years old; among countries in Central and Eastern Europe, only Hungary has less than 30 percent of the population under nineteen years of age. Second, the comparison is biased because countries with low family benefits, such as Albania and the former U.S.S.R. (below 1 percent of GDP), are usually excluded from the averages for Central and Eastern Europe. Finally, the outcome is influenced by factors other than the generosity of family benefits. For instance, between 1960 and 1984, expenditure on family benefits in the OECD countries remained a broadly constant fraction of GDP but fell sharply as a fraction of social spending, from 17.3 to 8.9 percent, because other programs, notably pensions, were expanded.

3. Although long lines and misery played a role in creating dissatisfaction in the communist regimes, rising poverty alone cannot explain the sudden demise of the system. A more likely explanation is the widening gap between expectations and reality for a large majority of people. The economic stagnation and downturn in the 1980s made it clear that the regime would not be able to deliver its promises and that the economy was entering a downward spiral.

4. The sharp decline in real wages in Poland in 1990 should be seen in the context of the explosion in nominal wages in 1988–89 when prices were still largely controlled, leading to massive shortages and finally hyperinflation. It is argued that the higher 1989 real wages were not translated into higher welfare because of shortages and forced savings; these forced savings were then washed away by inflation which ensued from wage increases (see Balcerowicz 1993a; Lipton and Sachs 1990).

5. To illustrate with a simple example, suppose that the income guarantee is $1,000 a year per person in the family and that all other income is taxed at 50 percent. A family of four with zero earnings thus has a total annual net income of $4,000. A family of four with earnings of $5,000, receives a guaranteed income of $4,000 and pays a tax of $2,500 (that is, 50 percent of its earnings of $5,000); its total annual net income is therefore $7,500. A family of four with earnings of $10,000, receives a guaranteed income of $4,000 and pays a tax of $5,000 (50 percent of its earnings of $10,000); its total annual net income is therefore $9,000. For a further discussion of negative income tax and details of the large literature on the topic, see Barr 1993b, chap. 11.

6. The definition and different characteristics of (a) child tax allowances and (b) tax-free child benefits are discussed below in the section on regulating market forces.

7. In 1993, Lithuania introduced a system of social assistance based on a negative income tax. Predictably, the scheme required high tax rates.

8. In almost all Central and Eastern Europe, restitution of church property has caused problems for existing social welfare or educational institutions. Many properties which used to belong to the church now house institutions which have nowhere else to go; their relocation would require considerable public expenditure which, during the current fiscal crisis, is not a realistic possibility.

9. This has been done in the new Social Law in Hungary, which offers a benefit for mothers with a work history who raise three or more children for a five-year period when the youngest child is between three and eight years of age (see Magyar Közlöny 1993).

10. The history of the British Poor Law, particularly during the period between the first and second world wars, is instructive as a case study of the problems of financing social assistance locally; the history of the U.S. 1935 Social Security Act is an example of how to reimpose some central control on a local system (see Barr 1993b, chap. 2).

Eleven

EDUCATION AND TRAINING

BRUNO LAPORTE • JULIAN SCHWEITZER

VIRTUALLY ALL COUNTRIES IN CENTRAL AND EASTERN EUROPE have mature education and training systems, often with a rich scholarly heritage which predates communism.[1] Scholars in the region have often demonstrated their excellence, particularly in the basic sciences. It could be argued, therefore, that reforming the education and training system is not a high priority because so many urgent matters are demanding the attention of cash-strapped policymakers and that restructuring can therefore wait. Why would such a policy of neglect be a critical mistake? At least since the development of human capital theory in the 1960s (Becker 1964), economists have realized that high levels of education and training are crucial determinants of a country's economic success. This has been confirmed most recently by two studies (World Bank 1991, 1994). The latter examined the determinants of success of the East Asian "tigers" and identified their high levels of education, from basic education to research, as one of the key factors distinguishing them from their peers. But quite apart from economic competitiveness, there is another reason to emphasize education and training now. Failure to invest in the education of young people will generate large downstream costs for the state and the individual because the poorly educated make up a large proportion of persons who are unemployed and living in poverty, and they consume a disproportionate share of public services and public expenditures (World Bank 1990a).

This is not to downplay the importance of the social and cultural purposes of education. The education and training system clearly plays a vital role in supporting the evolution of a social contract which emphasizes freedom of expression, equity, and the responsibilities, rights, and obligations of individuals. The study of art, music, and culture supports the preservation and evolution of cultural values and artistic life. Indeed, it can be argued that the whole idea of a nation rests on a bedrock of common values, the transmission, evolution, and protection of which are important purposes of the education system. A key function of education is social integration. This is unusually important in countries in Central and Eastern Europe, many of which have significant ethnic and religious minorities and communal tensions. The resulting benefits constitute an important reason why virtually all countries provide free primary schooling. The economic, social, and cultural purposes of education, moreover, are mutually supporting. For example, active participation in economic development presupposes that individuals understand societal values. An understanding and appreciation of the arts and humanities may help young people to develop value systems which are essential for equitable social and economic development. More efficient management of the education and training system (an economic objective) may dictate a degree of decentralization, which in turn supports creativity and diversity (a social objective).

Given the focus of this book, this chapter is concerned mainly with the economic objectives and challenges facing the education and training systems of countries in Central and Eastern Europe. From an economic perspective, the function of education and training is to equip people to participate actively in the formation of wealth. The success of the education and training system can, in principle, although subject to major measurement problems, be measured by the rates of return to investment in human capital.

Of course, the countries of Central and Eastern Europe have different cultures and social structures, and no single blueprint could possibly match these varied conditions. A discussion of reforms in education and training must therefore be approached with some humility.

The Inheritance

The postcommunist governments of the countries of Central and Eastern Europe inherited mature education and training systems which are a legacy of central planning and political control (see table 11-1 for enrollment data). Largely transformed after the Second World War into models of the Soviet system, education and training systems continued throughout the communist period to stress basic literacy and vocational training for the so-called productive sectors at the expense of creativity, diversity, and "unproductive" and ideologically unsafe topics in the social sciences. With some variations, the communist education and training system had the following components:

• The system of kindergartens and preschool education was well developed.

• Near-universal basic education was provided to both boys and girls up to grade eight (ages thirteen and fourteen) in state-run elementary schools.

• Secondary education in academic, technical, and vocational streams was provided to fourteen- to eighteen-year-olds. Academic secondary education prepared a small cohort of students for higher education. Vocational schools, often specialized into narrow occupational streams and frequently attached to enterprises, prepared skilled and semi-skilled workers for industry and agriculture.

• A relatively undeveloped system of adult education and training was largely confined to on-the-job training in enterprises.

• Higher education comprised university education and some professional training for a relatively small postsecondary cohort. Research was not carried

Table 11-1. *Gross Enrollment Ratios in Central and Eastern Europe,*
1985–89

Country and level of education	1985	1986	1987	1988	1989
Albania					
Primary	103	101	100	99	99
Secondary	72	73	76	77	80
Higher	7	7	8	8	8
Bulgaria					
Primary	102	104	104	104	97
Secondary	102	75	76	75	75
Higher	19	21	23	25	26
Czechoslovakia					
Primary	99	98	96	94	92
Secondary	84	81	82	85	87
Higher	16	16	16	18	18
Hungary					
Primary	98	98	97	96	94
Secondary	72	70	70	71	76
Higher	16	15	15	15	15
Poland					
Primary	101	101	101	100	99
Secondary	78	80	80	81	81
Higher	17	17	18	20	20
Romania					
Primary	98	97	97	96	95
Secondary	75	74	79	85	88
Higher	11	11	10	9	9

Note: The series on primary school enrollment are estimates of children of all ages enrolled in primary school. Figures are expressed as the ratio of pupils to the population of children in the country's school-age group. Many countries, but not all, consider the age for primary school to be six to eleven years. For some countries with universal primary education, the gross enrollment ratio may exceed 100 percent because some pupils are younger or older than the country's standard age for primary school. The data on secondary school enrollment are calculated in the same way. The definition of secondary school age also differs among countries; it is most commonly considered to be twelve to seventeen years. The higher education enrollment ratio is calculated by dividing the number of students enrolled in all postsecondary schools and universities by the population twenty to twenty-four years old. Pupils attending vocational schools, adult education programs, two-year colleges, and distance education are included.

Source: World Bank 1992a.

out on any scale in universities, being generally the preserve of research institutes controlled by industrial ministries or academies of science.

Strengths

The most positive bequest of the communist education and training system was the high quality of basic education. Even taking into consideration some distortion in the data, countries in Central and Eastern Europe achieved high levels of literacy compared with countries of similar gross domestic product (GDP) and even compared with some countries in the Organization for Economic Cooperation and Development (OECD). Despite similar caveats concerning the comparability of data, evidence from Hungary and Slovenia shows relatively high levels of achievement in school mathematics and science.[2] Close central control and standardization of teacher training, school curricula, and textbooks may have ensured reasonable equity of access, near-universal literacy and basic numeracy, and high standards for a limited range of cognitive objectives in a few subjects, such as mathematics and science, which the state considered economically important. Special schools for gifted children produced very high standards in mathematics, sciences, languages, sports, and the performing arts. Standards in some medical specialties were at world-class levels.

Social values may also have played a part in promoting a culture of high educational standards. In general, girls had equal access to education at all levels. Many countries enjoyed a tradition of scholarship dating back to the period before communism. Universities such as Charles (Prague) and Jagellonian (Cracow) were created in medieval times, and academies of science and humanities were founded in the nineteenth century.

Weaknesses

Among the negative bequests, central planning and political control of professors and teachers often stifled cultural and academic freedom, limited individual choice and diversity, encouraged dogmatic teaching, and limited the flow of information, particularly in the humanities and social sciences. The education and training system was an ideological and economic tool of the state—ideological to ensure conformity with Marxism and to stifle local culture and national identity, and economic to ensure a match between economic requirements and trained staff.

The major problems which resulted were political interference in education, excessive specialization, deficient incentives to efficiency, duplication of facilities, early decisions about children's specialist study, unequal access to the best schools and universities, poorly organized research, and a neglect of adult education.

Political interference was pervasive. At all levels in the system, but particularly in universities, many teachers were appointed for their political beliefs, or at least their lack of opposition to the system, rather than for their academic achievements. Academic standards suffered. The social sciences and the humanities endured particular neglect and distortion. Subjects like economics, management sciences, law, sociology, and psychology were either ignored or devalued. The consequences are serious for the economic reforms. Privatization is held up by the lack of accountants and lawyers; the scarcity of skilled managers hampers the growth of the private sector. Central and commercial banking is constrained by shortages of financial specialists. Since social scientists were discouraged from studying poverty, crime, and other "negative" social issues, institutions to assist disadvantaged sections of society are now weak at a time of great social stress.

Central planning led to excessive specialization and limited choice for the individual. Even though central planning in its extreme forms was abandoned from the 1960s onward in many countries, its legacy—manpower planning—survived. Industrial ministries determined the so-called requirements for skilled and unskilled workers based on production targets for their enterprises. As discussed in chapter 7, distortions in the labor market encouraged enterprises to hoard labor, while at the same time discouraging workers from entering "unproductive" sectors such as teaching or medicine.

Incentives to efficiency were absent in the institutions of education and training, as in other economic units. Assessment and examination systems did not provide information which could be used to improve the quality of education. The emphasis was on testing factual knowledge, using unreliable techniques which did not permit longitudinal analysis (comparison of standards year by year) or comparison between schools. Nor were there any incentives to improve financial efficiency. Financing at all levels came entirely from the state and was based on rigid cost-plus norms which allocated funds for teachers, books, and equipment based on historical data. Finally, given virtually full employment and distorted wage structures, there was no labor market measure of the success of different training institutions or university programs.

Fragmentation and duplication of scarce resources were widespread as enterprises and ministries competed to develop their own facilities. The curriculum of technical and vocational secondary schools was usually controlled by the relevant sectoral ministry, and these schools were often attached to enterprises. In contrast, and with inevitable attendant confusion, the academic curriculum was controlled by the Ministry of Education. Highly specialized universities reported to different government agencies, with little cross fertilization among academic disciplines or flexibility to introduce new subjects. For example, although closely located in the same town, the three universities in Debrecen, Hungary, reported to different ministries and did not share resources (they have since joined together into a larger *universitas*).

The practice of placing children into specialized vocational and technical training at an early age—around fourteen, before they are mature enough to choose their own career—was pervasive. In Hungary and Poland, about 75 percent of secondary schoolchildren were in vocational or technical streams in 1990–91, and the proportion was even higher in Romania. In practice, many of these children received limited general education beyond grade eight. By contrast, in Western Europe, specialization rarely begins before sixteen years of age, and continuing education designed to reinforce cognitive skills and communication is increasingly emphasized. The pattern in East Asia is similar. Specialization in Central and Eastern Europe was excessive throughout the system. In Poland, for example, some 300 occupational skills were taught until recently in secondary technical schools, reflecting the typical specialization of a command economy. By contrast, in Germany, about sixteen broad occupational programs are offered at the apprentice level for sixteen- to eighteen-year-olds.

Access to postsecondary and higher education was inequitable. As in other sectors, the proclaimed intent of the communist planners—that education and training should improve the lot of the ordinary worker and peasant—was not realized. One measure of the failure was the limited access to postsecondary and higher education. Students were all too often the sons and daughters of party officials and intellectuals, who used their influence to secure a place in elite academic secondary schools and universities. Students who were tracked into secondary vocational education found it almost impossible to enter university and a white-collar occupation. In addition, middle-level technicians and managers seeking training had few alternatives to traditional university programs. A separate source of inequity was that the socialist education and training system often failed children with special needs. Children at the bottom of the range of ability, with disabilities or from disadvantaged backgrounds, rarely received the specialized assistance they required.

Research was poorly organized. Although standards of science and mathematics were quite high at the school level, this did not necessarily translate into excellence in research and development or increased productivity. Few universities had a capacity for research, which was often isolated from the teaching process in research institutes, academies, or enterprises controlled by sectoral ministries.[3] Research and development were distorted in favor of heavy industry and the military; potentially more productive research in basic sciences, agriculture, and the environment was neglected. A cardinal weakness was the failure to offer new interdisciplinary subjects such as materials science and biotechnology.

Adult education, essential for mobility in a market economy, was neglected because workers were expected to remain in their first occupation throughout their working lives. Individuals were rarely, if ever, encouraged to take responsibility for their training needs and career development. Enterprises provided on-the-job training for their workers, but individuals who wished or needed to retrain for a different occupation found few opportunities to do so. In addition, much on-the-job training had little purpose. Wage levels in enterprises were often rigidly determined by qualifications and training, inducing firms to provide much spurious training so that workers could meet the required qualifications.

The Forces Driving Change

Economic, political, social, and cultural forces are all driving the reform process. On the one hand, countries desire to eradicate all vestiges of the old ideology and to reassert national and cultural values. On the other hand, however, the emerging market economy and the perilous state of government finances are exerting enormous pressures on the system of education and training.

Political, Social, and Cultural Forces

Changes in political ideology and a resurgence of nationalism and ethnic tensions are having a major impact on Central and Eastern Europe. Perhaps the first imperative is to eradicate Marxist dogma from school and university curricula and to reassert local cultural values over Soviet hegemony. After a half century of living in a closed society, there is a desire to travel, to learn about other cultures and economic systems, and to participate in the economic success of Western Europe. It is no coincidence, for example, that a fund set up by the Hungarian government to assist innovation in higher education is called the Catching up with European Higher Education Fund.[4]

The initial euphoria after the revolutions of the late 1980s has given way to the realization that the task of reform is immense and costly. Replacing teaching materials may take years because finances are so scarce that countries such as Albania and Romania cannot even heat school buildings in winter or undertake basic maintenance. Preparing new curricula and assessment systems requires new skills and additional resources; retraining and replacing teachers are long and costly processes. In many countries, authorities must take thousands of teachers of Russian and retrain them to teach Western languages. Purges of communist appointees have taken place at universities in the Czech Republic and in Eastern Germany, among other countries. However, hiring new university teachers, particularly in topics which are now in demand in the labor market, is proving difficult because of the shortage of qualified professionals, low academic salaries, and poor conditions for research. In Hungary, grants are being offered for young scientists to carry out research in an attempt to prevent a brain drain of the next generation of university academics.

The resurgence of nationalism and regionalism throughout Central and Eastern Europe is placing additional strain on the education and training system. Throughout the region, minorities are demanding more educational autonomy, particularly in the use of minority languages. Administrative and financial burdens are growing as authorities struggle to provide the additional teachers and educational materials which bilingual educational systems require. In addition, decentralization of school education produces stress as local governments take on new financial, administrative, and technical responsibilities with limited prior expertise and resources.

Economic Forces

The main thrust of this chapter is to consider the economic forces and their consequences. For the education and training system, as elsewhere, the move to a market economy involves the liberalization of (a) demand, (b) supply, and (c) prices and wages.

The liberalization of demand is leading to calls for more relevant education and skills training:

- *Individuals* are starting to respond to market signals. As salary differentials widen, individuals are beginning to seek skills which allow them to move into higher-paying jobs. For workers displaced by enterprise restructuring, new skills may increase their chances of finding alternative employment.
- *Enterprises* are beginning to operate in a competitive environment. Investing in human capital through the provision of pre-service and in-service training may enhance productivity and profits, although hard budget constraints are initially producing contradictory pressures.
- *The state* needs to respond to the demand for new knowledge and skills critical to the transition. For instance, emphasis needs to be placed on topics such as the environment and entrepreneurship, which were previously completely ignored.

The liberalization of supply is generating more diverse forms of education, ownership of educational institutions, and supply of educational materials:

- *Local governments* are progressively assuming more responsibility for the provision and financing of preschool and basic education. Given the limited basis of local taxation, however, the central government continues to play a significant financing role in redistributing tax revenues from richer to poorer districts, as well as a regulatory role in setting norms and standards to ensure quality and equity of access.
- *Institutions,* in particular at the level of higher education, are being granted more autonomy. New legal frameworks were introduced early in the transition which preempted any return to more centralized administration and allowed institutions more control over policies and resources.
- *Private providers* are starting to emerge, as offering education and training services becomes profitable. Skills training provided by private organizations and nongovernmental associations is beginning to expand in some countries.[5] In addition, private schools are being established.

The liberalization of prices and wages has led initially to declining output and a widening distribution of income, which is affecting the capacity of individuals and the state to finance educational services:

- Prices of educational inputs are increasing. This is particularly the case for textbooks whose price has increased significantly in real terms.[6]

- The fiscal situation has precipitated a crisis in the public financing of education and training in some countries (see table 11-2). For example, while GDP was itself falling in Poland, the share of education expenditures in GDP also declined, suggesting a steep decrease in real spending between 1990 and 1992. In Bulgaria, Hungary, and Romania, education expenditures appear to have been protected to some extent, although spending is falling in real terms. With the exception of Hungary, however, the share of education spending as a percentage of GDP is low in comparison with developing countries of similar incomes and OECD countries. Training expenditures have also been affected as enterprises, increasingly faced with hard budget constraints, move quickly to phase out their education and training programs.
- At a time of sharply declining income per capita, families are having to contribute increasing amounts for education. Parent-teacher associations are raising significant resources to cover items such as textbooks and educational materials. Private training centers and some institutes of higher education are charging fees.

Policy

In discussing the reform strategy, it is necessary first to consider the very different economic function which the system now has to perform in a market economy. As discussed in chapters 7 and 8, the education and training system now has to prepare young people for lifetime occupational mobility, since workers can expect to move between jobs and occupations as market conditions change (OECD 1990c). This will place a premium on workers with marketable skills, who can troubleshoot, anticipate problems, work in teams, and develop new production processes and services, that is, workers with good analytical and communications skills (Berryman and Bailey 1992). These workers will also be in demand in the new industries which countries hope to attract.

The Strategy

Thus a principal strategic objective of the education and training system in transitional economies is to prepare trainable, adaptable, and innovative peo-

Table 11-2. *Expenditure on Education as a Percentage of GDP in Selected Countries in Central and Eastern Europe, 1990–92*

Country	1990	1991	1992
Bulgaria	5.2	5.7	5.4
Hungary	5.6	6.2	6.4
Poland	4.6	3.8	3.1
Romania	3.2	3.7	3.9

Source: Unpublished data from ministries of finance and the World Bank.

ple who have the flexibility to shift occupations. To ensure an intelligent and innovative work force, the education and training system must encourage innovation, questioning, and the development of higher-order cognitive skills, rather than obedience, conformity, and memorization skills. This will require flexible and responsive institutions. The emphasis in the curriculum needs to shift from teaching facts and skills in an essentially static workplace with little mobility between occupations to teaching students how to think and use knowledge in a dynamic labor market. This signals a fundamental move from a teaching system designed to serve the needs of the state to one based on the developmental needs of the individual.

Preparing trainable, adaptable, and innovative people will require fundamental changes in the behavior of the main actors in the system: the state, the institutions, the teachers, and the consumers (students, parents, and employers). As in other sectors, this will require a careful balancing of policies aimed at preserving macroeconomic stability through targeting resources, containing costs, and mobilizing additional financial resources; building markets by promoting consumer choice, improving incentives, and introducing competition among providers of education and training services; regulating markets by maintaining educational standards and ensuring equitable access to services; and strengthening institutional capacity to implement the reforms.

Maintaining Macroeconomic Balance

Restoring macroeconomic balance has two important implications for the education and training system: public resources need to be used more efficiently, in part by concentrating them on priority areas; and the state can no longer assume the entire financial burden of education and training, implying a need to mobilize additional private resources.

CONTAINING COSTS WHILE IMPROVING EFFECTIVENESS. During the fiscal crisis, public finances need to be tightly targeted on the basis of explicit priorities. This section argues that the priority areas should include policies to maintain standards in basic education, restructure secondary education, improve adult training, and intervene selectively in postsecondary and higher education. These priorities should be buttressed by policies aimed at using educational inputs more efficiently.

Maintaining standards in basic education. This objective, though easily overlooked, is essential. Extensive research has revealed the high rates of return to basic education (Psacharopoulos 1993). Indeed, public investment in this area has some of the highest returns of any investment, public or private. The yields are long term, however, and politicians are tempted to downplay their importance. This would be a major error since basic education provides the foundation for all skills development.

Protecting basic education implies (a) defending poor districts against inequitable financing mechanisms arising out of decentralization; (b) ensuring adequate public financing for educational inputs which have been shown to be highly efficient in enhancing learning, particularly books, materials, student assessment, and teacher training; and (c) resisting any pressures for expenditure on policies, such as reducing class sizes, which have a much lower impact on student learning (World Bank 1990a). Indeed, in some instances, class sizes could be increased without risking quality. At the same time, it may be possible to reduce the size of the teaching force, thereby freeing resources for paying higher salaries or investing in teaching materials.

Restructuring secondary education. This is perhaps the major structural challenge for the education and training system. Much secondary education is now provided by a quasi-dual system, which unfortunately bears little resemblance to the parent German model. The large majority of secondary students who undertake occupational training, usually from the age of fourteen onward, are in apprentice training programs in which the school provides academic education and enterprises provide occupational training (this is the dual system). The remainder of the secondary vocational cohort is enrolled in secondary industrial or agricultural technical schools which are intended to provide a higher level of technical training.

Reforms currently under way in a number of countries in Central and Eastern Europe seek to address the following issues: early and narrow specialization; weak general education because specialized training crowds out essential components of the school curriculum; inflexibility of school-based vocational training systems in dealing with rapidly changing labor markets (Adams, Middleton, and Ziderman 1992); and the poor financial state or even bankruptcy of many enterprises, which renders much training by enterprises irrelevant or impossible.

The strategy for these reform efforts is twofold: (a) to establish a secondary curriculum for the age group up to sixteen years focusing on science and technology, broad occupational disciplines, languages, and social studies and (b) to develop a model for the sixteen-plus age group which allows for choice between academic and technical streams.[7] The technical stream covers the study of broad occupational clusters, including business studies and services, with continuing emphasis on communications and mathematical and computer skills.

Expanding adult education. Adult training—defined here to include school leavers—includes further education, on-the-job training, training for workers who wish to upgrade their skills to improve career prospects, and training for displaced workers. Because adults will need to change jobs and upgrade their skills frequently during their working life, adult training assumes considerable importance. The state has an important role in providing incentives to encourage the provision of training, but the state should not be the sole source of funding. The enterprise and individuals themselves should be encouraged to take responsibility for their own training needs, and the state can play an important role in encouraging training through appropriate tax incentives. Tax incentives will not be sufficient, however: small firms rarely have the capacity to carry out training even though training may be crucial to their survival in a competitive market, and the private sector cannot be expected to provide much training for disadvantaged groups in the labor market, for instance, the unemployed, school dropouts, and the handicapped.

The capacity to undertake occupational training will be needed in both the public and private sectors. Countries such as Hungary and Romania are creating national and regional training boards to fund and coordinate training by public and private institutions. In Hungary, national and regional training boards comprising representatives of government, trade unions, and employers, and jointly funded by the state and the private sector, contract with public and private training institutions to deliver training for dislocated workers, school leavers, and other groups. Development of these new institutions has been complicated by difficulties in strengthening the appropriate relations between the groups represented on training boards, including trade unions, which are highly fragmented. Despite these initial difficulties, results are quite promising.[8]

In addition to encouraging cooperation among the social partners (government, labor, and employer organizations), the state has an important role to play in (a) developing occupational training standards and accreditation; (b) ensuring an appropriate mix of financing by the state, individuals, and firms which reflects their new roles in the market economy and protects weaker groups such as persons with disabilities; and (c) ensuring standards and efficiency by evaluating and providing information.

Restructuring higher education. Universities and other institutions of higher education have a critical role. First, they must develop the higher-level skills and knowledge to assist the transition to a market economy. Second, they play a central role in the production and dissemination of scientific knowledge and innovation through high-level research and training. Finally, they provide a forum for debating ideas and promoting social and cultural values which are particularly important in the context of countries undergoing fundamental change. Many countries in Central and Eastern Europe have an ancient tradition of higher education and scholarship, which provides a solid foundation on which to build. Resistance to reforms may, however, alter the traditional structures on which the system is founded. This resistance may come both from supporters of the communist model of higher education or from those who are nostalgic about the elitist system prior to communism. These contradictory currents have already manifested themselves in debates over the duration and structure of education (five-year diploma courses similar to those in most of Western Europe versus shorter modular programs modeled on North American systems).

Higher education is now being restructured in many countries to improve access, choice, and relevance and to diversify sources of funding. The main priorities for reform include expanding enrollment (currently as low as 10 percent of the age cohort in Romania compared with 20–30 percent in OECD countries), introducing new programs, and diversifying ownership and finance. In Poland, changes in admissions policies in some university departments have been introduced to increase enrollment. In Hungary, the previously mentioned Catching up with European Higher Education Fund is being used to finance new university programs on a competitive basis in accordance with government priorities. In many countries, private universities and specialized institutes (for example, in management) have been created. This is an encouraging trend, since the private returns to education are high, and state subsidies should be directed first and foremost to basic education.

Decisions to invest public funds in efforts to expand the higher education system would need to be accompanied by significant reforms:

• Improve efficiency (student/teacher ratios as low as 3:1 were common in 1990). Methods under consideration include compulsory retirement (the Czech Republic) and experiments with normative financing based on student/teacher ratios (Hungary).

• Increase diversity, choice, and relevance by adding new subjects in science, technology, and the social sciences and by developing links with the evolving private sector. This is the primary objective of the Hungarian Catching up with European Higher Education Fund. In addition, it is essential to increase access for older students, which requires designing alternative pathways which allow school leavers and adults without the necessary formal entrance qualifications to enter higher education.

• Improve flexibility by introducing shorter, modular programs and developing vocational education at the postsecondary level. The restructuring and modernization of industries, the development of small- and medium-scale enterprises, and the emergence of a service sector are generating a demand for highly qualified technicians and middle-level managers which cannot be met at the secondary level and is not being met through traditional university programs. In countries like Poland, the long tradition of higher-level vocational training has been considerably weakened over the last forty years,[9] and new initiatives aimed at creating technician training institutions outside the university are under way.

• Develop productive relationships in research and development between higher education, research, and industry. Under socialism, much basic and applied research was carried out in research institutes under the control of industrial ministries, enterprises, or academies of sciences. The universities were mostly confined to teaching. Many of these research institutes are now obsolete or bankrupt because their parent ministries can no longer afford to fund them, and many have already closed. At the same time, universities have difficulty recruiting the best researchers because of salary constraints and poor access to research funds and facilities.

Maintaining the quality of inputs. Reforming the structure of the different components of the education and training system is only part of the story. Action to maintain the quality of inputs is critical to the success of the structural reform.

Teachers are key agents in bringing about reforms of the education and training system, since without their cooperation, nothing will change. The role of the teacher and the school principal is changing dramatically from deliverer of received wisdom to manager of resources, guide, and tutor. This change will require far-reaching reforms of the teaching profession. The successful introduction of new curricula and teaching materials, new methods of assessment, and decentralized school management all depend entirely on the quality and motivation of the teacher. The challenges include introducing new subject areas into the curricula, which may result in unemployment for some teachers whose skills are no longer required; rewriting existing curricula, which will require extensive retraining of teachers so that they can teach the new materials; direct access to libraries and information technology, which will need new skills; and decentralized management, which will require new skills of school principals, among others. Thus resources should be targeted toward reforming pre-service training to reorient new teachers to changes in curriculum and teaching practice and toward providing extensive retraining for existing teachers. Retraining of teachers is particularly crucial since the goal is to change long-entrenched attitudes as well as to introduce new techniques.

The quality and availability of books and educational materials are equally important. Given the magnitude of the change required in the system and the time it takes to retrain teachers and influence their behavior, ensuring access to appropriate teaching materials might, in fact, be the highest priority. During times of fiscal crisis, cutting public recurrent expenditures on education other than salaries is tempting. The effect is usually felt in the classroom as textbooks and materials disappear. This has an immediate impact on the quality of education, particularly for the poor, who have little disposable income with which to make up any imbalances. It is a truism to state that children will not become properly literate unless they have access to a diverse supply of books. Yet this is exactly the case even in middle-income countries, such as Brazil, which have neglected to supply this essential commodity in the classroom.

Most OECD countries assign the state responsibility for providing free of charge basic school textbooks for primary and secondary education. The main justification for this approach is that textbooks are considered essential to the educational process: without them, equality of opportunity is not possible, and targeting free materials specifically toward poor children may be administratively difficult. In Central and Eastern Europe, the decline in budgetary resources has led some education ministries to default on their obligation to provide free access to textbooks. In Poland, for instance, parents must purchase all textbooks with the exception of books for special schools for handicapped children and ethnic minorities. Given the significant decline in personal income during the early 1990s and the increase in the price of textbooks, this policy should be closely monitored. At a minimum, the state should provide financial support for poor parents to purchase books, as is the case, for example, in France.

Using inputs of physical and human capital efficiently. Finally, available data suggest that in most countries, educational inputs of human resources and physical capacity are not being used efficiently. Student/teacher ratios in the early transition were very low in all countries of Central and Eastern Europe.[10] Since expenditures on teachers' salaries make up a large proportion of the education budget in all countries, improved use of teachers can result in significant savings which can be redeployed elsewhere in the education system. The remedy would include altering three main contributing factors: light teaching loads,[11] undue specialization of subjects offered, and single-subject teachers. Reforming the teaching profession as well as streamlining the curricula, in particular in technical and vocational education, will lead to changes in the demand for teachers with particular skills. This calls for careful analysis of possible alternatives offered to teachers, such as professional upgrading, retraining for another assignment in teaching, retirement (including compensated early retirement), and compensation for moving into alternative employment.

So far as the use of physical capacity is concerned, higher education in Central and Eastern Europe is characterized by a multiplicity of poorly funded, poorly equipped, and often unproductive teaching and research facilities.[12] Certain institutions could be closed, when there is no demand for the programs offered or when quality is not up to standard. In Hungary, incentives are being offered to encourage small, specialized institutions to consolidate into larger and more comprehensive establishments offering a wide range of disciplines.

MOBILIZING RESOURCES. Public financing of education and training has many economic justifications (see, for instance, Barr 1993b, chap. 13). The main arguments may be summarized as follows. First, benefits flow not only to individuals but also to society at large. A highly educated and trained work force contributes to improved productivity and economic performance. Lack of education is closely correlated with poverty (World Bank 1990a). Second, economies of scale can be achieved through public provision of education for large segments of the population. Third, in the absence of comprehensive information about the quality and cost of education, consumers cannot always make rational decisions about their education, let alone about the value of their education in the future.

For these reasons, the state always plays a significant role in the financing of education. Given the competing claims on tax revenues, however, the state cannot and should not assume the whole financial burden. A new alliance is needed between the state, the employer, and the individual to share the financial burden of education and training (Laporte 1993). Since education is not homogeneous, the role of the state will be different for different levels of education.

Both primary and secondary education have large externalities (explained in chapter 2), and the public has an interest in ensuring that the right quantity and quality of education is produced at the right price. The government needs to ensure an adequate supply of education (either directly or through the private sector), control quality, and enforce mandatory school attendance. Since imperfect information leads to underconsumption, particularly by the lowest socioeconomic groups, a subsidy needs to be applied either to prices (free education) or to incomes (education vouchers).

So far as higher education is concerned, it is widely acknowledged that the proportion of benefits internalized by individuals rises with the level of education. In market economies, where the structure of wages is at least partially related to worker productivity, the discrepancy between the private and social returns to investment in higher education is large. This is particularly the case in publicly funded systems, where a large fraction of the cost is borne by the state and a large fraction of the benefits accrues to individuals in the form of higher wages. It is therefore desirable to shift part of the financial burden to individuals and private employers. As the structure of wages is liberalized, it should be possible to introduce fees in higher education. Countries including Bulgaria, the Czech Republic, Hungary, and Poland are moving in this direction. Cost recovery in higher education should be introduced alongside student loan schemes. Substantial reliance on tax funding will lead either to rationing of higher education and training through limited access or to a poor-quality system. At the same time, limited taxpayer funding, if not carefully targeted, will constrain access to higher education by lower-income students.

In the case of skill-specific training, the costs (which can be large) could be borne to a great extent by enterprises, which enjoy substantial benefits in the form of a more productive labor force, and by individuals, who are rewarded with higher wages. Although they benefit from investment in human capital, firms may underinvest for one of two reasons. First, the new owners and managers may not be aware of the benefits of having a skilled and flexible work force in their enterprises. Second, they may believe that skilled labor can be acquired more cheaply by poaching staff from other enterprises than by training their current staff (an example of the free-rider problem discussed in chapter 2). Correcting this market failure requires state intervention, and many countries (Brazil and Korea, for example) therefore use payroll taxes to mobilize resources for training.

One of the strongest arguments in favor of a payroll tax is that it can be used as an instrument for restructuring the labor force.[13] However, payroll levies tend to increase the price of labor and thus may inhibit employment growth and international competitiveness. In the case of countries in Central and Eastern Europe, the tax burden on enterprises is already high. It might therefore be desirable to introduce a training tax along the French model, which is paid only by enterprises which do not provide training at the same monetary level as the tax.

Of equal importance, individuals will invest in training if they perceive substantial economic benefits from doing so. In the German apprenticeship system, for instance, the apprentice accepts lower wages in return for training provided by enterprises, which leads to higher wages in the future.

Building Markets

Market forces can be used to improve the performance of the education and training system through increased consumer choice, through a more diverse supply of educational services, and through improved incentives to use resources efficiently.

INCREASING CONSUMER CHOICE. Education and training institutions need to become responsive to consumers by increasing their diversity and flexibility. In higher education, for instance, a wider variety of courses is needed to meet diverse cultural, social, and educational objectives. Similarly, vocational training institutions catering to the needs of adult populations (age eighteen and over) must provide flexible entry and exit programs tailored to satisfy the needs of particular target groups.

Increasing choice and diversity also requires fundamental changes in the way curricula and textbooks are organized. Although the state would normally determine the core curriculum and set standards, part of the curriculum should be left to the discretion of regional or local authorities, to take account of regional and ethnic diversity. Similarly, schools should be allowed to choose from a list of textbooks authorized by the Ministry of Education.

Increased choice will also necessitate changes in funding arrangements. A survey of financing in higher education (OECD 1990a) concludes that diversity and variety are the most important characteristics of higher education at the end of the twentieth century. The concept of global funding for higher education as a whole may have been appropriate in circumstances where universities and colleges were engaged in a relatively narrow range of activities evaluated on the basis of similar criteria. Since this is no longer the case, the higher education budget should now be divided into separate categories and different allocation criteria established for each program item. Ensuring maximum flexibility in the use of the budget at the institutional level is critical.

Finally, the role of information in the strategy to reform the education and training system is also crucial. Information must be available at all levels: to parents so that they can make informed choices about schooling and training; to students so that they can select higher education, training, and careers; to employers so that they can participate in the formation of policy and the delivery of education and training; to the unemployed so that they can make informed choices about training and career opportunities; and to the public at large, since the education and training system will continue to be largely financed through taxes.

DIVERSIFYING SUPPLY. Although education services partially or totally *financed* by the state may be justified for many economic reasons, such services do not need to be entirely *produced* in the public sector. (The central distinction between public funding and public production was discussed in chapter 2.) In many countries of Central and Eastern Europe, education and training are being provided increasingly by the private sector. Private institutions operate under different arrangements, with varying degrees of state subsidy. In Poland, for example, many institutions classified as "non-public" schools offer primary and secondary education.[14] Private training institutions for languages, management, informatics, and so forth have emerged in all Central and Eastern European countries. Providing public subsidies for private schools offers some advantages. It increases the number of educational services available; promotes competition, thereby reducing costs; gives parents and students a wider choice of schools and programs; and gives the state a measure of influence to regulate quality.

Diversified ownership of the production and distribution of textbooks and educational materials is particularly important to ensure wide choice and high quality. This in turn requires far-reaching reforms. For efficiency reasons, the traditional monopoly of the state needs to be challenged, and a market characterized by competition and free choice, with a wide variety of textbooks and materials available for purchase, should be established. For equity and efficiency reasons, however, the state may need to continue to subsidize primary and secondary texts and materials, but these subsidies should be transferred from the producers to the consumers (schools and individuals).

IMPROVING INCENTIVES. Education and training are a service which has a cost, transmits a benefit to its recipients through enhanced knowledge and productivity, and can be bought and sold. No matter how they are financed, education and training institutions must respond to the needs of their clients, maintain standards, and deliver services cost-effectively. This implies greater institutional autonomy, improved management practices, and the creation of an effective vetting mechanism with means to ensure that improvements happen when needed. The state also plays a crucial role here in creating a policy environment which favors improved management and efficiency. Barnes and Barr (1988) suggest that the government should allow higher education institutions to manage themselves within a framework which safeguards academic freedom but subjects them to scrutiny to prevent concealment of poor teaching or low academic standards.

Incentives are important, first, for the institutions which provide educational services, on whose behavior funding sources and mechanisms have a powerful influence. The OECD (1990a, p. 55) survey of financing in higher education notes the growing interest in the introduction of market incentives: "The move towards market mechanisms is taking two main forms: one is that public funding agencies are becoming more selective, sometimes taking the form of buying services from higher education institutions; the other is that universities and colleges are being encouraged to seek an increasing proportion of their finance from non-traditional sources."

In Central and Eastern European countries, supply-driven block grants were the norm until recently.[15] Over time, shifting a greater proportion of public funding toward demand-driven formulae should encourage greater market responsiveness. Ideas being tried out in a number of countries include capitation formulae which pay institutions by the number of students enrolled, formulae which penalize institutions for allowing excessive student repetition, and norms which ensure efficient student/teacher ratios. In addition, contracting arrangements between funding bodies and institutions of higher education should be encouraged not only for research contracts but also for mainstream teaching activities. This has started to happen. In 1991, the Polish National Committee for Scientific Research initiated a competitive peer review system for allocating funds for research, under the guidance of the commissions for basic and applied research. Hungary has adopted a similar approach for basic research under the Hungarian Research Fund (OTKA) and introduced competition for applied research under the auspices of the Office of Technology Development. Hungary has also introduced competition in the funding of higher education. This should assist institutions to be more responsive to the needs of the economy, as expressed by funding bodies, and to provide better value for money.

Incentives are also relevant to education staff. Teacher salaries are relatively low in Central and Eastern Europe, and teachers often depend on a complementary job to survive.[16] The living standards of primary and secondary education teachers should be protected to the extent possible, since they should not have to moonlight or engage extensively in private sector activities to survive. Protecting living standards is feasible only if efficiency is maximized in the ways described earlier. As wages are liberalized, the level of salaries for professionals in the education system needs gradually to be restored to attract and retain a well-qualified and highly motivated teaching force. Salary increases could be contingent on teachers participating in retraining and achieving certain standards of performance.

Regulating Market Forces

Where the state is no longer the sole provider of education and training services, it has two key regulatory roles: setting and maintaining national standards and ensuring equitable access to basic education services.

INTRODUCING A REGULATORY STRUCTURE. The state has a key role in monitoring the performance of institutions and maintaining standards of academic achievement. Crighton (1993, p. 1) argues that, "as educational systems in Central and Eastern Europe decentralise and diversify, it becomes paradoxically more necessary for the State to set, monitor, and if necessary insist on the maintenance of standards." A key aspect is the introduction of national systems of assessment and examinations which monitor quality, permit comparison between equivalent institutions, and deliver feedback to institutions and consumers. A national system of assessment will increase accountability: if educational results are poor, the public will demand reform.

Alongside these measures aimed at maintaining overall academic standards, it is necessary also to regulate private providers of education and training. The proliferation of such providers is a welcome development which assists the reform process considerably. However, the state needs to protect citizens against abuse. In order to ensure that the training provided is worthwhile, governments should encourage the accreditation of training providers and certification of training programs against national standards. National certification systems reduce uncertainty among prospective students and their subsequent employers about qualifications, permit mobility within the country, encourage personal development, and discourage the provision of low-quality programs. Participation by employers in these schemes should be voluntary to prevent additional rigidities from being introduced into the labor market. Importantly, certification and accreditation schemes do not need to be run by the government. Private associations can set up their own accreditation systems to ensure standards, as is the case, for example, for private English language schools in the United Kingdom.

ENSURING ACCESS. One of the most important functions of the state is to ensure equitable access to education, which is a fundamental right of individuals. This, in turn, requires state intervention to ensure that the costs and benefits of education are equally distributed among regions, socioeconomic groups, and ethnic groups. Two interventions are particularly relevant for countries in Central and Eastern Europe.

The move toward decentralization of education services carries the risk of unequal expenditure across educational districts. Local governments clearly have a comparative advantage in providing basic social services such as kindergarten and primary education. Given the significant differences in the tax base of individual localities, however, central government has a vital role in funding to ensure reasonable equity and minimum standards across the country.

So far as higher education is concerned, the shift toward greater private financing advocated in this chapter implies the introduction of loan schemes and measures such as scholarships to ensure access for the economically disadvantaged. An approach increasingly adopted in the West (see Barr 1993a) is to have income-contingent loan schemes, in which repayment takes the form of a given percentage of the student's subsequent earnings. One way of implementing such a scheme is to add the loan repayment to the graduate's social insurance contributions. Payments stop once the loan has been repaid. This strategy is particularly appealing for economies in Central and Eastern Europe. First, students pay part of the cost of their education themselves, which is both efficient and fair, and replaces parental contributions. Second, it circumvents the unwillingness of commercial banks to make long-term loans without a government guarantee to individuals lacking collateral. Third, it reduces the public cost of higher education, making it easier to expand the system and improve its quality.

Implementing the Reforms

Although the major task of implementing educational reforms involves building institutional capacity, the area is also politically highly charged.

BUILDING INSTITUTIONAL CAPACITY. Education and training cover a wide range of activities, and reform is needed of both their structure and finance. For both reasons, the demands on management and administration are heavy.

Managing complex reforms. The reforms suggested in this chapter require strong leadership and management at the central, regional, and institutional levels. Staff in the education sector in Central and Eastern Europe need to acquire specific skills and knowledge to work in a mixed, decentralized environment, particularly planning, management, financial, and computer skills. To promote greater effectiveness and flexibility, training should be provided to both senior managerial staff and to middle management in administrative offices with broad institutional responsibility.

Political reforms during the early transition dismantled many institutions which had had a prominent role under central planning. Several important functions have all but disappeared, and this undermines the capacity of institutions to function in a new economic and social context. First, the concept of planning in education and training should be rehabilitated and redefined. Few units at the central or regional levels now have the capacity or authority to collect, synthesize, and present data about the education system. Basic computer planning models, which can simulate the impact of alternative policies on enrollment, staffing, and infrastructure, are not readily available, and this limits the capacity of policymakers to make decisions. Second, educational research needs to be redeveloped and strengthened. As a result of political changes since the fall of the communist regimes, many education research institutes have closed. Policy development in education must be supported by a strong research capacity, and countries in Central and Eastern Europe need to organize this crucial function.

It is also essential to improve the information on which to base decisions. Management information systems comprising financial, material, and human inputs and, especially, outcomes such as student learning and success in the labor market need to be improved. Educational management information systems should be accessible at the national, regional, and institutional levels to assist education policymakers and managers to make more informed decisions, to monitor the impact of policies, and to direct resources in ways which improve the quality of education. They would also facilitate communication with education systems in other European countries, which in turn could encourage innovation and change.

Finally, objective criteria and transparent processes for allocating resources should be introduced. In most countries in Central and Eastern Europe, historical budget allocations, expenditure patterns, and crude norms for staffing and physical resources are the dominant factors in allocation decisions, which are best described as the result of elaborate bargaining. Managers have been conditioned to believe that the more resources garnered during the year, the better off the institution will be in next year's budget, regardless of the quality of output achieved or the efficiency of service delivered. The main task for most countries is to formulate budgets in the context of the macroeconomic framework where the purpose of the outlays and the expected results are explicitly understood.

Financing and managing decentralization. Most countries are moving very fast toward decentralizing certain education services. In Hungary, local governments are responsible for kindergarten and primary education, but the financing is shared with the central government, which provides per student grants depending upon the level of education. In Poland, preschool education is already the responsibility of local governments, and primary education is progressively being transferred to local governments. Because local governments have limited capacity to raise revenues, the financial resources devoted to important services such as preschool education and child care have decreased. This is a serious setback for a previously well-developed system. Attending preschool has an important positive effect on later school achievement, particularly for disadvantaged children (Meyers 1992), and the lack of child care negatively affects the participation of women in the labor force.

Some important principles need to be applied to decentralization policies. First, local governments clearly have a comparative advantage over the central government in providing basic social services such as kindergarten and primary education. They are in a better position to assess preferences and target resources. Second, as just discussed, differences in the taxable capacity of individual local governments to raise revenue give the central government a vital funding role to play in ensuring reasonable access and minimum standards across the country. Third, many governments will wish to maintain a core curriculum throughout the country to assist social integration and to ensure educational standards. Fourth, the central government needs to address the weak technical and implementation capacity of many local governments by providing training and technical assistance.

Measuring efficiency. Implementation of these policies will depend to a considerable extent on the efficiency of resource use. The education and training system cannot afford to waste precious resources on ineffective or irrelevant programs. Particularly during a fiscal crisis, it will be well-nigh impossible to persuade ministries of finance to provide financial resources for the sector without demonstrable improvements in efficiency. Measuring efficiency in education and training is problematic, however, since measuring inputs, outputs, and the relation between them is difficult. Inputs can be quantified in measures such as the number of students and teachers, curriculum hours, teaching loads, and units of materials. But these are only proxies for what goes on in the classroom and do not provide information about the quality of the teaching and learning process, which is much more difficult to measure. Output measures of student achievement are technically difficult and have to be interpreted carefully if used to compare institutions and programs; it is even more difficult to relate teaching inputs to educational attainment, given the different social, psychological, and educational profiles of students in the classroom. Finally, even though the causal link between investment in education and economic growth is now well established at the macroeconomic level (World Bank 1990a), it is methodologically extremely difficult to establish such links at the microeconomic level.

These problems are particularly acute in the countries of Central and Eastern Europe, which lack management information systems to measure educational inputs properly or assessment and evaluation systems to measure student achievement. Neither do they have sufficient historical data on the labor market to measure the relationships between earnings and education. Notwithstanding the technical problems and institutional difficulties associated with measuring efficiency, quantum advances are essential to track the impact of the reform effort.

BUILDING POLITICAL CAPACITY. Contentious issues include the development of new curricula, the provision of new textbooks, and teachers' pay. On the first, the political imperative of "out with the old" applies to education as to other parts of the system. One of the problems facing policymakers, therefore, is how to protect those parts of the old system which worked well and, at the same time, to head off demands for reforms which experience elsewhere suggests are unlikely to work well.

The provision of new, non-Marxist textbooks has already begun and should be left largely to the private sector, both as a signal of political freedom and for reasons of economic efficiency. Education authorities should buy books for those parts of the system for which schoolbooks should be provided to children free. A difficult political decision is how to divide the education budget so as to prevent the demands of teachers' salaries from crowding out spending on textbooks. The issue of teachers' pay during a time of fiscal crisis has no simple answer.

In facing the political dimensions of all these issues, governments should remember that in a democracy (in sharp contrast with the old system), parents are a powerful part of the electorate. Education ministers should therefore be prepared to communicate directly with parents (for example, using television) to explain the problems and the government's proposed policies to address them. In the matter of teachers' pay, for example, the government's greatest allies are likely to be parents.

Conclusions: Priorities and Sequencing

The central message of this chapter is that improving the quality of education and training is vital for the future economic health of countries in Central and Eastern Europe. This will require involvement and investment from the different actors in a market economy: the state, private business, and individuals. No single blueprint exists for restructuring the education and training system, and countries in Central and Eastern Europe are by no means alone in facing the need for reform. Similar debates—for example, about the nature of the relationship between education and the labor market—are occurring in many market economies. The following represents a cross section of short- and medium-term policies which should be appropriate in most of Central and Eastern Europe.

Short-term Policies

Standards in preschool and basic education should be maintained as a matter of high priority. Basic education is the foundation on which skills and knowledge for the workplace are built. A failure to invest in basic human capital will result in major costs to the state later on, as the poorly educated consume disproportionately more state services. In addition, a semi-literate work force cannot compete in a global market. Since the population in primary education is static or declining in much of Central and Eastern Europe, quality can be maintained in basic education without any significant increase in expenditure. Areas of focus include:

- Reforming curricula
- Ensuring the availability of in-service teacher training
- Providing an adequate supply of books and educational materials.

Secondary education needs to be restructured to ensure a flexible and trainable work force, capable of responding to the rapid changes taking place in the labor market. The emphasis should be on:

- Increasing enrollment in academic and technical streams which lead to matriculation
- Reorganizing technical and vocational education around broad clusters of occupations
- Delaying specialization and transferring pre-employment training outside the formal school system
- Developing appropriate curricula and associated teaching materials and books
- Redeploying and retraining teachers.

Reforms should be initiated at the postsecondary level. A diversified system of postsecondary and higher education, capable of providing opportunities to a wide range of abilities and levels and able to provide the economy with high-level knowledge and skills, is an essential component of every successful market economy. Reforms include:

- Creating alternatives to long-duration university programs
- Giving universities autonomy, while at the same time providing them with incentives to introduce new specializations and to improve the efficiency with which they use resources
- Modernizing libraries and information services.

Medium-term Policies

Policies with a medium-term dimension are of three broad types: those concerned mainly with improving the quality and mix of education and training, those aimed at improving the efficiency with which that mix is delivered, and those concerned mainly with institutional capacity.

Adult education and training should be reformed to make it easier for individuals to upgrade or change their skills so that they can compete in a changing labor market. The emphasis should be on:

- Developing a partnership between public and private financing and the provision of training
- Decentralizing the system to ensure that it responds to local needs
- Ensuring standards through regulation and provision of information.

Postsecondary and higher education should be expanded to meet demand. The focus should be on:

- Changing curricula and admissions procedures to allow for more flexible access
- Recovering costs through fees and other income-generating activities
- Introducing student loan systems
- Providing incentives to promote private sector involvement.

Teacher education should be reorganized. This is crucial since the teacher is the key player in the education and training system. Reforms include:

- Consolidating teacher education institutions and strengthening education departments in the universities
- Developing an education inspectorate
- Developing institutions and incentives to ensure an adequate and continuing provision of in-service teacher training.

The education and training system should be decentralized in the medium term. Without accompanying reforms of local government and the tax system, however, rapid decentralization may backfire. In the absence of appropriate mechanisms to equalize funding across regions and to ensure minimum standards, decentralization may lead to a decline in the quantity or quality of services. Preschool education in many countries has already suffered in this way. The focus should therefore be on:

- Ensuring equity of public funding across regions
- Developing and enforcing norms and standards
- Providing technical assistance and training to local government.

Policy should support the emergence of private providers, not least in order to diversify the finance of the education and training system during the fiscal crisis. This can be accomplished by:

- Removing obstacles to licensing private providers
- Providing incentives through the tax system and in other ways such as offering access to idle facilities and equipment
- Subcontracting training and retraining services for groups such as the unemployed.

Regulatory and information systems require development to protect consumers and promote informed choice by consumers and employers. Priorities include establishing:

- A national system of assessment and examinations
- Accreditation of private providers
- An information system on providers and their performance.

Notes

1. The discussion in this chapter focuses on those countries which were not part of the U.S.S.R. Readers may, however, be able to draw some parallels with the situation in the former U.S.S.R.

2. Educational Testing Service 1992. Recent Soviet research on the use of mathematical and scientific knowledge among thirteen-year-olds in nineteen OECD and former communist countries quoted by Heyneman (1993) highlights an interesting difference between the two groups. The three former communist countries in the sample (Hungary, Slovenia, and the U.S.S.R.) rank highest on the awareness of facts but significantly lower on the integration and application of those facts to new situations. The ranking in the OECD countries is typically the reverse. These differences do not necessarily reflect a good or bad education, but deep differences in the role played by education in market and centrally planned economies.

3. Anecdotal evidence for the split between education and research comes from one of the authors, who in 1989 asked a prominent biomedical researcher from a Hungarian research institute whether he taught students. The researcher replied that yes, he had taught students, but only in the United States, where he had been a visiting fellow. He had never taught students in Hungary.

4. This name, chosen by Hungarian officials, connotes a system which was separated from and is now trying to rejoin Western Europe.

5. Poland maintained a tradition of private training throughout the communist period, and private training has flourished since. In Poland in 1992, at least forty enterprises of reasonable size competed in this market. Organizations such as the Chamber of Craftsmen, the Skill Development Centre (ZZDZ), and the Federation of Scientific and Technical Associations (NOT) managed hundreds of training programs.

6. Between 1988 and 1991, the price of textbooks in Poland increased by a factor of 26 at the primary and secondary levels and by a factor of 30 to 40 at the tertiary level, with similar increases occurring in all the countries of Central and Eastern Europe.

7. A relatively new trend in some OECD countries is to introduce technology education into the curriculum. The intent is to enable students to acquire a general understanding of design, production, and marketing processes.

8. For example, initial results in Hungary for ERAK (Regional Labor Development Center) show a placement rate of 60 percent for graduates in a region with a 20 percent unemployment rate. The unit cost is only 25 percent higher than unemployment benefits for the corresponding period.

9. As illustrated by the closing in 1973 of the Wavelberg and Wotwand High Technical School, which was founded before the Second World War.

10. In 1990–91, student/teacher ratios in secondary education were 18:1 in Poland, 17:1 in Bulgaria, and 14:1 in Hungary.

11. In 1990–91, the number of teaching hours was eighteen hours per week in Poland, which is low compared with other European countries, where twenty-five hours per week is more normal.

12. In 1992, Bulgaria (a country of 9 million people) had thirty higher education institutions and forty-five postsecondary education institutions spread over twenty-five different cities. In addition, some 175 academy-based research centers varied in size from 4 to more than 1,000 employees.

13. In 1992, enterprises paid the following: in Poland, a 2 percent payroll tax toward the Labor Fund, part of which is dedicated to retraining the unemployed; in

Romania, a 4 percent payroll tax, which is used partly for unemployment compensation and partly for retraining the unemployed; in Hungary, a 1.5 percent payroll tax earmarked for training and another 5 percent payroll tax earmarked for the Solidarity Fund, which finances both unemployment compensation and proactive employment programs. However, expenditures on unemployment benefits have crowded out expenditures on training.

14. "Non-public" schools in Poland can be divided in two groups, depending on whether or not they receive state subsidies: private schools, which receive no subsidy, and social schools, which receive subsidies equivalent to 50 percent of the average unit cost in public schools.

15. In 1990, funding formulae were based on salaries in the case of Poland and courses offered in the case of Czechoslovakia.

16. In Poland in 1988, average wages and salaries in the education sector were the lowest in the entire socialized economy. The index was 71.3 in education, 98.3 in agriculture, 103.6 in manufacturing, and 179.1 in mining and energy (Poland, GUS 1989).

Twelve

HEALTH AND HEALTH CARE

ALEXANDER S. PREKER
RICHARD G. A. FEACHEM

To REVITALIZE THEIR HEALTH SERVICES, governments in the former socialist states of Central and Eastern Europe are experimenting with a new wonder drug called market mechanisms. This is rather like the doctor who gives penicillin to a patient who has a known allergy to it but will die without it. It is necessary to understand the associated dangers so that appropriate measures may be taken to prevent the treatment from killing the patient.

Market forces have of course existed in Central and Eastern Europe for a long time in the form of the underground economy, and they are a normal part of the health sector in most of the highly industrialized countries. So by themselves market forces in the health sector are nothing new. In most Western countries, however, the trend over the past century has been toward an increasing role for government because of known market imperfections in the health sector. In Central and Eastern Europe, by contrast, market forces are being introduced to deal with the systemic failure of excessive rather than insufficient government involvement.

This chapter presents a conceptual framework for understanding why the health sector performs poorly in Central and Eastern Europe (excessive state intervention), the prescribed treatment (market forces), the allergy (market imperfections), and the remedy (a new public/private partnership in the financing and provision of health care). The chapter looks at policies relating to health and, a closely related but distinct topic, at the organization of health care.

The Inheritance

It would be senseless for governments in Central and Eastern Europe to attempt to fix things which are not broken or to focus excessively on problems for which there are no known solutions (for studies of the different countries, see Preker 1990; Feachem and Preker 1991; World Bank 1991a; World Health Organization 1991; UNICEF/WHO 1992). They should, therefore, conduct a rapid assessment of the strengths, weaknesses, and perennial problems which their health sectors inherited, so that they can begin to redress the most urgent needs. Although each country has its own unique history, some of the common threads presented below are readily apparent.

Strengths

The first major lesson to be learned from the socialist experience is that, under the right set of circumstances, state involvement in the health sector is desirable and may at times be essential. Rapid economic growth, expansion in the social sectors (income transfers, health, education, and culture), and more readily available food, shelter, and employment led to significant improvements in living standards and health status in many of the socialist states during the early years of central planning.[1] In comparison with earlier periods of physical destruction, economic hardship, starvation, and homelessness in countries like the Baltic states, Poland, and the U.S.S.R., the 1950s were characterized by improvements in human well-being and health status despite the obvious price of political oppression and loss of personal freedom.

Entitlement by the entire regional population to a full range of health services was one of the remarkable achievements of the socialist regime and its health care system. No other region in the world, not even Europe or China, has ever succeeded in providing such extensive coverage of comprehensive health care to a population of similar size. Since equity in access to affordable health services appears to be one of the most important determinants of health status at upper income levels, preserving this positive legacy should be a high priority.

By the early 1970s, countries such as Czechoslovakia and Hungary had very mature health systems. Compared with developing countries with similar per capita gross domestic product, or even compared with the highly industrialized countries, the health sectors of Central and Eastern Europe were well endowed in basic physical infrastructure, trained staff, and education programs. Structurally integrated networks of hospitals, clinics, and other clinical facilities, based on the Soviet health care model, secured universal access to curative health services throughout the region. Patients had their first point of contact with the lower tiers of the health system through individual outpatient departments of hospitals, polyclinics, diagnostic departments, emergency services, community health centers, rural health centers, and industrial health services. Doctors working in these settings acted as gatekeepers, treating what they could and referring more difficult cases to higher levels of care. University hospitals and national specialist institutes capable of providing advanced technological interventions formed the tip of this pyramid. A highly structured system of hygiene and epidemiology stations formed a similarly sophisticated and integrated network of public health services which concentrated on the control of infectious, occupational, and environmentally related diseases. These positive bequests from the former socialist regime are summarized in table 12-1.

Because the previous system failed to produce many of its predicted economic and social benefits, radical reformers during the transition have been

Table 12-1. *Positive Bequests from the Socialist Regimes in Central and Eastern Europe*

Bequest	Characteristic
Entitlement	Entire regional population
Burden of financing	Evenly distributed
Access	Few financial barriers
Range of services	Comprehensive
Network of facilities	Structurally integrated
Resource base	Extensive infrastructure

quick to condemn, and even have been tempted to discard, almost everything that existed during the past. As a result, there is now a serious risk that, at least in the health sector, some of the countries in Central and Eastern Europe may throw out the baby with the bathwater. Not only would this waste valuable assets such as excellent vaccination programs and maternal and child health services, but it could also deprive a significant part of the population of basic health care at a time when unemployment and deteriorating standards of living are having a negative impact on health.

Weaknesses

Unfortunately, despite these positive attributes, the national health services in Central and Eastern Europe were remarkably ineffective in promoting good health or preventing illness and disability from known and avoidable causes. The second major lesson to be learned from the socialist experience is, therefore, that relying too heavily on a state monopoly in a centrally planned and supply-driven health sector lowers the efficiency and quality of care. This must be quickly corrected if health services in the region are to be successfully integrated into the emerging market economies.

Interestingly, the health sectors in Western countries such as the United Kingdom, where governments have assumed a similar monopolistic and centralized dominance in the financing, ownership, and provision of health services, suffer from many of the same problems as those observed in Central and Eastern Europe. Thus, even though government involvement may be necessary to securing some positive attributes in the health sector, it is not in itself a sufficient ingredient in securing others, such as good health. There are two major sets of problems: poor health status and low-quality health services, as summarized in table 12-2.

POOR HEALTH STATUS. The single strongest predictor of a nation's health status is not the character of its health services but its per capita GDP (Schieber 1989; World Bank 1993d). The relationship between GDP per capita and life expectancy at birth is strong, especially at lower levels of income (figure 12-1).

Both the northern and southern countries of Central and Eastern Europe, together with the Central Asian republics of the former U.S.S.R., occupy positions in figure 12-1 which suggest that overall, they are a little more healthy than their income level would predict. This is especially the case for the southern countries of Central and Eastern Europe. If, however, a measure of adult health is taken, such as risk of death between fifteen and sixty years for males, a different picture emerges (see figure 12-2). The southern countries of Central and Eastern Europe and the Central Asian republics have male adult mortality risks close to those predicted by their GDP per capita. The northern countries of Central and Eastern Europe, by contrast, are more wealthy and less healthy. They have a male adult mortality risk of around 29 percent when, at their level of income, the predicted figure is 21 percent. In both figures 12-1 and 12-2, the national wealth and health status of Central and Eastern Europe is broadly similar to that of Latin America and the Caribbean.

At the same time, the relationship between life expectancy and national wealth is weak at higher levels of income. Wilkinson (1992) suggests that among industrial countries income *differentials*, not the *level* of income, are a better predictor of health status, with more equitably distributed wealth being

Table 12-2. *Legacies of Poor Health and Inefficient Health Services in Central and Eastern Europe*

Key problem	Issues
Health status	High mortality, especially in adult men; high morbidity; unhealthy life-styles and environment
Policymaking and management	Ineffective intersectoral coordination; low priority given to health and good health care; lack of responsiveness to local needs; weak management, tracking, and evaluation
Structure	Rigid and overly centralized structure; too much emphasis placed on institutional care; neglect of public health and primary care; distortions in the mix of public and private financing
Function	Lack of functional integration; ineffective, inefficient, and low-quality services
Resources	Arbitrary statistical norms (physical and human); imbalances with surpluses and shortages; excessive use
Training and research and development	Narrow and excessive specialization and isolation; graduate education isolated from universities; research isolated from teaching; noncompetitive funding
Financing	Underfinancing compared with capitalization; adverse incentives

Figure 12-1. *Life Expectancy by GDP per Capita and Region, 1990*

Life expectancy at birth (years)

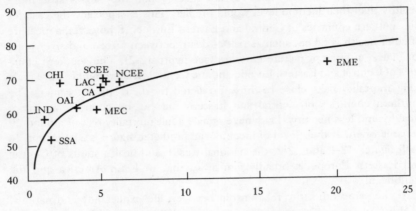

GDP per capita (thousands of U.S. dollars)

Note: GDP per capita is in 1991 U.S. dollars converted from local currencies at purchasing power parity rather than the exchange rate. The line is drawn from the data points for every individual country; the regional pooled means are shown as a single point for each region.

The regions are defined as follows: CA, Central Asian republics of the former U.S.S.R. (Azerbaijan, Kyrgistan, Tajikistan, Turkmenistan, and Uzbekistan); CHI, China; EME, Established market economies (Australia, Canada, Japan, New Zealand, United States, and Western Europe); IND, India; LAC, Latin America and the Caribbean; MEC, Middle Eastern crescent (from Morocco in the west to Pakistan in the east, including North Africa and the Middle East); NCEE, Northern Central and Eastern Europe (Belarus, Czech Republic, Estonia, Hungary, Kazakhstan, Latvia, Lithuania, Moldova, Poland, Russian Federation, Slovak Republic, Ukraine); OAI, Other Asia and islands (all of Asia except for China, India, Japan, and Pakistan, plus the Pacific Islands); SCEE, Southern Central and Eastern Europe (Armenia, Bulgaria, Georgia, Romania, and Yugoslavia); and SSA, Sub-Saharan Africa.

Source: Data supplied by Christopher Murray and colleagues, Center for Population and Development Studies, Harvard University, Cambridge, Mass.

associated with greater longevity. In light of this, the poor status of adult health in the northern countries of Central and Eastern Europe is noteworthy in view of their history of a relatively compressed income distribution and supposed absence of poverty. There is, therefore, a real risk that the move to a market economy will make health status decline in some countries as poverty increases among some subgroups of the population.

Although health status in Central and Eastern Europe today is reasonable in relation to the wealth of the region (except for that of adult males in the northern countries), an examination of trends over the last five decades tells a very different story and poses a considerable challenge for policymakers (Fea-

Figure 12-2. *Risk of Death for Males between Fifteen and Fifty-nine Years of Age, by GDP per Capita and Region, 1990*

Risk of mortality (percent)

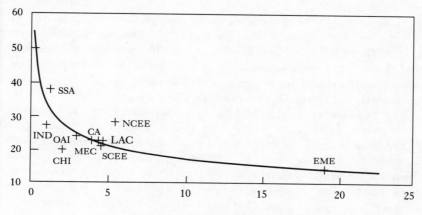

GDP per capita (thousands of U.S. dollars)

Note: The risk of death measures, for every 100 males who are alive at fifteen years of age, the number who will be dead by sixty. The horizontal axis is in 1991 U.S. dollars converted from local currencies at purchasing power parity rather than the exchange rate. The line is drawn from the data points for every individual country. The regional pooled means are shown as a single point for each region.

The regions are defined as follows: CA, Central Asian republics of the former U.S.S.R. (Azerbaijan, Kyrgistan, Tajikistan, Turkmenistan, and Uzbekistan); CHI, China; EME, Established market economies (Australia, Canada, Japan, New Zealand, United States, and Western Europe); IND, India; LAC, Latin America and the Caribbean; MEC, Middle Eastern crescent (from Morocco in the west to Pakistan in the east, including North Africa and the Middle East); NCEE, Northern Central and Eastern Europe (Belarus, Czech Republic, Estonia, Hungary, Kazakhstan, Latvia, Lithuania, Moldova, Poland, Russian Federation, Slovak Republic, Ukraine); OAI, Other Asia and islands (all of Asia except for China, India, Japan, and Pakistan, plus the Pacific Islands); SCEE, Southern Central and Eastern Europe (Armenia, Bulgaria, Georgia, Romania, and Yugoslavia); and SSA, Sub-Saharan Africa.

Source: Data supplied by Christopher Murray and colleagues, Center for Population and Development Studies, Harvard University, Cambridge, Mass.

chem 1994). After recovering from the war years, the region enjoyed a health status which overlapped that of the West. Life expectancy at birth was higher in Czechoslovakia than in Austria throughout the 1950s (Bobak and Feachem 1992). Furthermore, in the decade from the early 1950s, improvements in health status outpaced those in most Western countries except Japan. Infant mortality rates fell by nearly half in the communist countries, and life expectancy at birth increased by around five years. By contrast, life expectancy over this period increased by two and a half years in the Federal Republic of

Germany and one year in the United States (Eberstadt 1993). By the mid-1960s, only one or two years separated average life expectancy in Central and Eastern Europe from that in the advanced capitalist countries, and the gap was closing.

From the mid-1960s, the relative trends changed dramatically. Health status in Central and Eastern Europe stagnated or deteriorated, while in the highly industrialized countries, it improved steadily. Between the mid-1960s and the late 1980s, life expectancy at age one fell for males throughout Central and Eastern Europe (the greatest fall—three and a half years—was in Hungary) and rose by less than one and a half years for females.[2] Over the same period, age-standardized male mortality rates in Bulgaria, Czechoslovakia, Hungary, and Poland rose 2–13 percent, while in the Netherlands, Sweden, Switzerland, and the United Kingdom, they fell 12–27 percent. The widening gap in life expectancy and mortality between East and West was particularly striking in middle-aged adults. In the communist countries, death rates for males aged forty-five to forty-nine years increased between 7 percent (German Democratic Republic) and 131 percent (Hungary) between 1965 and 1989, while they decreased in the highly industrialized countries (Eberstadt 1993). The risk of death for men between fifteen and fifty-nine years of age in the late 1980s was higher in Hungary than in Zimbabwe and higher in Czechoslovakia than in Viet Nam (Feachem and others 1992). A substantial proportion of this divergence is attributable to an epidemic of cardiovascular disease and, particularly, of ischemic heart disease in middle-aged males. (Other leading causes of death and morbidity include cancer, respiratory disease, and accidents.) By the mid-1980s, mortality rates from ischemic heart disease in men aged forty-five to fifty-four years were twice as high in Czechoslovakia as in Austria, while in the 1950s they had been the same (Bobak and Feachem 1992).

Superimposed on these longer-term trends of declining health are the effects of the transition. Although the data are somewhat fragmented and the precise causal relationships difficult to establish, the results in some countries are dramatic. The increase in crude death rates in the Russian Federation "assumes truly apocalyptic connotations. [It] increased by 33 percent . . . between 1989 and the first seven months of 1993" (UNICEF/ICDC 1993, p. 20, fig. 7).

It is no consolation to the countries in Central and Eastern Europe to know that their overall health status is similar to that in Malaysia, Sri Lanka, Uruguay, and Venezuela, since it is not with these countries that they must compete. They need human capital capable of competing effectively with the countries of the European Union. At present, the gap in health status between the former socialist economies and the highly industrialized countries is wide and growing, especially in the cohorts of working age. The gap in health status shown in table 12-3 is due only to differences in mortality. These mortality

Table 12-3. *Health Status in Central and Eastern Europe and Established Market Economies, 1990*

Health measure	Central and Eastern Europe[a]	Established market economies[b]
Life expectancy (years)		
At birth		
Average	72	76
Range	69–73	74–79
At fifteen years	59	62
Risk of death (percent)		
Between birth and five years		
Average	2.2	1.1
Range	1.3–3.6	0.6–1.3
Between fifteen and fifty-nine years		
Males	28	15
Females	11	7

a. Albania, Belarus, Bulgaria, Czech Republic, Hungary, Lithuania, Moldova, Poland, Romania, Russian Federation, Slovak Republic, Ukraine, former Yugoslavia.

b. Australia, Austria, Belgium, Canada, Denmark, Finland, France, Federal Republic of Germany, Greece, Ireland, Italy, Japan, Netherlands, New Zealand, Norway, Portugal, Spain, Sweden, Switzerland, United Kingdom, United States.

Source: Calculated from World Bank 1993d, table A4.

excesses place a burden on the economies of Central and Eastern Europe because of the lost investment in human capital (individuals who die in middle age have received publicly funded education and other services), medical expenditure made prior to their deaths, and the more general opportunity cost of lost lives. An even greater burden is likely to be imposed by the huge morbidity differential which underlies the mortality differential. Lost productivity due to high rates of sickness in the labor force and lost investment due to the high costs of caring for the chronically ill and their families put the formerly socialist economies at a striking disadvantage when competing with their healthier rivals.

Repairing the damage caused by the last three decades of communist rule and closing the gap in health status must, therefore, be a central objective for human resources policy throughout the region. This will be neither easy nor quick. If they could achieve rates of overall decline in mortality similar to that of Chile over the past two decades (roughly 2 percent a year), it would take eastern Germany twelve years to catch up with western Germany, and Hungary would need twenty-three years to rejoin Austria (Eberstadt 1993).

For many years, either these trends went undetected or information about them was suppressed for political reasons. The general philosophy—that citizens are passive recipients of state-run health services rather than active participants in a process of improving life-styles and reducing environmental risks—contributed to the problem.

LOW-QUALITY HEALTH SERVICES. During the early years of communism, the health sector (as part of the so-called non-productive sector) was accorded a lower priority than the industrial sectors. As a result, the enormous potential benefits of an intersectoral approach to health and health services planning under the former five-year plans were largely lost. These problems in policymaking at the central level were compounded by poor flows of information in the tracking and evaluation of health trends and lack of authority in decisionmaking at local levels (a result of the inability to decentralize). The main resulting problems were poorly targeted investment, inadequate integration between the different parts of the system, excessive specialization, a narrow base for financing, and grossly deficient incentives to efficiency.

Poorly targeted investment had, by the early 1970s, led to a massive but lopsided buildup in acute care hospitals and excessive specialization at the expense of public health services and primary care. The rigid and over-centralized Soviet-style national health services which evolved from this process contained both significant overlaps and significant gaps because different ministries all tried to provide services for their own enterprise-based workers. The private sector was excluded from nearly all activities.

The different components of a complex network of services were not functionally integrated. Standards were determined by arbitrary statistical norms, leading to many imbalances. Surpluses and shortages developed due both to variations in local patterns of use and to political patronage. The compulsory catchment areas and the role of primary care doctors as gatekeepers were unpopular with patients who rightly felt that they were wasting their time waiting in line to be told that the services they needed were not available at that level of care. Not surprisingly, they were willing to pay substantial gratuities to be referred quickly up the line or to use services outside their official catchment area. In the poorer countries, such as Romania and across the southern crescent of the former U.S.S.R., this shopping around was often unproductive because neighboring areas with higher levels of care, more often than not, had similar shortages. Instead of providing comprehensive and functionally integrated services, the system created many indirect barriers to access through corruption and adverse incentives.

Excessive specialization led to the absence of a broad education and to the development of narrow skills which are now difficult to adapt to the more complex demands of a market economy. For instance, girls would enter technical training in nursing at the age of fourteen, depriving them of a broader educational base. Since medicine was considered a technical skill, doctors and other health care providers received training provided by ministries of health. Medical education was, therefore, isolated from the education provided in general universities by ministries of education. Training of public health specialists, health economists, health service managers, general practitioners, nurses, and many other health care personnel was seriously neglected. Under the Ceaucescu regime in Romania, the nursing profession was virtually abolished, leaving the country with a serious deficit of trained nurses by Western European standards. Research was isolated from teaching and financed through noncompetitive grants. International isolation in training, research, and technology was pandemic throughout the region, especially in Albania, Bulgaria, and Romania. These problems of the health care system echo those of the rest of the educational system, and the similarities are striking.

Health services were financed almost exclusively through general revenues. Although an equitable and possibly efficient source of financing, such dependence left the health sector vulnerable to political agenda. It also failed to send a clear signal to patients and health care providers that health care is not free even when provided without direct charge at the point of delivery. Under-the-table user fees in the form of gratuities or "black money" became widespread throughout the region. In countries like Hungary, up to 20 percent of health care expenditure is estimated to have taken the form of out-of-pocket payments. The recurrent budget required to operate the resulting overcapitalized and underfunded health services outstripped the financial resources of most countries.

Incentives to efficiency were virtually nonexistent either to motivate patients to maintain good health and use scarce resources judiciously or to encourage health care workers and institutions to provide high-quality care. Doctors working on salaries had an incentive to minimize their work load by referring complex problems up the line. Directors of hospitals, similarly, had a strong incentive to avoid performing expensive diagnoses and treatments themselves and to refer difficult cases to higher levels of care. The less work, the less strain on their global budgets. As a result, queues and waiting lists were common even though the number of doctors and beds per capita was higher than in most Western countries.

Perennial Problems

The third major lesson to be learned from the past is that some problems are an unavoidable part of the human predicament and may simply have to be accepted. No health sector reforms or expenditure will entirely eliminate the problems of aging, biological defects, poverty, and social misery. These existed both before and during the former socialist regimes, just as they do in market economies today. Universal entitlement and free access to health services may have alleviated some of the associated hardships, but they did not eliminate them altogether.

The Forces Driving Change

The current reforms in Central and Eastern Europe fundamentally redefine the role of the state in the health sector, as elsewhere. This includes a call for more efficient allocation of resources through market mechanisms, greater individual freedom through democratic processes, and stronger institutional capacity of health systems through decentralized devolution of responsibility and management. The polar extremes of those driving forces which are most relevant to the health sector are summarized in table 12-4. The transition from a centrally planned command economy and socialist political system to a market-oriented economy and democratic political system is accompanied by a swing of the pendulum of policy choices.

Some countries, such as Poland and the Czech Republic, have opted for a "big bang" approach to restructuring, while others are following a more gradual path. In all instances, however, an elixir of rapid liberalization of demand, supply, prices, and wages, combined with the fiscal constraints of stabilization programs, provides a potent, and at times toxic, brew for ailing health services. In most countries, the health sector was not prepared for the resulting problems discussed below: an overshoot in economic liberalization, the rejection of constraints on freedom, and a resulting institutional collapse. The mounting backlash against these negative effects of the transition is now threatening to destabilize the whole reform process. This presents a particularly difficult short-term problem for policymakers in the region, who have to

Table 12-4. *Driving Forces in the Reform Process*

	Dichotomies in policy options	
Driving force	*For centrally planned and socialist political systems*	*For market economies and democratic political systems*
Political factors		
Ideology	Collectivist	Individualist
Political process	Autocratic	Pluralist
Governance	Totalitarian	Democratic
Economic factors		
Economic model	Socialist	Capitalist
	Command	Market
	State	Laissez-faire
Ownership and financing	Public	Private
Prices and wages	Predetermined	Competitive
Production	Supply driven	Demand driven
Labor markets	Restricted	Mobile
Incentive structures	Bureaucratic	Meritocratic
	Normative	Performance
Economic equilibrium	Static	Dynamic
Institutional factors		
Policy and legal	Five-year plans	Incremental
Structure	Centralized	Decentralized
Function	Simple and uniform	Complex and diverse
Personnel	Super specialized	Broad skills

address the strengths, weaknesses, and perennial problems of the past as well as those of the transition.

The first major driving force is the tendency for economic liberalization in the health sector to go too far too fast. The removal of centralized state control over the health sector and the rapid introduction of unregulated competitive markets have already led to significant market failure. In the Czech Republic, ownership of most health care facilities was quickly transferred to local communities. Substantial parts of the former national health service have been privatized, especially in the case of pharmaceutical products, medical equipment, supplies, ancillary services, and ambulatory health services (private offices, clinics, pharmacies, and diagnostic centers). Many health care providers, such as general practitioners, consultant specialists, dentists, and pharmacists, working outside the hospital subsector no longer see themselves as public employees but rather as entrepreneurs in private practice. Patients see themselves as consumers of health care, demanding services in return for their taxes or social insurance contributions. Unfortunately, instead of engendering a partnership between the public and new private sector, unrestrained privatization within an excessively relaxed regulatory framework has led to unscrupulous profiteering and pillaging by health care providers and unchecked use by patients (both problems are caused in part, though not wholly, by the third-party payment incentives discussed in chapter 2).

With the sudden liberalization of prices and wages, the cost of critical supplies and pharmaceuticals has exploded. In Poland, for example, public expenditure on pharmaceuticals increased from 12 percent of health care expenditure to more than 30 percent almost overnight. The unbalanced shift in incentives from a normative-based system to a performance-based system has exacerbated the problem of cost containment.[3] In the Czech Republic, the original annual budget of the newly created health insurance fund was more or less exhausted within six months of its inception at the beginning of 1992 and had to be replenished; expenditure in the first six months of 1993 was almost as much as in the whole of 1992. In Hungary, the proposed replacement of global hospital budgets with performance-related, diagnostic-related groups (see the glossary) is likely to have the same effect if not accompanied by strict mechanisms to cap the budget.

In countries like Albania, large-scale layoffs in the industrial sectors have led to significant dislocations, housing shortages, family disruption, and unemployment as firms begin to shed labor to become more efficient and try to avoid bankruptcy during the recession. These social disruptions have increased the demand for health care at a time when resources are extremely limited.

A second driving force, the sudden rejection of constraints on freedom of choice, has led patients and health care providers throughout the region to shop around indiscriminately, wasting valuable resources and increasing the pressure on an already overused health service. In Poland, the obsession with steering committees and popular consultation has led to ineffective policy-making and management. Critical pieces of legislation remain deadlocked in Parliament for months, even though many of the proposed reforms could easily have been introduced through regulation. In Hungary and surrounding countries, rejection of past socialist ideals and excessive confidence in individual autonomy and self-sufficiency have almost completely replaced all concern for collective protection and equity. Gypsies, migrant populations such as immigrant workers, and minority groups such as AIDS patients are now threatened with loss of entitlement and of access to adequate health care.

The third driving force, the rejection of the communist model, has led to a massive collapse in the prevailing centrally planned institutional framework for health services which had previously been provided through the public sector and which has not been replaced by private or nongovernmental sectors. The abolition of five-year plans has left many ministries without a clear strategy for the future and without an institutional capacity to develop new policies.

Purges of the nomenclature and a witch hunt for previous card-carrying communists have decapitated the ministries of health and many health care institutions in the Czech Republic, Poland, and the Slovak Republic, leaving a dearth of experienced senior staff. Medical doctors with little training or experience in management have assumed key administrative posts in health ministries, hospitals, and other institutions, creating a policy vacuum in which poorly prepared and contradictory reform proposals are often presented to Parliament at the same time. Important decisions about the budget allocation process and major capital investments are now heavily influenced by vested interests such as the medical profession, pharmaceutical companies, and equipment manufacturers and importers. As a result, the investment needs of hospitals and clinics may be increasing unnecessarily, not least because maintenance and repairs are being neglected.

Excessive decentralization and a breakdown in referral networks in countries such as Bulgaria, the Czech Republic, Hungary, and the Slovak Republic have led to potentially expensive overlaps, as every community hospital wants its own neurosurgical unit and the latest technological equipment. A splintering of the former centrally organized wholesale and retail distribution systems has added to the crisis caused by rapidly rising prices. The result has been serious shortages of pharmaceuticals and critical supplies in much of the former U.S.S.R., where many of the distributional links which formerly led to Moscow have now been cut. Severe supply shortages are, of course, experienced also in countries ravaged by war.

Policy

Governments in Central and Eastern Europe are finding that reforming their health sectors during a period of major socioeconomic adjustment is vastly different from reforming them in a more stable environment. Frequent changes in leadership in Poland during the early years of the reforms, hyperinflation in Latvia, and purges in the civil service in the Slovak Republic make it difficult for government to introduce sustainable health care reforms. Reforming the health sector is also different from reforming other parts of the economy because health services cannot shut their doors while they are undergoing significant restructuring and reform. As a result, the massive layoffs and liquidation of assets in eastern Germany following reunification did not extend to the health sector.

Health sector reforms, therefore, carry with them special opportunities and risks. Policymakers in Central and Eastern Europe have an opportunity to redesign their health services and health financing systems, while drawing on the best of Western experience and avoiding known pitfalls such as the escalation of costs in the United States, waiting lists in the United Kingdom, and excessive prescription of drugs in Germany. Yet, they run the risk that lower standards of living, weaker commitment by governments to maintaining essential services, and restrictions in public expenditure on health care will weaken the health sector's role as a critical link in the social safety net. For instance, the privatization of family doctors' offices in Russia, under consideration in 1993, could worsen the shortage of doctors in regions outside Moscow. Continued underfunding of immunization programs or maternal and child clinics could worsen the already high infant and maternal mortality rates.

The Strategy

In an effort to be consistent with the underlying thrust of the transformation process, many governments in Central and Eastern Europe are trying to introduce health and social policies which simultaneously support the broader aims of associated political, economic, and institutional reforms, as well as objectives which are specific to the health sector itself. The underlying thrust of the transformation is to increase the overall well-being of the population by restoring individual freedom, increasing living standards, and creating a strong safety net for those who are unable to take advantage of these benefits without additional assistance. In the case of health and health care, as elsewhere, this includes incentives for individuals to become more self-sufficient, competitive markets to improve efficiency, and decentralization of management to improve the responsiveness of institutions to local needs.

Illness limits people's autonomy, reduces their participation in employment, and increases their dependence on health services. Thus poor health has a direct negative impact on labor mobility, productivity, and public expenditure. Unrestrained demand for health services and explosive increases in prices, if left unchecked, could trigger a vicious cycle of poor economic performance and deteriorating standards of living which would jeopardize other measures taken to improve health. The current sense of urgency, therefore, stems not only from ethical motives to improve health and the quality of human capital but also from a desire to underpin the emerging market economies with sound fiscal policies and institutional reforms.

Most countries in Central and Eastern Europe have begun to formulate policies which address these broad objectives. They can be grouped under three major categories:

- The first set of policies is directed at improving health, without which self-sufficiency and individual freedom cannot be restored. This includes measures to raise standards of living, to promote healthier life-styles, to protect the environment, and to improve the effectiveness of preventive and curative health services.
- The second set of policies focuses on health care and is directed at restoring macroeconomic balance, building new and more efficient markets, and controlling for market failure, without which economic growth and productivity will not be restored. In the health sector, this includes measures (a) to maintain macroeconomic balance through cost-containment policies and the introduction of non-budgetary sources of health care financing; (b) to build markets so as to increase consumer choice, diversify supply, and improve labor productivity; and (c) to regulate market forces to improve risk pooling, and ensure quality control.
- The third set of policies is directed at improving the political and institutional capacity of the health sector to implement relevant reforms. This includes measures to introduce new legislation, to strengthen the physical and administrative infrastructure of institutions, and to reorient the training of personnel.

Variations in the evolving health systems which are observed throughout the region can be explained largely on the basis of differences in how policymakers combine these options (World Bank 1993d).

Improving Health

The first and most important reforms being introduced in the health sector in Central and Eastern Europe, like the Close the Gap Program in Hungary, aim to reduce the difference in health status between the countries in the region and the highly industrialized countries with which they will compete in the future.

The determinants of the health gap between Central and Eastern Europe and Western Europe are not well understood. A plausible breakdown would attribute 30 percent of the gap to differences in wealth and associated socio-economic factors, 50 percent to known life-style risk factors, 10 percent to environmental pollution and occupational risks, and 10 percent to deficiencies in the preventive and curative health care services.

Good health depends not only on income per head but also on the distribution of that income. The risk factors causing disease are experienced particularly by the poor and less educated, and it is they who must be reached by incentives and educational programs. Unhealthy jobs are taken by workers with no other options, typically the least skilled and least educated. It is the poorest who live in the shadows of belching chimneys. Closing the gap in health status, therefore, requires a concern for equity and an emphasis on reaching the most disadvantaged sections of the population. Fortunately, so far, no country in Central and Eastern Europe has withdrawn universal entitlement or equity in access to services, although the health insurance programs recently introduced in the Czech Republic, Hungary, Russia, and the Slovak Republic could alter this situation.

Although policies to increase standards of living and economic growth are extremely important, they lie, for the most part, outside the health sector. Policymakers in Central and Eastern Europe are, nevertheless, trying to avoid any actions in the health sector which would contribute unnecessarily to the fiscal deficit or to inflationary pressures during the transition, since this would indirectly hurt health by retarding economic recovery and improvements in living standards. At the same time, they are concentrating their attention on those factors which have a more direct impact on health: life-styles, pollution and occupational risks, and preventive health care.

CHANGING LIFE-STYLES. The largest single contribution to the health gap is the high and rising rate of cardiovascular disease among adults, especially adult men. Risk factors for cardiovascular disease include excess consumption of alcohol, smoking, obesity, lack of physical exercise, and poor diet (high in animal fat, salt, and cholesterol; low in fruits and fresh vegetables). All these factors are more prevalent in Central and Eastern Europe than in the highly industrialized countries, and the most important factor, smoking, is much more prevalent. Surprisingly, no country in the region has introduced effective policies to reduce these risk factors. Such policies would include:

- Taxation-based disincentives to consumption of alcohol, tobacco, and unhealthy foods
- Legislation on alcohol, tobacco advertising, and food labeling
- Public education programs to inform and sensitize the population about diet, physical exercise, and dangerous behavior.

REDUCING POLLUTION AND OCCUPATIONAL RISKS. Environmental pollution is widespread in Central and Eastern Europe, and bringing it down to an efficient level, which takes account of both the costs of pollution and the benefits of the polluting activity, is a priority for government action. This is for more reasons than the negative impact on health. The most serious problems are dust and gases in the air, lead in the water and soil, and nitrates and heavy metals in the water. Air pollution is particularly damaging. Black spots are prevalent throughout the region, and at the junction of the Czech Republic,

Germany, and Poland is the so-called Black Triangle, where about 6.5 million people are exposed to extremely polluted air. In the Czech Republic, air pollution may explain about 9 percent of the gap in health status with Austria (World Bank 1993d). Occupational risks are widespread and varied. Under the communist regime, worker organizations, management, and the state conspired to allow appallingly unsafe and unhealthy working conditions to be maintained. This conspiracy was shrouded in an elaborate system of inspection, certification, and regulation which was corrupt and ineffective.[4]

Most governments in Central and Eastern Europe realize that rectifying this legacy will not be cheap or easy. As a result, effective inspection and regulation, and incentives for investing in anti-pollution technology and occupational safety, have not yet been introduced. In the future, such measures must be applied evenly, and preferably throughout the region, to prevent compliant industries from being disadvantaged in the marketplace. Worker organizations, such as Solidarity in Poland, could play a central role in helping to develop a wider public consensus.

SUPPORTING PREVENTIVE HEALTH SERVICES. In some respects, preventive health services were performing well under the previous regimes. In particular, most countries offered good services for pregnant women and babies, high rates of immunization among children, and effective programs for controlling infectious diseases, such as typhoid and tuberculosis. Immunization coverage rates in countries such as the Baltic states, Bulgaria, Czechoslovakia, Hungary, and Poland were among the highest in the world, and diseases such as poliomyelitis and measles were reduced to very low levels.

Maintaining these achievements, and building upon them by adding new vaccines such as hepatitis B, has not yet become a high enough priority for many countries. Control of some communicable diseases is now threatened in the former U.S.S.R. by problems in the production, purchase, and delivery of vaccines. Without more effective action, illnesses such as tuberculosis, which is on the rise in Western Europe, will pose a great threat to the poorer countries in Central and Eastern Europe.

Improving education and services for women and their babies is an effective way to increase health status and avoid unnecessary medical expenditure. In contrast with past achievements in providing effective immunization services, family planning under the communist regime was grossly inadequate throughout Central and Eastern Europe, leading to many unwanted pregnancies, frequent and expensive hospital abortions, and dangerous back-street terminations of pregnancies (Johnson, Horga, and Andronache 1993).

IMPROVING THE QUALITY OF CURATIVE HEALTH SERVICES. Although the direct impact of curative health services on life expectancy and morbidity may be much less than the effects of public health measures such as those described above, policies in this area are nevertheless important for several reasons.

Well-being can be measured in terms of quality as well as length of life. Since modern medicine is constantly finding new ways to relieve suffering, the curative services cannot be ignored simply because they do relatively little to increase life expectancy. For instance, victims of accidents may suffer permanent disabilities and loss of productive employment when their fractures are not set properly.

Furthermore, the way curative health services are financed has an important effect on the financial resources available to other areas of the health sector, such as preventive services, and to income support programs, which, in turn, indirectly affect standards of living and health. Although international guidelines exist for the cost-effectiveness of some basic interventions, such as the treatment of diarrhea or various forms of immunization (World Bank 1993d), policymakers in Central and Eastern Europe are finding that they must often develop solutions specifically for their own country. Hungary and Poland are already reviewing the costs and outcomes of such policies. Based on these reviews, they are drawing up "baskets" of affordable health care which match the financial resources available to the health sector.

Maintaining Macroeconomic Balance

Given that perhaps one-third of the health gap between Central and Eastern Europe and Western Europe is due to differences in wealth and associated socioeconomic factors, many governments are according very high priority to policies in the health sector which support measures to restore macroeconomic balance. These policies are of two broad forms: (1) policies to contain public expenditure on health care in an effort to relieve the fiscal imbalance and avoid crowding out the newly growing, but fragile, private sector and (2) policies to mobilize non-budgetary resources for health care financing to rejuvenate the sector. In both cases, various techniques are being used successfully.

CONTAINING COSTS. In the past, it was relatively easy for governments to maintain tight control over health care expenditure. Since the sector was financed exclusively through the state budget, expenditure caps could be enforced by indexing budgets at or below the rate of inflation (global budgets in the case of hospitals and clinics; salaries or capitation fees in the case of doctors and other health care workers). Hard budget caps, such as global budgets for hospitals and budget envelopes for doctor services, continue to be used throughout most of the region as well as in many Western European countries (Culyer 1989). This has markedly reduced health care expenditure both in real terms and relative to GDP in some countries. For example, in Poland caps led to an annual drop in health care expenditure of approximately 10 percent in real terms between 1990 and 1992.

As countries begin to introduce diversified sources of health financing and more complex performance-based reimbursement for health care providers, controlling health care expenditure through budget caps is becoming more and more difficult. For instance, different variants of the decentralized German health insurance model (sickness funds), under which doctors and institutions are paid according to the number of examinations and procedures they perform, are becoming extremely popular throughout the region. German policymakers control health care expenditure under this model through a complex process of managed price fixing. Negotiated agreements between the government, the sickness funds, and the medical profession, rather than market forces, determine prices. Undesired increases in the number of examinations or procedures are, in this way, offset by downward adjustments in prices. Even using such complex methods, many Western European countries have difficulty controlling health care expenditure (Abel Smith 1992). The Czech Republic introduced a similar model in 1992, which replaced salaries by carte blanche fee-for-service payments, before having adequate information systems to track costs or establishing a process for negotiating prices. The entirely predictable result was that within a few months, the authorities had completely lost control of health expenditure.

MOBILIZING RESOURCES. During the early stages of the transition, the health sector in most countries had enough internal reserves to withstand significant budget cuts without seriously compromising the quality of care provided. In the face of continued budgetary restraint and explosive increases in prices, especially in the case of imported pharmaceuticals and equipment, this is no longer true. Countries like Bulgaria, the Czech Republic, Hungary, the Slovak Republic, and several countries of the former U.S.S.R. have, therefore, turned to public—and to a small extent private—health insurance and direct charges to shore up the dwindling state budget.

This silver bullet solution of health insurance did not mobilize financial resources for the health sector for two reasons: (1) the health insurance contributions in countries like Hungary usually replaced rather than supplemented budgetary sources of health finance, and (2) rising unemployment and a growing informal sector in countries like the Slovak Republic reduced compliance and eroded the original contributions base. As a substitute for the state budget, social insurance for health care is, therefore, no panacea as an additional source of finance. When the government is forced to bail out an insolvent health insurance fund, as in the Czech Republic, shifting health care financing from the state budget to national health insurance may even damage other efforts to contain public expenditure.

Medical doctors in countries which have not introduced health insurance are looking with envy at the incomes of their Western European neighbors, while ignoring the negative experiences of other countries in Central and Eastern Europe. They continue to lobby their Parliaments for German-style health insurance. In the poorer countries of the region, such as Bulgaria, the

Slovak Republic, and the southern crescent of the former U.S.S.R., a rapid collapse in the collection of contributions would have a devastating effect on equity and the quality of care provided. In many of these countries, governments are realizing that state budget support for the health sector is unavoidable and that giving up cost controls under the assumption that health insurance will regulate itself is courting disaster.

Building Markets

Governments in Central and Eastern Europe are finding that it is neither desirable nor possible to insulate the health sector from the market forces sweeping the region. Many are turning to radically new health policy paradigms which can be grouped under three headings: (1) policies to liberalize demand by increasing consumer choice and redefining the role of governments, patients, health care providers, and enterprises as active partners in the health sector; (2) policies to diversify supply by establishing a new enabling environment for nongovernmental and private sector activities in the ownership, production, and financing of health care; and (3) policies to improve incentives in a new competitive market for health care. The individual elements of these policies are summarized in table 12-5 and described in greater detail in the following sections.

The strategy to use market forces in the health sector is double edged: on the one hand, it offers a chance for new approaches to improving health and the performance of health services; on the other, it carries an immense risk of reopening the Pandora's box of well-known market imperfections discussed in chapter 2. The budgetary crisis which occurred in Poland in 1992 because of a

Table 12-5. *Policy Options to Improve Market Forces in the Health Sector*

Policy area	Extreme policy options	Health paradigms
Consumer choice	State vs. individual	Partnership
Governments	Monopolist vs. minimalist	Stewardship
Patients	Recipients vs. consumers	Active participants
Providers	Employees vs. vested interests	Preferred providers
Enterprises	Mini-states vs. capitalists	Responsible agents
Diversified supply	Planning vs. marketplace	Managed competition
Economic goods	Public vs. private	Mixed goods
Ownership	State vs. providers	Public/private mix
Production	Planned vs. supply and demand	Public/private mix
Financing	Taxation vs. insurance	Public/private mix
Incentives	Normative vs. input and output	Outcome-based
Incentive structure	Bureaucracy vs. personal gain	Efficiency
		Cost-effectiveness
		Quality and satisfaction

sudden increase in the price of pharmaceuticals and the recent breakdown in the supply network in Russia are only the tip of the iceberg.

Because of such market failures, many Western countries have recently begun to use managed competition, or managed care, to create a restrained, but competitive, environment in which the advantages of market forces can be exploited, while the undesirable effects are controlled (Enthoven 1988). Governments in Central and Eastern Europe are discovering that market forces are an excellent way to improve efficiency, but only if accompanied by regulation, some public production, and targeted subsidies to minimize the associated abuses and deficiencies (Evans 1984).

INCREASING CONSUMER CHOICE. As a first step, many countries are improving consumer choice by redefining the role of governments, patients, health care providers, and enterprises in the health sector. In the past, both state and individualist extremes led to many unsatisfactory outcomes. The communist era provides ample evidence that well-being and health are poorly served when consumer choice is totally suppressed. At the same time, experience from Western countries over the past century provides equally good evidence that excessive reliance on consumer sovereignty has its own shortcomings. The changes require redefinition of the role of government, of patients, of health care professionals, and of enterprises.

The central government in the Czech Republic, Hungary, and Poland has already relinquished much of its previous monopoly by transferring the ownership of most health care facilities and the responsibility for providing services to local authorities and the private sector. Likewise, responsibility for financing health care is being transferred from the state budget of the central government to decentralized and semi-autonomous health insurance funds. Instead of withdrawing completely from the health sector, however, governments in these countries are assuming a new and important role as regulators or stewards of the emerging nongovernmental and private sector activities. Whatever the policy developments in a particular country, governments in all cases remain responsible for putting policies and programs into place in order to protect equity and the quality of care as patients are becoming more active participants in securing health. This is especially true since patients are not well informed about the cost-effectiveness of most treatments, and supply shortages would cause severe distortions in the market value of treatment without such intervention.

All the key players are adapting quickly to the new context by responding to price signals in a competitive environment. Patients are becoming active consumers of health care, in contrast with their passive role under the old system. In the Czech Republic, patients have gained increased freedom to choose their doctor and mode of treatment. The remaining barriers are created by geographical and resource constraints.

Health professionals, too, are responding to price signals. In Hungary, health care providers are negotiating the place of work or mode of practice with local governments and the health insurance fund. In most countries, enterprises and governments are choosing the extent to which they participate in health care activities. Although many doctors sit in Parliament and on influential legislative committees, there has been little self-regulation of the health care professions. Even in Poland, where syndicate activities started early under Solidarity during the 1980s, self-regulation is not effective because responsibility for conducting professional activities and enforcing professional standards is vested in single professional organizations such as the Physicians Chamber, Nurses Chamber, or Pharmacists Chamber. The strict internal controls necessary to prevent abuses such as supplier-induced demand, especially where payment is on a fee-for-service basis, are often missing.[5]

During the socialist years, the boundaries between the enterprise and the state were blurred: both were responsible for the welfare of workers, and both provided extensive and elaborate networks of health services, social programs, and education. In some countries, such as the former Soviet republics, industrial subsectors like mines, railways, police, and military provided health services for up to 10–15 percent of the population, financed largely through direct or indirect public subsidies. As enterprises become preoccupied with making a profit and avoiding bankruptcy, they are seeking to withdraw from their position as mini-states within states. The potential, sudden collapse in health services of these sectors would require massive and expensive restructuring; this is a time bomb which few governments have taken adequate steps to defuse. Nor have they sought to ensure that enterprises assume a new role as responsible employers who provide safe workplaces and protect the environment in line with international standards.

DIVERSIFYING SUPPLY. Governments are finding that diversification of supply requires major changes in the way they treat health care as an economic good. It also requires major changes in ownership and financing.

Health care as an economic good. As discussed in chapter 2, private goods exhibit excludability (an individual can be prevented from consuming a good until he or she has paid for it), rivalness (consumption by one individual prevents consumption by another), and rejectability (an individual can choose to forgo consumption). True public goods have significant elements of non-excludability, non-rivalness, and non-rejectability. Mixed goods have some but not all of the characteristics of private goods. A breakdown occurs in efficiency, equity, and sustainability when public goods or goods with significant externalities are allocated through competitive markets. The reverse problem—when private goods are allocated by central planning—is one of which Central and Eastern Europe needs little reminding: all have experienced low-quality health care, black money, queues, and the like.

Governments in Central and Eastern Europe are finding that health care is not a homogeneous private good which can be uniformly submitted to market forces. Policymakers have to answer a number of major questions:

- Which sector, public or private, should *produce* the service in question?
- Which sector should *finance* the service? In particular, what charges, if any, should be made to patients?
- Should patient choice be constrained, for example, by requiring family doctors to act as gatekeepers for treatment by specialists or by restricting drug subsidies to a range of basic drugs?

Different types of health activity involve different packages of these elements.

Public health services, such as sanitation services, control and prevention of communicable diseases, and health promotion, and other activities, such as research and development and professional education, have significant public goods characteristics. Yet some of the elements which make up these services—such as clean water, collection of refuse, immunizations, public health campaigns, individual research projects, and postgraduate training—may have sufficient characteristics of private or mixed goods to be subjected to market forces. Some of these elements are, therefore, usually sold as public utilities rather than given away as public goods. Hungary, for example, has recently introduced a controlled internal market by establishing a competitive process for funding research. Nongovernmental organizations are being contracted to conduct public health campaigns in Poland. And private pharmaceutical companies are being provided with incentives to produce desired vaccines in Russia. In the case of such public or near-public goods, production can thus be public or private, consumers typically do not pay charges, and consumption is relatively unconstrained.

Hospital care, including expensive diagnostic and therapeutic care, although in principle a private good and hence marketable, has not been privatized anywhere in the region. Such action would lead to politically unacceptable inequity and allocative inefficiency if left to market forces alone. In this case, production was historically mainly public, although that is changing; consumer charges historically were mainly unofficial, though again that is changing; and consumer choice was constrained by ambulatory clinics acting as gatekeepers.

Ambulatory care and long-term residential care were generally treated as private goods even under communism because it was difficult, if not impossible, to prevent their sale in the informal economy. Following the collapse of the socialist regimes in the Czech Republic, Hungary, and Poland, the new democratic governments were quick to introduce legislation which allowed the private production of medical drugs and supplies and the services of general practitioners and pharmacists. In such cases, supply is becoming increasingly private, and the role of consumer charges is growing.

One of the great challenges facing governments in Central and Eastern Europe is to decide where to draw the boundary between different types of health care goods and services, especially in the case of mixed goods, which are open to a fair amount of interpretation. The choice made in the Czech Republic, which now charges fees for access to private clinics and hospitals, is quite different from that of the Slovak Republic, which still provides free access to public clinics and hospitals. A second challenge is to develop effective policies for dealing with the various types of abuse as they arise. Governments throughout the region have so far been unsuccessful in introducing effective policies to deal with gratuities, bribes, and the emerging informal health sector economy.

Ownership and finance. Many countries in Central and Eastern Europe are also introducing major changes in the ownership and sources of financing for health care. In fact, much of the striking variability in the health systems in both Western countries and Central and Eastern Europe can be explained by the mix of public and private production of health care. Table 12-6 presents this mix: the columns show different types of financing of health care, and the rows show the different types of ownership.

Interestingly, none of the health systems in the highly industrialized countries or in Central and Eastern Europe falls wholly into a single cell. For example, the systems in the United Kingdom and the Nordic countries are primarily in cell A, but also have components of B and C, and J to L, in their public sectors. Their private sectors include a range from D to L. Canada and Australia are mainly in B and C, but likewise include D to L in their private sectors. In Canada, J to L are prohibited by law for standard services in the public sector, but direct charges are levied for above-standard services, pharmaceuticals, and many other goods and services provided through ambulatory care. The systems in some continental European countries are mainly in cells D to F, but include A to C through the public sector for targeted services

Table 12-6. *Typology of Ownership and Sources of Financing for Health Care*

	Source of financing			
	Public financing		Private financing	
Production	Prepayment	Prepayment	Prepayment	Direct charge
	Government General revenue	Statutory Social insurance	Regulated Private insurance	Competitive Out-of-pocket user fees
Purely public	A	D	G	J
Private nonprofit	B	E	H	K
Private for-profit	C	F	I	L

Source: Adapted from Reinhardt 1989.

and populations. Australia, Canada, France, Germany, and the United States are similar in their use of health insurance but very different in the ownership of health care services. Although health care in most of Central and Eastern Europe is currently concentrated in cell A, about 10 to 15 percent of the private sector and black market activity takes place in cells J, K, and L. As the countries move toward health care systems with a more diverse public/private mix, they are already beginning to occupy other cells in this schema, especially cells D, E, and F, given the tendency to move from general revenue financing to that of national social insurance.

The source of financing for the health sector in the highly industrialized countries and Central and Eastern Europe falls simultaneously into two broad categories: direct charges and third-party prepayments. Direct charges are made in the form of official user fees or unofficial gratuities paid directly to health care providers when services are rendered.[6] Third-party prepayments are made through taxes, social insurance payroll levies, and earmarked health insurance contributions paid by individuals, families, and employers to intermediaries (the state budget, social insurance funds, or private health insurance companies) before services are rendered by health care providers (hospitals, clinics, doctors, and pharmacies). In the West, most prepayment schemes are subsidized either directly through deemed contributions covering the noncontributing population (the poor, unemployed, and elderly) or indirectly through tax credits. Likewise, the new health insurance schemes in the Czech Republic, Hungary, and Russia also rely on direct subsidies to cover the nonactive parts of their population.

IMPROVING INCENTIVES. Governments are quickly discovering how to let the genie out of the bottle in a market economy by improving the structure of incentives. Incentive structures in the health sector are, of course, nothing new in Western countries. Positive incentives include profits, subsidies, professional recognition, and special status. Negative incentives include losses, fines, professional disapproval, and exclusion. As parts of the health sector are being privatized, and wage differentials are increasing, such incentives are beginning to have an effect.

The incentive effects of third-party reimbursement of services supplied on a fee-for-service basis are highly conducive to exploding costs. Western countries such as the United States, where similar problems are observed, have tried to deal with the problem through regulation and countervailing incentives, for instance, by encouraging health care providers to join together in preferred provider organizations, which offer a predetermined basket of care for a fixed price (see Sandier 1989 and, for general discussion of the underlying principles, see Barr 1992, sec. VI[B]).

In Hungary, recent changes in the mechanism used to pay health care providers (capitation payments for family doctors, fee-for-service payments for specialists, and diagnosis-related groups for hospitals; see the glossary) have had a much greater impact on the character of service delivery than earlier changes in ownership and the introduction of contributory health insurance. A well-run prepayment scheme can offer good coverage and comprehensive care, as well as low administrative costs and good control of expenditure. A poorly run scheme can be expensive to manage while providing poor coverage and incomplete care. There is, therefore, no ideal model for the countries in Central and Eastern Europe to follow.

The underlying principle for governments which want to introduce such incentives is, therefore, simple: use them carefully because they will almost certainly be more powerful than expected. Governments which use incentives skillfully as thermostats to turn desired behavior on and off can greatly improve both efficiency and equity. For instance, outcome-based incentives reward providers who improve the health of their patients, not just those who see more patients each hour. The idea is that the reimbursement regime should be based on outcomes not on inputs, giving doctors an incentive to improve the health of their patients instead of merely rewarding them for the amount of treatment provided irrespective of its cost-effectiveness.[7]

Governments which use incentives poorly will quickly find themselves facing escalating costs due to supplier-induced demand and reduced efficiency as resources are moved into narrowly defined areas of low priority. The cost explosion in the Czech Republic was entirely predictable and could have been prevented though the introduction of appropriate incentives such as preferred provider organizations and of adequate mechanisms to cap the budget. The incentive structure certainly increased the productivity and income of doctors but led to few gains in the quality and cost-effectiveness of their interventions.

Regulating Market Forces

Market imperfections in health care are pervasive:

- Consumer information is highly imperfect, creating serious problems of quality control.
- Workers have only limited power to insist on safe working conditions, and, without appropriate sanctions or regulations, employers have little incentive to improve safety.
- Major problems with externalities mean that, although environmental cleanup is in the interests of society as a whole, individual firms have no incentive to produce in a less polluting way.
- Private insurance is unable to cover some important medical risks.
- Third-party reimbursement, particularly in a fee-for-service environment, creates incentives which can lead to uncontrolled escalation of costs.

In addition, and separately, in a market system the poor are likely to be excluded from medical care because of its cost.

Examples of the ill effects when these problems are ignored are legion. At least 35 million Americans have inadequate medical insurance coverage; costs are difficult to contain; and the massive dumping of low-quality drugs and the importing of defective medical equipment have already occurred in Poland and other countries in Central and Eastern Europe because the government does not—or cannot—enforce adequate standards for quality control.

INTRODUCING A REGULATORY STRUCTURE. Because the problems are so pervasive, policies to regulate market forces have to be an integral part of measures used to harness market forces. Thus they emerged repeatedly in the discussion of the previous sections and are recapitulated here only briefly. State intervention, apart from income transfers, is of three types: regulation, finance, and public production.

Regulation is necessary in a variety of forms:

- Quality control is made necessary by imperfect information. The government has a stewardship role which includes controlling the quality of medical services generally, ensuring that professional standards are upheld, and maintaining regulatory regimes for the testing, production, and sale of drugs.
- Containing medical expenditure in the ways described is necessary because of the third-party payment problem. Regulation of medical spending can be imposed at the level of the total system, as in the United Kingdom, or at the level of the hospital, as in Canada, or at the level of the individual provider, as in Canada or Germany.
- Environmental action is needed because of the externality problem; such policies need to be stimulated both by appropriate incentives and by appropriate regulatory structures.
- Safety at work requires regulation *and* enforcement to ensure minimum standards.

Public funding of medical care is necessary for two very different sets of reasons: to ensure the provision of public goods and to ensure wide-ranging access:

- Public health activities such as vaccination programs and public health education are generally largely, if not wholly, publicly funded.
- The non-active populations with low income require subsidies, particularly the elderly.
- The poor also require subsidies.
- Drugs, particularly limited drug formularies, are generally subsidized.
- Health finance, more generally, is frequently government run (tax funding) or government mandated (social insurance), to address failures in the private health insurance market and to ensure adequate access for the poor.

Public production has various aspects:

- Public education is needed to counter imperfect information. Examples include programs to inform the population about diet, physical exercise, and the ill effects of alcohol, narcotic drugs, smoking, and the like.
- Other public health activities, such as maintaining clean water, need to be sponsored by the government, even if some or all of the production is in the private sector.
- Medical education and medical research, having significant public good attributes, also involve some government activity.
- Public clinics may be necessary to ensure that health care is available in rural areas and slums.

EXPERIENCE ELSEWHERE. Three additional observations from the highly industrialized countries have a direct bearing on future health policies in Central and Eastern Europe.

Public funding, whether through taxation or social insurance, is the major source of health financing in the OECD nations. Countries like the United States which rely to any great extent on private risk-rated health insurance (cells G, H, and I in table 12-6) appear to have much greater problems (a) with ensuring equality of access—because of uninsurable risks and the poor—and (b) with containing costs—because of the third-party payment problem—than countries which rely on other sources of health finance. At the same time, countries like Sweden and the United Kingdom which rely heavily on public ownership (cells A, D, G, and J) appear to have much greater problems with efficiency and productivity than countries which rely more heavily on nonprofit and private ownership. Health care financing mechanisms which rely on general revenues or national health insurance, combined with some direct charges and nonprofit or private ownership, would appear to be a much better choice for Central and Eastern Europe than private health insurance and public ownership.

Direct charges on their own are not effective as a major source of finance. Providing financial protection against the unpredictable risk of illness and the high cost of modern health is the main reason why pooling risk through third-party prepayment schemes has become the cornerstone of health finance over the course of the twentieth century. Direct charges, although they may play an important role in financing less expensive care and discretionary services, provide neither adequate protection against the risk of catastrophic illness nor a sufficient source of financing for expensive health care, typical of modern health systems. As a result of these factors, direct charges contribute less than 20 percent of the financial resources available to the health sector in most Western countries. Governments in Central and Eastern Europe must balance the expected benefits of introducing copayments as a disincentive against excessive use and as a source of additional revenue against their negative impact on vulnerable populations during the transition. So far, none has formalized the gratuities or black payments which are characteristic throughout the region.

No system is perfect. Significant tradeoffs between efficiency and equity are associated with each of these mechanisms. Greater equity may be achieved through the risk-pooling characteristic of prepayment schemes. Greater efficiency may be achieved through the market forces associated with direct charges. Governments in Central and Eastern Europe, like those in the highly industrialized countries, are finding that neither model by itself is ideal.

The main message is that in the case of medical care, a carefully designed blend of market forces and government intervention is needed. As market forces are introduced in Central and Eastern Europe, problems with cost containment, quality assurance, and equity are beginning to appear. To counter these effects, many governments are now rewriting the social contract for the health sector based on policies which (a) protect highly vulnerable populations, such as mothers, children, and the elderly, through targeted entitlement to specially designed services; (b) secure affordable access to a basic basket of cost-effective health services for the whole population; and (c) distribute the financial burden of illness across the population as a whole.

Implementing the Reforms

Alongside policies to improve health and those to increase the cost-effectiveness and equitable distribution of health care, a third set of policies seeks to improve the political and institutional capacity of the health sector to implement relevant reforms. This effort includes introducing new legislation, strengthening the physical and administrative infrastructure of institutions, and reorienting the training of personnel.

Most governments have introduced a new legal framework for the legitimate entry of the private sector (private ownership and provision of health services), syndicate activities by the medical profession and other professional groups, and decentralized decisionmaking. Poland has passed the Health Care Institutions Bill legalizing private practice, Hungary and the Czech Republic have transferred ownership from the central government to local communities, and Russia has vested Oblasts with almost autonomous power over their health services, similar to the Canadian provinces or the states in Australia.[8] The Physicians Chambers in Poland was the first professional body to gain the status of a self-regulating medical syndicate, with responsibility for setting professional standards and negotiating collectively. Many countries are passing new regulations to control the quality of pharmaceuticals and medical equipment.

Although countries such as Russia and other former Soviet republics are still struggling with critical imports of pharmaceutical products and other nondurable goods, the Czech Republic, Hungary, and Poland are concentrating on building an infrastructure for the health sector. This includes new diagnostic and therapeutic equipment to improve clinical interventions, computer systems to improve cost control and management, and limited public works to consolidate outmoded and inefficient physical facilities. Since catching up with standards in Western Europe would cost billions of dollars, governments are finding that it is critical to establish clear priorities. Without such priorities, every community hospital will attract gifts of CAT scanners or other expensive high-technology equipment from bilateral donors without having the recurrent budget to operate them effectively.

As many countries in Central and Eastern Europe are decentralizing and privatizing their health services, they are finding that clinical and nonclinical personnel need much broader training in such fields as health policy, management, health economics, chronic disease epidemiology, computer science, and medical sociology than was either necessary or possible in the past. Modern management skills and data analysis are required to cope with the complexity of many of the new health insurance systems. The success of many of these reforms will lie in a fundamental new orientation in the education system and applied research in the health sector.

Conclusions: Priorities and Sequencing

In most of Central and Eastern Europe, a collapse in the public enterprise sector, inflation, and unemployment have reduced the real income of large segments of the population and created a fiscal crisis. The resulting poverty deprives the population of exactly the healthy living arrangements, diets, and life-styles which are necessary for good health. At the same time, reduced tax revenues make it impossible for governments to maintain their historical commitment to public expenditure programs like health care, just at the time when they are most needed. Of necessity, therefore, most governments are cautiously introducing reform in the health sector in two phases.

Short-term Policies

The first phase (up to two years) concentrates on urgent measures which need to be taken or avoided to survive the early transition. The following actions are not intended to be exhaustive but rather to focus on critical measures which must be taken by government to avoid a collapse of the health sector and to strengthen its role as part of the social safety net.

Policy should ensure continuing delivery of basic health services and cost-effective acute critical care. This includes:

- Ensuring that the share of GDP devoted to the health sector is maintained during the transition, either through the state budget or through some form of national health insurance.

- Providing immediate relief of critical shortages in areas where the health system is collapsing. Of particular importance is action to guarantee the continuing availability of essential vaccines, drugs, and supplies.
- Allocating an adequate budget for services designed to protect vulnerable populations. These include maternal and child clinics, immunization programs, and social services for the elderly.
- Selecting and implementing cost-effective corrective interventions, especially for cardio/cerebrovascular and related diseases, accidents, and pulmonary diseases. This includes maintaining properly equipped and staffed ambulance services, emergency rooms, and intensive care units.

Measures to contain costs should focus particularly on avoiding actions likely to lead to a cost explosion. This entails:

- Introducing hard budget caps, such as global budgets for hospitals and salaries, or capitation payments, for doctors and other health care workers. Such action is vital to contain costs during the transition. Other forms of reimbursement, such as fee-for-service payments and itemized retrospective payments, are difficult to control and, unless accompanied by wide-ranging and sophisticated regulation, will lead to exploding costs.
- Avoiding the fool's gold of using health insurance as a way to mobilize additional financial resources for the health sector. The introduction of such a mechanism would be premature. During the early transition, the potential contributions base was shrinking because of rising unemployment and a growing informal sector; in addition, an adequate mechanism for collecting contributions suitable for a system of health care finance based on social insurance has yet to be put in place in most countries.

Medium-term Policies

Policies with a medium-term dimension could be started immediately, or in the near future, if the short-term issues just described are addressed at the same time. Since their impact will be mainly in the medium term, policymakers must ensure that measures taken to restructure health services do not destabilize and lead to a collapse of the sector during the transition. Excessively rapid decentralization and changes in the financing mechanisms could have just such an effect, particularly if they are introduced before the institutional capacity necessary to implement such changes has been built up. Many of these, therefore, are being left for a second phase of the reform process. The actions listed below are of four broad sorts: those which relate mainly to improving the quality and mix of actions to improve health, those aimed at improving the efficiency with which those services are delivered, those which relate mainly to health finance, and those intended to improve institutional capacity.

New and more effective approaches to public health and disease prevention should be introduced. Actions include:

- Mobilizing greater community participation in health promotion and prevention programs
- Launching national campaigns to promote healthier eating habits, to reduce alcohol and substance abuse, to discourage smoking, and to encourage greater physical activity
- Providing safe alternatives to abortion through improved family planning programs
- Coordinating an intersectoral approach to occupational and behavior-related illness and accidents.

Imbalances between public health, primary care, institutional care, and community services require correction. The emphasis should be on:

- Setting up or improving training programs in primary care, public health, and community services
- Increasing the intake of students into these programs and restricting the intake into specialist training
- Raising the income of primary care, public health, and community service workers relative to other clinical specialties
- Increasing the relative weight of investment in primary care, public health, and community service facilities and programs relative to hospital-based care
- Setting up professional bodies to improve the status of individuals who choose a career path in these areas.

Efficiency should be improved with particular focus on:

- Containing costs through regulatory mechanisms and hard budget caps
- Stimulating productivity through performance-based reimbursement for the health sector, such as capitation payments for general practitioners, case-mix adjusted budgets for hospitals, outcome-based reimbursement for specialist services, and changes in ownership (transfers to local government or privatization)
- Encouraging competition among providers.

A new public/private mix should be established in the provision of health services. This would involve:

- Creating a legal framework which facilitates appropriate private sector activities in the health sector
- Setting up accessible loan facilities for doctors, dentists, laboratory technicians, and other health workers who want to set up private practice or private clinics
- Removing subsidies from the public production of pharmaceutical products, medical equipment, and supplies

- Introducing publicly mandated financing of privately owned health care facilities.

The sources of health care financing should be diversified by:

- Introducing national or government-mandated contributory health insurance
- Adding copayments to some goods and services
- Excluding above-standard services from publicly financed programs
- Eliminating public subsidies such as tax credits from nonessential services and private health insurance.

Institutional capacity should be strengthened through legislative reform; consolidation, rehabilitation, and renewal of basic infrastructure; modernization of equipment; upgrading the training of personnel; and improved quality control systems.

Again, this list is not intended to be exhaustive but rather to focus on early action which can be taken to strengthen and reconstruct the health sector. Even in the medium term, few governments are planning to expand dramatically the physical infrastructure of their health sectors. Such investment is being postponed until there are signs of significant economic recovery, which, in some countries, may be several years away.

Notes

1. Being interrelated, it is extremely difficult to attribute improvements in health status to any one of these factors.
2. Life expectancy at age one rather than at birth is used because infant mortality (mortality before age one year) was not accurately measured in some countries in the region and because the relatively poor performance of the communist countries is particularly evident for mortality beyond one year of age.
3. A normative-based system would finance hospitals on the basis, for instance, of the number of beds; a performance-based system of reimbursement would pay hospitals in a way related to the number of patients treated. The incentives given by the two methods are clearly different: the former encourages hospitals to have lots of beds, but to keep them empty; the latter encourages hospitals to give large amounts of treatment, whether or not it improves health.
4. For example, as discussed in chapter 5, an adverse report by a safety inspector could be interpreted as sabotage of the plan.
5. As discussed in chapter 2, retrospective reimbursement for medical providers who operate on a fee-for-service basis creates incentives to oversupply; if uncontrolled they can easily lead to exploding medical costs.
6. User fees are also referred to loosely and inconsistently as tariffs, copayments, deductibles, *ticket moderateur* (France), and cost recovery. They are usually legal charges used to supplement income from third-party prepayment schemes in both industrial and developing countries. Gratuities are also referred to as black money, under-the-table bribes, tips, and tokens. They are often illegal charges used to supplement doctors' salaries.

7. Many Western countries are experimenting with funding regimes which give providers the incentive to keep patients healthy. Health maintenance organizations (see the glossary) are one example. Along similar lines, reforms in the United Kingdom in the early 1990s allowed some family doctors to act as the funding agency for certain types of hospital treatment, again giving them incentives to keep their patients healthy.

8. Russia is divided into eighty-four Oblasts.

Thirteen

DRIVING CHANGE: POLITICS AND ADMINISTRATION

IAIN CRAWFORD • ALAN THOMPSON

There is nothing more difficult to arrange, more doubtful of success, and more dangerous to carry through than initiating changes in a state's constitution. The innovator makes enemies of all those who prospered under the old order, and only lukewarm support is forthcoming from those who would prosper under the new. [Machiavelli 1513]

Stability is not everything, but without stability there is nothing. [Schiller]

SUCCESSFUL REFORMS ARE WHAT THE PEOPLE EXPECT. That was why they defied the tanks. It is now a question of how soon they can expect reforms to improve their individual welfare. The real problem was the popular assumption that their lives would be like those in the West within weeks, months, or at the worst a year or two. Now that these illusions have been punctured, it is important that they be replaced by realistic hopes and that the economies can be seen to improve measurably as quickly as possible. People must believe that the short-term sacrifices are just and that they will be worthwhile in the long run.

Suppose, then, that all the relevant ministers decide to adopt the recommendations in the previous chapters. Would success of the reforms be guaranteed? The answer is, not necessarily unless broader political and administrative problems have been solved.

The reforms will not succeed without sufficient support, not just for the general idea of reform but also for a significant proportion of the specific measures necessary to put that general idea into effect. The government, for instance, has to convince Parliament of the need for pension reform. In this, as in other matters, Parliament will be acutely conscious of the views of the electorate. The major political problem, as the quote from Machiavelli makes clear, is that government has to cope not only with individuals and groups who are actively hostile to the reforms, but also with the apathy of people who *would* support the reforms if they had any faith in their ultimate success. Without such support, government may fail to deliver a policy for political reasons, such as the refusal of Parliament to pass a stringent budget law. Even if government has the political support, it may fail to deliver the policy because of administrative shortcomings, such as an inability to implement a promised increase in pensions fast enough. The first is fundamentally a matter of political communication; the second is a matter of building an effective public administration.

In addition to convincing the electorate and Parliament about the rightness of their policies, departmental ministers have also to convince their ministerial colleagues. All the policies advocated in earlier chapters have major cost implications. Deliberately there has been little discussion of how a budget, enormously constrained by the fiscal crisis, should be divided among competing claims on resources. Those decisions will have to be taken by ministers in the countries concerned. Yet departmental policies do not operate in a vacuum. They are determined in large measure by interactions between departments and by competing political priorities:

- Each spending department will compete for scarce resources of finance, of administration, and of legislative time.
- The overall limit on public spending will be decided by the interaction of the spending departments and the Ministry of Finance. The total resources available will be a matter of economic circumstances and of government decisions about general levels of taxation, a high-taxing, high-spending economy (for example, the Scandinavian countries in table 1-3) or one with lower taxes and, consequently, a smaller public sector (for example, Japan).
- Both the total of public spending and its division among departments are affected by how departmental policies interact with overall government objectives. The government has to balance the demands of social policy with other demands such as infrastructure, the environment, industrial investment, and defense; it has to form views about the pace of reform (for example, about how much unemployment is politically tolerable); and it has to take account of its electoral popularity given the proximity or otherwise of the next election.

Alongside such questions about the *size* of the government sector are separate but related questions about its *structure*. Important decisions are needed about the dividing line between the market and the state. As discussed in chapter 2, market imperfections are a particular problem in the social sectors. Government thus has to make choices, which are political as well as economic, about such questions as whether the government should alleviate unemployment through selective subsidies to enterprises; how large the role of private pensions should be; the extent to which health care should be privately produced, even if it is largely publicly funded; and what the state's role should be in retraining displaced workers. In the social sectors, in particular, decisions have to be made not only about the volume of spending, but also about how and by whom the various services are to be produced. Some of these decisions will end up being made not for economic, but very much for political reasons.

These topics—political communication, strong public administration, and the competing claims of different spending priorities—though clearly far transcending the specifics of the social sectors, have a critical bearing on social policy. The success of social sector reform depends on the ability of government to garner sufficient political support for the overall reforms. It also rests on the ability of the public service to implement the administrative aspects of change. In addition, the success of reform in any particular part of the social sectors, such as health care, will depend on the success of the relevant department in its negotiations with other departments, particularly the Ministry of Finance.

The Objective: Political Stability

A primary prerequisite for all these tasks is political stability. Without it, none of them can be carried out effectively for long. Political stability is not, of course, an absolute condition, nor can all Western democracies claim to have consistently enjoyed it.[1] Stability, in this context, is not a condition where change cannot take place, but rather a situation in which the government and the governed are responsive to each other, and citizens have general confidence in the institutions of the state.

Given the speed and extent of the changes, the first transitional governments could not have hoped to achieve political stability immediately. The early period of transition, which combined economic and political reform, was (and in some places remains) revolutionary. Revolution is dangerous, unpredictable, often violent, and, as history tells us, uncertain of success. Balcerowicz (1993a) has described this as a period of "extraordinary politics," which he sees as a time when politicians have high political capital, and the electorate is ready to accept radical economic measures. This honeymoon, however, is short-lived and impossible to repeat, and the "normal politics" which follow in the second phase require normal political behavior. This may be seen as an argument for concentrating radical and dramatic change in the early days of reform, but that historic moment is over in most countries, and the daily grind of normal politics has arrived. This process may seem to the general public to be rather boring compared with the heady days of the early revolution, but this is the essence of stable, safe, democratic government.

In stable political systems change is usually incremental. Even the governments of Ronald Reagan in the United States and Margaret Thatcher in the United Kingdom, which were seen as fairly radical, were far from revolutionary. All the emerging democracies should aim to reach a plateau from which further development is incremental. Thus a prime objective of all policy advisers, international aid agencies, foreign governments, and politicians themselves should be to reinforce political stability, without which there can be no confidence in the institutions of the state. Without political stability, democracy itself is precarious. Without confidence, both domestic and international, the reforms are seriously hampered, investment is not forthcoming, and the potential for chaos cannot be ignored. Crick (1992) makes a powerful claim for the "science" of politics. He sees politics as *the* component essential to the organization of a stable society.[2] Without a system of political organization little else of an organized nature can take place.

Political stability is the prime objective, first, because of the potential consequences of its absence, both for the countries themselves and for the rest of the world. Some states possess considerable stocks of nuclear and chemical weapons, the accidental or deliberate use of which would have catastrophic global consequences. New disarmament agreements have been signed, but earlier agreements still have to be implemented. Many transitional states produce nuclear power with units of questionable safety. For the reasons discussed in chapter 5, environmental and pollution control was inadequate under previous regimes, and the consequences of an environmental disaster in a situation with little or no political authority could be catastrophic.

The replacement of the old regimes has already led to civil and international wars. Some have been the result of ethnic, religious, or nationalist disputes which had been bottled up by the previous totalitarian governments. They clearly illustrate some of the dangers inherent in the lack of strong government. Events in the states which formerly comprised Yugoslavia provide a terrible example of the consequences of instability. Nationalism is a clear threat to politicians seeking reform in Central and Eastern Europe, not least because the promotion of nationalist issues or the targeting of an "enemy within" is a tempting diversionary tactic for politicians unable to meet inflated expectations.

Political stability is of prime importance also because economic reform becomes much more difficult without it once the first flush of public enthusiasm has passed. The primary function of government is to govern and provide a safe and secure environment within which citizens can live their lives. In the new democracies governments must establish the rule of law and prove that they can enforce it. Reforms which establish property rights and depend on ownership as an economic motivation cannot work unless state agencies are both willing *and* able to protect the citizen and his or her property. Enterprise will not take place unless the entrepreneur can be sure that his property will be safe both from criminal threat and from future expropriation by the state. The same is true for foreign investment.

Constraints and Potential Pitfalls

Serious political and administrative constraints exist, some inherited from the communist regimes, others exposed by or caused by the transition process.

Political Constraints

Political problems can usefully be divided into four kinds: those resulting, directly or indirectly, from the inheritance; those resulting from political pressures; those resulting from a lack of clarity about the constitutional rules under which the political game is played; and those resulting from a lack of experienced professional players of the political game.

Alongside these, one additional (and almost universally underestimated) constraint is that politicians who entered politics committed to long-term economic and social reform in practice find themselves spending 90 percent of their time fire-fighting, that is, dealing with short-run crises. Many of the early reform policies were designed on the run. This is not unusual. In established democracies, politicians constantly have their carefully prepared plans blown apart by events beyond their control. Postwar British Prime Minister Harold Macmillan (known as Supermac for his skillful political leadership) identified "events" as the only aspect of politics which really frightened him.

THE INHERITANCE. The general inheritance was discussed in chapters 3 and 5. Two aspects of the political inheritance stand out: a top-down approach to government and single-party rule. A further aspect—a consequence of the transition—is the effects, particularly in Russia, of the loss of empire.

Top-down government. Government was an entirely top-down exercise. The supreme authority decreed, and the communist party machinery disseminated and implemented, policy at the local, national, and even international level. Two-way communication was neither necessary nor encouraged. Feedback is not a requirement of an authoritarian, top-down system. It is, however, a principal requirement of a government which is ultimately answerable to an electorate and dependent on public support. That said, some two-way communication did take place under the old system, at least at a local level, in that individuals with grievances, such as a missing pension payment, could often get help from their "elected" representative, who would ensure that his or her assistance received due recognition in the press. Party officials in rural areas, having to live with the effects of their policies, similarly responded to local needs. The individual representative's power, however, was very limited.

Single-party rule rendered any pretense of democratic elections pointless. The single-party system is the most potentially corruptive of genuine democracy. In Western countries, politicians compete for power within their parties, as they did in the old communist system, but, unlike that system, the parties themselves compete against each other for electoral support. This leads to the constant generation of new ideas and policies and to accountability to the electorate. Under single-party rule, such competition is nonexistent.

The end of empire. In the former U.S.S.R., and particularly in Russia, empire is an enormously important part of the political inheritance. Historically, large empires, notably the Roman and British, have gone into slow decline or have collapsed over a period of years. The collapse of the Soviet empire has been dramatic and, for its citizens, traumatic. If throwing off the imperialist power is one of the most compelling driving forces for change in Central and Eastern Europe, the Baltics, and the Asian republics, what of Russia? Soviet imperialism predates communism. The people of Russia have lived at the center of an empire for a very long time. Since the late 1940s they have been citizens of one of the two most powerful states in the history of the world. They have always had a strong sense of patriotism which was heavily reinforced by the experiences of the Second World (Great Patriotic) War. Until very recently, they had demonstrated a powerful sense of national pride, carefully fostered by propaganda. The shock waves emanating from the collapse of empire began with the loss of the European satellites and intensified with the breakup of the U.S.S.R. The fact that the old regime has not been replaced by strong government has left a tremendous void.

The situation in Russia is very different from that in postwar Germany and Japan, where a sense of national guilt contributed to a determination to bury the past and create a new future. In the circumstances of Russia, there is bound to be a sense of national grief—a crucial element of the political inheritance of current government and one which will have to be taken into account well into the future. This grief has at least two sets of political implications:

• Russia, while remaining a nuclear power with a permanent seat on the Security Council of the United Nations, has lost international prestige and influence and must find a new role on the world stage. Hence foreign policy will be an important factor in domestic politics, albeit not as important as economic and political reform. Foreign governments and international organizations should be particularly aware of this factor. The political signals will be confused: on the one hand, criticism of the West for not doing enough to help, and, on the other, resentment at being the recipients of charity.

• Whereas the populations of countries like Hungary, Poland, and the Czech and Slovak republics can comfort themselves with their new-found national self-determination, this is not true of the population of Russia, who may be much more inclined to look nostalgically to the days of the old regime. It may be that, although the change of regime is common to all, the lack of liberation effects will reduce the mass psychology which Balcerowicz cites as a factor in the level of readiness to accept radical economic measures. As Western politicians quickly discover, the public memory is often, but selectively, short. The recollection of being part of the powerful super state will remain with the population when memories of its shortcomings have faded, especially if the pain of transition persists for too long.

POLITICAL PRESSURES. Although the relative strengths of the effects differ, the politics of all the reforming countries are heavily influenced by four sets of forces: a distrust of government, pressures to decentralize, inflated expectations about the speed of change, and pressures to avoid excessive unemployment.

Distrust of government. Attitudes can be measured and will vary over time as issues come and go. Distrust of government is likely to be less severe in countries such as the Czech Republic which have a history of democratic culture or where repression was in large measure imposed by the U.S.S.R. In such cases, the new democratic governments should experience less difficulty in developing mutual trust with their citizens. Where the repression was severe and imposed by their own government, such as Romania and Russia, distrust will be harder to overcome. The problem will be even more serious in countries with substantial ethnic minorities, such as Azerbaijan, Georgia, Moldova, and the former Yugoslavia.

Pressures to decentralize. Centrifugal forces arise most acutely in the form of nationalist aspirations. The reaction against strong central control led to the demand for power to be devolved away from the central government to an extent which could make some states nearly ungovernable. National and ethnic minorities demand a degree of autonomy or even complete independence. Local regions, towns, and even villages are demanding powers over their own affairs. This is often a response to the breakdown of authority of the central government as much as a desire for local autonomy.

Unrealistic expectations. Governments run serious risks when they allow expectations to take root which cannot be fulfilled. This has been a problem since the first day, when the new leaders stood on the Berlin Wall and failed to acknowledge that things were going to get very much worse before they got better. Since then, politicians have had the impossible task of trying to satisfy vastly overblown public expectations.[3] While this is true for longer-term expectations, it is also true for expectations on a day-to-day basis. Western politicians who are about to announce some policy (an annual budget, for example) will encourage the press to speculate that the news is going to be very bad. This tends to ensure that the not-quite-so-bad news is better received than it otherwise would have been. Since public confidence is an important factor in producing economic upturns, this approach can be important for economic success. When democracy itself is very fragile, it is even more important.

The politics of unemployment. Given inherited attitudes toward employment, it was inevitable that expectations about unemployment would be disappointed and, in some countries, very badly disappointed. The move from a centrally planned to a market economy made unemployment inevitable in all the countries of Central and Eastern Europe. In Russia, the problem is aggravated by the existence of a large army.

The cultural inheritance discussed in chapter 3 is crucial in this context. It is useful to divide Central and Eastern Europe into two groups of countries: those which were industrialized only under communism, and hence had no previous experience of unemployment, and those, on whom communism was imposed only after 1945, whose prewar experience included industrialization and the interwar depression.

Countries like Hungary, Poland, and the Czech and Slovak republics had been industrial market economies before the Second World War and had experienced the interwar depression. Western levels of unemployment are therefore more likely to be acceptable to them. As well as the economic motivation for change and the desire for democracy, an additional motivation in the former Soviet satellites is the desire for national self-determination. The people of countries like the Baltic states, the Czech Republic, Hungary, and Poland are prepared to pay a price for freedom. Part of that price is acceptance of short- and perhaps even medium-term hardship, including levels of unemployment not experienced since the 1930s. Unemployment in these states is severe but, as discussed in chapters 7 and 8, is more likely to be susceptible to Western-style solutions than unemployment in the Russian Federation itself.

Russia, Belarus, and some other parts of the former U.S.S.R. did not experience the economic turmoil in the West of the 1920s and 1930s and have no history of mass unemployment. Thus there is an extreme reluctance to accept unemployment at the levels common in Western Europe, let alone those in prospect during the early transition. In those countries, unemployment is almost certainly a problem more to do with the "dignity of labor" than with short-term financial hardship. There is virtually no experience of idleness. Politicians should not assume that they can adopt policies leading to even low levels of unemployment by Western standards without running the risk of serious civil unrest; trade union leaders in Belarus, for example, spoke in 1993 of the unacceptability of unemployment levels of under 1 percent. Given that, in the new democratic conditions, public acceptance must be nurtured, unemployment will be a difficult area for policy in the foreseeable future. Self-determination is not a driving motivation for the population of Russia itself.

The relevance of Western experience in this context is very limited. Two policies, however, workfare and employment subsidies, should be mentioned as examples of the political aspects of unemployment. Workfare, which conditions unemployment benefits on compulsory training or public works, has been the subject of much debate in the highly industrialized countries. So far, because of its coercive nature, it has been politically unacceptable on anything but a small scale in the West. Such policies, however, should be considered in the context of the unique circumstances of Russia. Their absence may be politically problematic, to put it no stronger, for a number of reasons which, when taken together, are specific to Russia:

• Under the old system, everyone was given a job, so that people are not used to unemployment and are not likely to feel coerced by workfare.

- Reform will involve relocating perhaps tens of millions of people.
- The condition of the infrastructure and the environment is such that major public works programs could be productive.
- Large-scale demobilization of the armed forces is necessary.

The problem with the workfare solution is that public works programs on the scale required will need considerable organizational resources. It may be that in Russia and in one or two of the other countries of the former U.S.S.R., these organizational resources exist in the military and can be harnessed for the purpose. One way of gradually running down a large conscript army would be to demilitarize rather than demobilize units by allocating them to specific public works projects and reinforcing them with unemployed civilians from the locality. Such projects would thus have a core labor force which was organized, disciplined, and capable of building infrastructure and carrying out large-scale environmental cleanup. It might be argued that such a scheme could easily lead to the creation of the kind of non-jobs common under the old system. However, it may be better politically to run this risk than to allow a situation with millions of idle, disaffected, and potentially disruptive individuals and their families.

A clearly understood set of policies should be in place before a large-scale military demobilization or an industrial shakeout takes place. Mass unemployment is perhaps the most dangerous social phenomenon, outweighing even a short-term reduction of living standards. Even Western politicians tend to underestimate the dignity of labor aspect of unemployment. It is unlikely that unemployment benefits beyond the absolute minimum levels can be afforded, and this is unlikely to be sufficient to remove the political dangers. These political arguments should be set alongside discussion of the role of public works and retraining in chapter 8.

Employment subsidies, discussed in chapter 7, are another approach, potentially applicable in all the countries of Central and Eastern Europe just as they are used selectively in the West. Such subsidies are another policy which is just as likely to be adopted for political as for economic reasons. For the reasons just discussed, unemployment is felt even more keenly in former communist than in Western countries; and the problem is even worse in single-industry towns or in regions which are highly dependent on a single or closely related set of industries. Not only is such unemployment economically disastrous and, on that scale, unnecessary for restructuring; it also creates entire regions intensely hostile to the reform effort. The result is opposition to the central government and, in some circumstances, pressures to secession.

CONSTITUTIONAL ISSUES. Although constitutional issues lie, for the most part, outside the scope of this book, some points cannot be ignored, since the lack of appropriate instruments can cause serious problems of governance.

Issues which are particularly relevant concern electoral arrangements, the idea of a mandate, under which politicians are held accountable for past promises, and the arrangements which determine the relations between central and subsidiary levels of government.

First-past-the-post electoral systems (see the glossary) most often lead to governments with a majority sufficient to ensure the passage of legislation. The system usually prevents the proliferation of parties in Parliament. The main disadvantage is that the system makes it almost impossible for new parties to emerge and difficult for established third parties to break through to power. Proportional representation (see the glossary) can avoid these problems but may cause other difficulties, some of which have emerged in Central and Eastern Europe. If the system has no threshold mechanism to limit the number of small parties, the result (as in Poland before 1993) can be fragmented parliamentary representation and, consequently, difficulty maintaining a stable coalition.[4] Italy, which has had such a system since the Second World War, has decided to move toward a first-past-the-post system.[5]

A different set of problems arises where electoral arrangements allocate parliamentary seats to specific groups (in Belarus, for example, a significant number of seats was reserved for representatives of disabled groups and pensioners). Pressure groups are an important and inevitable component of a pluralistic society, but their place is *outside* Parliament. At least two major problems arise when parliamentary seats are reserved for the representatives of specific groups: these representatives will seek to block any constitutional reform which would deprive them of their seats, and they will try to prevent any legislation adversely affecting the group they represent. Thus, for example, a country might have a high pension and a low retirement age, which in the short run is fiscally unsustainable; the pensioners' representatives in Parliament could block, if necessary by logrolling, any legislation to reduce pensions or to increase pensionable age.[6]

A mandate is not a concept usually enshrined in the constitution, but in most Western states, it has quasi-constitutional status. The underlying idea is that a government is elected on the basis of a previously stated ideological position and on a menu of policy proposals which are at least broadly outlined at the time of the election. If no mandate exists, or none is adhered to, there can be no democratic accountability. This is especially true if elected politicians are not linked with a constituency. Where the concept of a mandate is absent, major swings in policy are frequent when a new minister takes office. Short-run swings in policy (as opposed to gradual evolution of policy during the lifetime of a government or a sharp change in policy upon the election of a new government) do not serve the reforms. Such swings are inevitable in the short term; in the longer term, they may engender distrust. As the early pace of change slows and nuances of policy positions become more important, greater stability of policy should become easier.

The relationship between central and subsidiary levels of government raises a large range of issues, including that of unitary versus federal states and that of the balance of power between central and local government. A key problem for countries in Central and Eastern Europe is that they have inherited an inability to decentralize. At least two problems arise. First are centrifugal pressures which can overshoot. A national policy framework is desirable in some areas, even if implementation allows for local discretion. Examples include systems of unemployment benefits which assist labor mobility and the setting of standards in health care and education. In the extreme, the inability to maintain a national framework can lead to political instability. Second, two very different types of organization are frequently confused: (a) local administration which is accountable to a local electorate and (b) local administration which is accountable to a central ministry and which is therefore simply a branch of the central government. Ample scope exists for both types of arrangement, but they are different from one another and are useful in different circumstances. Confusing them can lead to arbitrary differences in policy in different parts of the country.

THE DEMOCRATIC DEFICIT. The move to democracy has exposed a lack of relevant political experience and expertise.

A lack of experience of democratic politics. One of the most acute shortages at the start of the transition was the lack of professional politicians with experience of the day-to-day conduct of democratic politics. Given the nature of the previous regimes and the circumstances of change, this was hardly surprising. The people available to form postcommunist governments were academics and other intellectuals, trade union leaders, and people who had been purged by or had defected from the communist party. These people were not, in the main, experienced politicians. The former communists and trade union leaders were not familiar with the techniques of democratic institutions. Neither was the electorate. Even in countries with a history of democracy, virtually nobody was left who had any democratic experience.

The problem with academics is that, even though they may have considerable knowledge of the components of public policy or of political theory, they are more than most people inclined to be obsessed with policy rather than with politics or administration. In fairness, there was no choice. Someone had to get on with the job of running the country, and it must have taken immense courage to take on the task. In the West, academics frequently enter politics, but they usually do so only after they have served some sort of political apprenticeship by participating in local or single-issue politics or by being closely involved as advisers to working politicians. Western academics have the added advantage of living day to day in a democratic culture. Even so, if they had to take over the government of a country, they would not generally be successful without the support of professional political advisers.

Clearly, professional politicians cannot be produced overnight, although people with an aptitude for politics will learn on the job, largely by trial and error. In any case, there is no choice but to proceed in this way even though mistakes can be very dangerous for emerging democracies, which are so precariously balanced between success and failure. It makes sense to offer these countries advice on political, communications, party organizational skills, and, if requested, training in some of these areas. Poland must be governed by Poles and the Czech Republic by Czechs, but the best possible advice on all aspects of government and politics should be made available, not only by the aid agencies but also by Western trade unions, political parties, and pressure groups, as well as academic institutions and think tanks.

Mass communications. This is the area of political skill which has developed most in recent years in the West and one which, with rare exceptions, is lacking in Central and Eastern Europe. The use of modern media techniques is essential to success in modern politics and is the most rapidly developing area of specialist experts. Western political parties spend a high proportion of their, usually limited, finance in this area. Governments have greatly expanded and transformed their information services. A senior member of Mrs. Thatcher's government said of Bernard Ingham (her press secretary and head of the Government Information Service) that he would be remembered by historians long after most members of her cabinet had been forgotten. As press secretary, Ingham was one of the prime minister's closest advisers, which was precisely the reason for his success (it also made him a very controversial figure). In the United States, following a period of poor opinion poll results shortly after his election, President Clinton, the first Democratic president for many years, appointed as director of communications David Gergen, a media expert who had run the White House press office under two Republican presidents. This appointment was to strengthen a team which had been successful during the elections but was much less effective once in government.

The contrast with at least some of the reforming countries is sharp. A report of President Clinton's appearance on a television town meeting during his visit to Russia in early 1994 pointed out that

> no Russian politician has gone on television to explain to ordinary people why they are having to tighten their belts and what benefits they might expect in future. Had reformers regularly talked and listened to Russian people in a friendly . . . manner, they might have been better rewarded in December's elections. [*Independent (London)*, January 15, 1994, p. 1]

The same paper in a leading article on the economic reforms went on to argue that

> the fault lies not primarily in economics but in politics. The reformists have . . . failed so conspicuously to carry their message to the nation,

that elections intended to strengthen them have only weakened them. [*Independent (London)*, January 18, 1994, p. 15]

Party organization. This aspect of political institutions was a real strength of the old system and is perhaps the former communists' biggest advantage. The main, sometimes the only, factor holding together the "rainbow coalitions" which governed immediately after the fall of communism was the opposition to communism. These coalitions may be difficult to maintain in the post-revolutionary phase of reform. The new parties which have been most successful electorally have learned the importance of well-organized, grass-roots party machines. The failure of Civic Forum in the Czech Republic to construct itself as a political party with grass-roots membership and support led to its virtual elimination as an electoral force. For similar reasons, reformist parties in the Russian election of December 1993, performed poorly.

Without the development of mass membership and local organizations, parties will continue to lose elections and may be particularly vulnerable to the former communists, who have experience in sophisticated party mechanisms, or to nationalist pressures:

> One of the most valuable forms of help that Mr. Yeltsin could receive from the West would be advice and know-how specifically aimed at helping him and his followers to establish a reformist party. Without a grass-roots party, the president has no hope of transforming the reformist impetus into a cornerstone of multi-party democracy. [*Independent (London)*, March 11, 1993]

A separate problem is that discipline in Parliament is often difficult to maintain in systems where coalitions are necessary, especially where the number of small parties is large.

Administrative Constraints

The major administrative constraints derive fairly directly from the administration inherited from the old system. Two aspects stand out: the administration was not designed or called upon to give good impartial advice to help ministers to introduce policies capable of being administered, and it had little experience of rapid change, having neither the physical nor the technical capacity to cope with large-scale change.

LACK OF AN ADVISORY ROLE. The old administration had only one role, to implement policy handed down from government. The ability to question and advise on policy was not a requirement. The role of the administration was to manage implementation. In the West, in contrast, administration is not restricted to carrying out instructions; it also has a very positive role in advising ministers on matters of policy as well as implementation. In its second role, it ensures that policy proposals are linked to past policies and administrative experience.

In Central and Eastern Europe, the role of adviser did not exist under central planning, and many ministers during the early transition were academics and thus reasonably prepared to take on the role of policy development. But no professional administrators, nor indeed anyone else, were available for this task. The new politicians simply had to do the job themselves on a day-to-day basis as the problems of transition emerged. This differs from the Western model in important ways: more than is normal in the West, the minister in Central and Eastern Europe concentrates on policy rather than on politics; by concentrating on policy development, the minister generally fails to give administrators strategic guidance, and policy is developed without regard to its administrative implications.

LACK OF FLEXIBILITY. The administrations were highly centralized, with a heavily bureaucratic central power base. With only very limited delegation of power to the regional or local level, they retained control of a top-down style of management and implementation. Management style was limited to passing instructions down the organization, with little or no staff participation. The whole administrative structure was based on the individual holding power in a vertical chain of command. Status, and by implication salary, was based on how much control an individual had within the organization. Such a rigid top-to-bottom approach left little room for consultation with colleagues, resulting in considerable duplication of effort.

The incentive for the staff was to retain control of whatever small part of the administration was deemed to be their responsibility. There was no incentive for an individual to delegate or devolve work or to report up the organization anything other than what the manager wished to hear. Change was imposed from above, with little opportunity for staff to influence change. Individual initiative was not encouraged, and where it existed, it did so without the knowledge of management and therefore without any coordination.

Over time an administration based on power becomes dictatorial and loses its regard for the effects of the operation it was designed to administer. Government by decree, and the assumption that, once legislated, a particular action will take place, led to a lack of monitoring of the effects of policy. There was little understanding at the center, where decisions were made, on the actual effect of past policy or the degree to which that policy achieved its aim.

CONSTRAINTS ON THE SPEED OF CHANGE. It is possible to introduce reforms more quickly only if governments are able to change the basic structure of the administration. The new governments have to build on administrations in which power is heavily concentrated at the center. Staff outside the center are used to acting on instructions and not to making decisions, and across the administration, staff are not used to change and have had no incentive to become involved in delivering services to the public. In any other circumstances, this situation would suggest the need for a cautious approach to change, building a strong civil service at the center, and only then moving on to implement the large-scale changes envisaged. In the human resources sector at a time of transition, this "softly softly" approach is not possible. The public expects, and will very quickly demand, effective systems of education, health, and cash benefits, as well as effective action to ameliorate unemployment. The public distrust of administration will have to be addressed before real progress can be made in introducing good new policy initiatives. In the interim transitional period, change will have to take account of the constraints and not attempt to move to a radical new approach without addressing the very real effects of the inherited cultural problems.

Further problems have arisen from the shortage of administrators in some parts of the public service, aggravated by the fact that administration has received too few resources for many years, had little or no computer support, and relied heavily on a pencil-and-paper approach. As discussed in chapters 9 and 10, most short-term cash benefits and family allowance under central planning were administered by enterprises; health care, education, and other social services were often also administered at the enterprise level. The government apparatus will have to assume administrative responsibility in most of these areas, but currently too few administrators have the necessary skills.[7]

Driving Change

Various elements interact in the political and administrative cycle. Administrators should be involved at three discrete stages:

- They should assist in policy formation, not least to ensure that policy proposals are administratively feasible.
- They should implement policy in the light of legislative and regulatory enactments.
- They should check the impact of policy changes, feeding this information back into the policy formation process.

The political cycle, in contrast, is a perpetual process, going round and round the following stages:

- Planning policy
- Preparing and enacting legislation
- Checking when the policy first takes effect.

The second and third stages of the political cycle are public and high-profile periods, so that politicians will receive a lot of feedback. Politicians make speeches, appear on television, and write newspaper articles. The goal of these activities is to ensure that (by the end of the process, if not at the beginning) the electorate as far as possible understands the policy and generally approves of it. Where the policy and the views of the electorate are in conflict, the politician can either change the policy or seek to change the views of the electorate. Throughout, politicians will check with the public, with their major sources of political support, with party members, and with opinion formers. They will also receive feedback from the civil service. This process, in principle, applies to all policies at all times, though obviously politicians will give priority to policies which are particularly important or controversial.

Political Communications

In the short run, governments can impose political change from the top, provided that they obtain a parliamentary majority to pass the necessary legislation and persuade administrators to carry out the detailed work of implementation. Any change, however, creates winners and losers. As the discussion of emerging poverty and rising unemployment makes clear, changes as dramatic as those required to achieve the desired economic and social progress in Central and Eastern Europe will generate very large numbers of people whose conditions, at least in the short run, deteriorate seriously as a result of government policies. Many of the losers will be the very people whose cooperation is required to make the changes work, such as administrators appointed for political reasons under the old system, whose privileges and pay are seriously reduced by the reforms. Not all short-run losers will oppose policies if they can be convinced that the promised results can and will be delivered and that they, their children, and the nation will eventually enjoy better times.

In order for changes to be achieved with the maximum public support, politicians need to acquire skills in two related but separate areas: reading public opinion and shaping public opinion. Political communication has the twin tasks of informing and persuading the public about government policies and of giving politicians feedback about what the electorate thinks. Such information helps decisionmakers to present policy, to tailor policy, and, importantly, to time policy initiatives. Under the new democratic conditions, the opposition will exist and will also be campaigning for public support, so public opinion will be subject to competing pressures.

The first requirement for successful political communications is a strong and efficient government information service staffed by experienced professionals. In mature democracies, successful politicians rarely embark on policy initiatives without taking the presentational consequences into account. The public relations aspect must be considered as part of the policy formation process. Public relations are unlikely to be successful if they are merely tacked on as an afterthought and expected to make the best of a bad job by attempting to sell the unsalable. For public relations to work, the people responsible for communications advice must have access to ministers and policymakers, along with the responsible administrators and policy advisers, at the very highest level.

READING PUBLIC OPINION. Targeting is a sophisticated art, just as much in politics as in the areas discussed in earlier chapters. It is important for getting the best support out of scarce resources. Information about how expected support is holding up or about the intentions of potential supporters is vital. In multi-party systems, it is important to know how support is moving between the other parties. The first measurements, and often the most accurate, are gleaned from the party grass roots and, in constituency-based systems, from constituents. Politicians can easily get too caught up in Parliament and affairs of state and miss these signals. Experienced politicians do not ignore them. This is one of the advantages of a constituency-based system and of parties with good grass-roots organization and a good local government base.

Another important mechanism for collecting feedback is the public opinion survey. Feedback is a perpetual process, so that politicians are constantly aware of shifting public opinion. A distinction should be drawn between the general surveys published in the national press and the private polls commissioned by the parties themselves. The polls published in the press are important because they undoubtedly influence public opinion. Each party's own polls should pay particular attention to their target electorate and give the party information on how they and their policies are regarded by these groups.

INFLUENCING PUBLIC OPINION. Given the inherited distrust of government, politicians will have to convince even the groups which stand to gain from the reform that the proposed changes will deliver the promised benefits. A clear problem in this context is the extent to which many politicians in Central and Eastern Europe have allowed, and sometimes encouraged, inflated expectations. Promoting public understanding of the issues is a priority. The overall objectives of the reforms, to increase individual freedom and raise standards of living, can be achieved only if sufficient public support is generated for the necessary policy initiatives. To achieve public support, public debate must be promoted so that the main issues are understood and the benefits of change and, at least as important, the disbenefits of lack of change are clearly understood.

Winning the argument, for instance with the old guard, the new zealots, and the party, is a separate but linked aspect of shaping opinion. Some groups oppose the reforms, and opposition parties will be campaigning for public support. So long as conditions remain worse than they were before the previous system collapsed, the old guard will have a powerful platform. Even after improvements have been delivered, the long-run losers will be potential supporters of the unreformed communists, although this threat should diminish fairly quickly. Other groups will champion ideologies such as fascism and extreme nationalism, which could represent a very real threat.

Only a government with public opinion on its side will be capable of resisting the hostile reaction of groups within society. Disgruntled groups will always exist, and their discontent can manifest itself in different ways, including strikes, protest marches, and other acts of civil disobedience. These acts in themselves, although disruptive, can be survived, but government must win the arguments and prevent mass support from developing for the protesters and also more serious civil disruption from taking place. Public patience may be short-lived, and government will have continually to reinforce its message if the dispute runs for any length of time. Initial public support cannot be taken for granted, and government communicators must be involved at every stage of development. Strikes, marches, and protests are primarily publicity stunts, and government must therefore be prepared to fight and win the public relations battle.

Policy Formation

In the early stages of transition, the immediate objectives of economic and political reform were obvious and policy largely predetermined. Members of the first transitional governments were by and large self-selected, since they were often the people who had most volubly criticized the old regime. They assumed power to put through programs of reform. These people came from different backgrounds: they were academics and intellectuals, workers' representatives such as Solidarity in Poland, and in some cases members of previous communist regimes who had become disillusioned by the system's economic and political failures.

Beyond the initial objectives of reform, however, the new governments were all faced at a very early stage with the question of detailed policymaking. They had to do this in a very different way from mature liberal democracies. In the West, many complex interactions underlie the commitment of political parties and governments to particular policies. The governments in transition have had neither the time nor the machinery to do this and have had to cut corners and concertina months and years of careful political bargaining into weeks and sometimes days. While wholly understandable in its historical context, this process of policy formation is unsustainable in the long run, for reasons of political stability, democratic accountability, and rational policy formation.

Policy should be formed through the interaction of different groups: external agencies, the civil service, political and policy advisers, political parties, and opinion formers (including media commentators and academics). It may take some time before the pace of change slows sufficiently to allow this luxury, which nonetheless should remain a goal.

External influences are currently a major factor. These include the International Monetary Fund, World Bank, European Bank, and European Union, all of which influence policy, not least through conditions imposed on loans.

Influences of this sort, it is to be hoped, will decline as the transition proceeds and the capacity for forming policy internally strengthens. In the medium term, however, other external influences, for instance, contacts made through membership of the European Union, will become stronger. A separate external influence is exerted by Western investors who pressure governments to adopt free market industrial strategies before they are prepared to become involved in investment or joint ventures. External influences can also operate through domestic opinion. The populations of countries in Central and Eastern Europe view the West as a model and are demanding Western life-styles. There can be no doubt, for example, that the aspirations of citizens living in what is now Eastern Germany were greatly affected by television from Western Germany.

The civil service will have a growing input into policymaking. It is the organization which has to implement all policy, and it will be the first to realize when policy is not working or when problems are about to occur. The input of the civil service should be based not on ideology but on administrative practicality.

Political and policy advisers are almost essential in the West, both for governments and for leading politicians. While ministers are daily campaigning for government policy and trying to gain and retain broadly based public support, their policy advisers should be conducting the detailed analysis. The lack of professional politicians available to form transitional governments has encouraged the sort of people who should be policy advisers to become politicians. As the new democracies develop, these roles will have to be separated, and the different skills of politicians and advisers developed. In the West, think tanks are common; they can be set up by government, or attached to political parties, or to independent, usually academic, bodies. These groups are increasingly influential on detailed policy matters as government involvement in regulation becomes more complex and the pressures on politicians grow.

Political parties must develop their own internal machinery for making policy. The communist parties in Central and Eastern Europe had policymaking mechanisms in many ways similar to those in the West, where parties, through conferences, internal party elections, and general day-to-day response to events, constantly evolve and shape policy menus. In liberal democracies, however, this does not happen in a vacuum. Party competition leads to fairly sophisticated, detailed examination of policy (a luxury regularly available more to parties in opposition than to those in government). Being in government is more about campaigning for the things you need to do than about becoming involved in background policy work.

The concept of a mandate has not yet become important in most of the countries in Central and Eastern Europe, in part because there has not yet been time. The lack of a mandate can create problems, both internally and externally, in countries where policy changes dramatically between elections. These changes are sometimes unavoidable, but lending institutions which believe that a series of policies has been agreed upon become nervous when they detect regular and serious lurches in policy; foreign investors are frightened for the same reasons. For democracy to be meaningful, a party or parties elected to government should govern broadly on the basis of the policies on which they were elected. This should be a long-term objective, the achievement of which will become easier when competition leads to distinguishable differences in each party's position on policy. This will probably take two or three electoral cycles.

Political parties should take notice of opinion formers, including editors of newspapers, key radio and television commentators, academics, and foreign writers, whose opinions are increasingly accessible. There is a danger, however, that the politicians who have emerged in Central and Eastern Europe, having generally come from an intellectual elite, will not be sufficiently aware of the views of the wider public.

Finally, for particularly intractable policy problems, it is possible to establish a committee (such as a Royal Commission in the United Kingdom) composed of eminent non-political figures who consult widely and produce a set of recommendations. The advantage of such an approach at its best is its perceived impartiality, and hence its legitimacy; the disadvantage is that it can be time-consuming.

Designing the Law

In reforming the legal system, two very different sets of issues arise: the constitutional framework and the appropriate use of primary and secondary legislation.

THE CONSTITUTIONAL FRAMEWORK. As emerged in chapter 5, the legal inheritance is in important respects inadequate for a market economy. Filling gaps and amending inappropriate legislation are urgent tasks. Rushing into ill-considered new legislation can, however, create more problems than it solves. The approach to new legislation has to be structured if the new laws are to be acceptable and if the changes themselves are not to increase instability. The need constantly to change legislation shortly after its introduction is common in many parts of Central and Eastern Europe and does little to build confidence in the new governments. Legislation must be capable of withstanding the test of time; otherwise, it will fail to gain public confidence and respect, as occurred with previous legislation. The urgent need to produce vast amounts of new legislation must be balanced with the need to get it right the first time.

Three basic elements should be considered in the design of new laws:

• Any law should clearly and unambiguously set out the rights and entitlements of individuals, and these should be described in a way which leaves

no room for misinterpretation by the courts. Laws which are open to a range of interpretations will lead to protracted legal battles to establish whose interpretation is to be accepted.
- Laws should have a clear sanction for noncompliance or for exceeding the extent of those rights. There is little advantage in having legislation which seeks to impose standards on individuals without accompanying sanctions. The sanctions should be capable of being enforced.[8]
- The law should be capable of being administered effectively. There is no point in passing a law stating that all citizens should have a new passport within three months if the administration is not able to change all passports in less than a year, nor in announcing complex pension increases to be paid next month if the necessary recalculation requires six months.

With so much to reform, including in many cases the legal system itself, shortcuts are inevitable. The temptation to govern by executive or presidential decree is strong. In some cases, especially where parliamentary support for reform lags behind that of the executive, such a decree may be the only answer. In many instances, a better approach would be to use skeletal primary legislation, allowing ministers to use secondary legislation or statutory instruments to fine-tune policy. Once more, a tradeoff exists between perfect democratic scrutiny and the needs of urgent reform.

PRIMARY OR SECONDARY LEGISLATION. The way in which the legislative framework is constructed is crucial to protecting individual rights and maintaining a flexible administration. In the West, legislation is commonly split into two tiers, primary and secondary. Indeed, with the growth of government activity everywhere, secondary legislation is virtually unavoidable. Primary legislation devolves power from Parliament to a person (a minister) or a body. The power is expressed either as a requirement ("the minister will . . . ") or as an option ("the minister may . . . "). Where the power is given as an option, there should be a clear statement of the broad intention of the law, together with clearly defined, specific powers which may be exercised under secondary legislation. This statement should not contain the details required to implement the legislation nor any aspect which may be subject to frequent change. Secondary legislation, in contrast, governs the exercise of the powers devolved by Parliament. It should set out the precise details of how the primary legislation is to be implemented, and in the transition stage, it should give maximum flexibility to the administration consistent with the protection of democratic principles. Parliamentary oversight is generally much less detailed and time-consuming for secondary than for primary legislation.

Taking pensions policy as an example, primary legislation should specify the broad conditions, who will receive it, at what age, and with what restrictions (excluding, for example, persons while they are in prison or out of the country). Even these conditions may be better left to secondary legislation until the new economies settle down. The primary legislation must, however, state clearly what provisions will be contained in secondary legislation, such as conditions of entitlement, method of calculation, rates of payment, and method of making future increases. Leaving detailed conditions to secondary legislation in this way gives discretion to individual ministers but still retains a measure of parliamentary scrutiny.

To build public confidence, all aspects of primary and secondary legislation should be subject to the independent scrutiny of the judiciary. Any citizen should be able to challenge the application of the broad principles in primary legislation and the detail in secondary legislation outside the executive and administrative structure. Only by separating the powers of the executive from the powers of the judiciary to enforce and interpret can the citizen be fully protected. Again in the short run, this may have to be an aspiration rather than a reality, but politicians should clearly state their intention to move toward this condition as rapidly as practically possible.

Strengthening the Administration

The discussion which follows should not be thought of as narrowly administrative. Administration has major political implications. The point of delivery of any government service brings the public into direct contact with the government. Indeed, for most citizens, it is their only direct contact with government. The public will therefore judge the effectiveness of policy by the effectiveness of its delivery. Designing an acceptable and efficient administration is therefore essential for political as well as for policy reasons.

OBJECTIVES OF ADMINISTRATION. The function of administration is twofold: to help the government to deliver its policy in a cost-effective way, and to serve the public. It is this concept of service which in the West has given rise to the term "civil service." In order to build a central administration capable of giving reliable and trustworthy advice, several requirements must be met:

• Ministers and their administrators need to understand clearly their respective roles and responsibilities. Thus ministers should recognize the role of administrators in policy development, and administrators should accept the resulting responsibility.

• A relationship of trust and understanding must exist between ministers and administrators.

• Objectives must be clearly agreed upon and understood. In most instances, the administration exists to deliver government policy accurately, on time, and at an acceptable cost. This will most likely be achieved if the prime objective is to keep the administration simple. When designing an administration to implement a piece of policy, the tendency is to become involved with the detail and to lose sight of the primary aim. This, unfortunately, was particularly true during the early transition, when countries were changing from one policy to another. Whatever the administration is designed to do, the question to be asked is always, "Is this as simple as it could be?"[9] The great additional advantage of simple systems is that they are much easier to monitor and therefore more difficult to abuse. The more specific objectives of accuracy, speed, flexibility, and cost can be designed and tested against the prime objective of simplicity.

The design of the administration will be greatly influenced and constrained by the policy to be delivered and the legislation which empowers it. If administration is to be effective, consideration should be given to administrative needs *at the time the policy is designed and legislation drafted*.

THE STRUCTURE OF ADMINISTRATION. Building an administration where none existed is difficult; changing an administration while trying to maintain the services of the old one is even more difficult. In times of rapid change, it is rarely possible to move directly to the ideal system, which may take a number of years, and an administration may have to be planned for a lengthy interim period. It may even be necessary to have interim arrangements which appear to be in direct conflict with the longer-term aim. For example, the protection of standards in education might in the short run involve continued, centrally organized arrangements, even though the ultimate aim is to decentralize and to provide support for the emergence of private providers. In these circumstances, it is even more important to have clear long-term objectives when planning the short-term administration.

A key question in designing an administration is whether it should be (a) centrally accountable (though the administration itself could be central or local) or (b) locally accountable. Local offices in the first case are *not* local administration, but local arms of central administration. The difference between the two types is not always understood, and a local office is often wrongly thought to be local administration.

The crucial factor in deciding whether an administration should be national, regional, or local is deciding where the decisionmaking power and accountability should rest. In many cases, this decision will be contained in legislation, and if the administrative needs have not been taken into account when drafting the legislation, the administrative design will be constrained. Four factors are relevant in deciding on central, regional, or local administration: the consistency of policy delivery and standardization of the service provided, the size of the administration and the economies to be made by the scale of individual units, the financial need for cross-subsidy between regions, and the need to finance the organization from local sources.

The relative values of these factors are seen in the two basic administrative models:

- A central administration offers a prescriptive, standardized approach, ensuring uniformity of implementation. It also offers the best mechanism for financial cross-subsidy between regions, together with reduced cost resulting from the opportunity to combine some tasks at a central point. It can allow for local variations, but only those decided on at the center. Central administration can have local delivery. An example is the provision of cash benefits, where for reasons of universal implementation, financial cross-subsidy, and central record keeping, a centralized administration is necessary. To maintain a large network of local contact offices, a sizable local organization is also required.

• A local administration reduces the opportunity for cross-subsidy but increases the options for regional variations to be decided at the local level. An example might be regional training opportunities.

In deciding whether or not to have a locally accountable administration (elected local authorities), governments may prefer to start with what they already have. If structures of local or municipal government exist and can be democratized, they should be used. Locally elected tiers of government can be useful: they can take pressure off the central government by dealing with politically contentious local issues which would otherwise divert the national leadership; they are a useful breeding ground for much-needed national political talent (much of the previous leadership came to national politics via this route); and they are valuable in building grass-roots party mechanisms.

It is a mistake to think of these issues as narrowly administrative. Inappropriate administrative structures can cause political instability. This is another area where administration can have major political implications. Regardless of whether or not a structure of local government is in existence, the laws passed by the central government should be interpreted and implemented uniformly across the whole country. The legislature adopts and enacts laws. The task of the executive is to implement those laws; to do so, it must inform enforcement officers about the precise interpretation of the law and its supporting regulations; that guidance has to be uniform and should normally be overturned only by an adverse legal decision or by subsequent legislation. In particular, central laws should not be subject to local interpretation, except where explicitly permitted.

The spread of computerization offers an important example of how this can go wrong. Hardware is increasingly organized as a network of local computers. It is, however, vital that the software which implements central government laws should be produced centrally and revised by the center whenever the law is amended. In some countries, localities tend to develop their own software, which implies that both the original program and any updates (for example, to take account of an increase in pensions) will incorporate local interpretation of the law. This may not cause greater problems than wasteful and inconsistent administration. However, where laws are repeatedly amended—a frequent occurrence in Central and Eastern Europe—local interpretations will, in all likelihood and with important consequences, diverge increasingly. The logical result will be loss of control from the center, increased fragmentation, implicit or explicit rejection of the power of the center to impose legislation, and, ultimately, perhaps the breakup of the country concerned.

STAFF AND SUPPORT SERVICES. Once the structure has been decided, the next step is to bring together the other elements which will make the administration work: the staffing and the support services, including computer assistance. One of the main messages of chapter 11 was that educational policy should emphasize the teacher. Similarly, staff are the most valuable asset in any administration. The shortage of suitable candidates is critical, and a major role exists for training and educating civil service staff—an activity which has, perhaps, not been given as much emphasis as it should.

The correct numbers, selection, and training of staff are what make an administration effective. Assessing the number of staff required varies with the policy being implemented. In the social sectors, where policy is directed at specific groups in the population, staffing levels based on the size of the group served is more appropriate than staff levels based on the total population. For example, payment of unemployment benefits, where most of the work is at the local level, requires flexible levels of staffing related to the number of unemployed people in the area, not the total population of that area. In addition, the social sectors will make increasing demands on the administrative capacity of the various levels of government as enterprises cease to deliver benefits.

Selection of staff is more than usually important when the quality of service has to be maintained in the face of increasing fiscal constraint and rising demand for the service. In the past, when a professional qualification led to a higher wage for the employee and possibly also for the manager, the incentive was to employ overqualified staff for some administrative work. Well-designed legislation keeps the administration simple (the primary objective) and minimizes the need for discretion in implementation. This, in turn, reduces many tasks to clerical work, without the need for professional expertise. An example is the distinction in chapter 10 between highly skilled social workers and clerical benefit officers for social assistance. Professionals have a role to play in all aspects of social sector administration, but the organization should be designed to ensure that their professional expertise is required for most of the work they perform, not just a small part of it. Good administration requires the right mix of administrators and professionals.

Training takes resources but should be given a high priority. Chapter 8 emphasized the importance of targeting training. This is precisely such an area. A variety of training needs have to be met. In many areas, the expansion of services requires significant numbers of new staff. Existing staff need to be trained to implement new and revised legislation and to learn new methods of working; managers will require training in different management styles and practices. Training is a good example of the need for uniformity; no matter what administrative structure is in place, the training should be designed at the center, even if the delivery is devolved to the local level. Senior professional staff, familiar with the requirements of a modern mixed economy and the role of the state, represent one of the greatest shortages in Central and Eastern Europe. Such staff should be rewarded as well as possible, with job security and good pension rights as well as salaries, to keep them in the government service; fast-stream programs should be instituted to attract a proportion of the best graduates to the civil service; and resources should be devoted to educating and training potential high flyers in the civil service and the brightest and best graduates. As the private sector grows, these people will be in great demand.

COMPUTERIZATION. It is a fundamental error to think that computerization is the answer to bad administration. It never is. Computerization greatly assists good administration, however, improving it or allowing it to provide better service with greater accuracy. For bad administration, computers may improve accuracy, but they could also burden the administration in ways which make it more staff intensive, with little or no improvement in service. Before deciding which computer system is best suited to the administration, a clear view is needed of why a computer system is considered necessary. What is the computer intended to do (keep records, perform calculations, provide statistics, or offer a word processing service)? Why is a computer essential for this task (greater speed, great accuracy, calculations or data manipulation which would be impossible without support, or improved use of staff time)? Where are computers required in the administration (at the local level, at the regional level, or at the national level, and what links should be provided between all three)?

Only when the administrator has a clear idea of the answers to these questions can he or she talk to experts about which system would best meet those needs. It should also be remembered that staff require time and training. The full value of a computer system will be achieved only when the staff who operate it are convinced of its value. Once convinced, however, the staff themselves will push for bigger and better systems. A number of models in the West could be adapted with good results. However, no one administration from outside Central and Eastern Europe will provide all the answers for any one country's particular administrative needs. A key message is that the best administration is achieved when it is designed alongside the policy it seeks to administer.

Checking the Outcome

The process of initiating policies and checking their outcomes is perpetual. While the necessary legislation is being passed, both political advisers and administrators constantly interact with politicians and adjust and amend their proposals slightly in response to problems which become apparent during the process. The political outcome therefore should be obvious during the process of enactment. Policy change, however, is rarely free of unintended consequences. These can generally be rectified given some flexibility in implementation, but further legislation may become necessary.

Virtually all Western democracies can show laws which had to be scrapped after passing through the whole parliamentary process simply because insufficient account had been taken of the obvious political hostility and administrative difficulties during the policymaking and law-making processes. The infamous poll tax in the United Kingdom is a perfect example. This measure to finance local government was passed through the legislative procedure by a government with a substantial parliamentary majority and yet was abandoned shortly afterwards, before the subsequent election. The problem, in part, was the attempt to introduce unpopular legislation without ensuring that policy design was robust enough to protect the government from the parliamentary opposition. This failure was a significant factor in Margaret Thatcher's loss of office as prime minister. It is often the case that the general public is unaware of the likely consequences of legislation and understand the effects only when implementation begins. Ministers must use all the powers at their disposal to assess the national feeling—grass-roots party opinion, general public opinion, press reaction, and private party opinion—to ascertain that minorities of political importance to the party in government are not disproportionately dissatisfied.

Democratically elected governments are formally accountable, in that they have to face election from time to time. The executive, however, should be accountable to Parliament for its actions between elections. There are many possible mechanisms for accomplishing this. The most common is some form of committee structure in which the executive is publicly examined on the detailed day-to-day work of government. This is particularly important where the executive, as in the United States, does not sit in Parliament.

The administration should also be accountable. One of the great failings of the previous communist administrations in Central and Eastern Europe was that once they put a policy into place, they assumed that all would be well. No provision was made for collecting the feedback necessary to check the impact of the policy or the performance of the administration in implementing it. The result was a complete lack of accountability on the part of administrators. The style of the past was to implement policy without having to answer for the effects of that implementation. Whichever style of administration is chosen, administrators should have to account to the government for its effects and costs, and the government should be accountable to the general public. For that reason, senior civil servants are commonly subject to parliamentary scrutiny for administrative aspects of government.

Management must be held accountable for the accurate application of policy if administration is to be improved. This requires centralized information. Part of the administrative design should therefore concentrate on flows of information, keeping the information up to date and ensuring its accuracy. In addition, policymakers require accurate information on the application of the policy to shape changes and future policy: Is the policy working as intended? Should it be changed? Is it being administered accurately and on time? Are there local or regional issues which have to be addressed? Is the financing correct? What trends are evident for the development of policy in the future?

For all this to be possible, a great deal of information must be passed back to administrators and policymakers at the center. Continuous information is essential and provides the link which ensures that the policy is developed and improved, not just changed without considering what has gone before. This should be a continuing process and is vital if administration and policy are to be developed in a structured way during the transition. It is also vital for the future if the administration is to keep pace with, and take advantage of, developments in management as they occur. No administration in the West is a static structure; all are subject to constant change. The new administrations should allow for this constant change and development.

Conclusions

It is helpful to separate conclusions about the political process from political and administrative policy actions.

The Political Process

During the revolutionary phase of the early transition, much policy, of necessity, was very short run. It was concerned (a) with trying to make change irreversible and (b) with trying to prevent a complete breakdown of the system, by ensuring the continuing payment of unemployment benefits and pensions and of doctors' and teachers' salaries. In most countries, however, the revolutionary period is past.

Normal politics should follow the revolutionary phase, and normal political rules will apply. The process is cyclical, generally following two cycles: the formal and the electoral. The first is annual, the fiscal and parliamentary year. Budgets are set, and legislation planned and enacted. Each stage must be explained and feedback noted. Politicians need to "sell" their policies and explain their actions, particularly in the later stages of the electoral cycle. It is better to make the most radical changes early in this cycle so that the benefits can become clear before the electorate is asked to judge them.

The combined political and administrative cyclical process is as follows:

- Publicly float the idea and monitor the response, which may involve a formal consultative process

- Prepare detailed policy proposals, where possible taking into account the views of interest groups
- Campaign for the policy, get ministers and prominent supporters on the radio and television and in the press, and win the argument with opponents (this is not easy if, for example, the issue is a refusal to increase teachers' salaries, but government must explain why this is necessary; not everyone will agree, but it is important to measure the strength of the opposition)
- Test at all times that support exists within the party for the policy, since splits are often more damaging in themselves than the issue which initiated them
- Enact the legislation; widespread public debate will continue during the parliamentary debates, triggered by the enactment procedure
- Implement the policy
- Listen to the feedback.

The feedback will come from many sources, very importantly the civil service, party supporters, and the electorate. This is true at *all* stages, from the floating of ideas, throughout the consultative process, and in checking the final outcome. The feedback from checking the outcome is often the starting point for a new policy initiative. Thus the process is cyclical and perpetual.

Short-term Policies

The following policies will have an impact in the short run:

- Because of institutional capacity constraints, policy design should be as simple as possible.
- Administrative needs should be given considerable weight. In particular, individuals with implementation skills should be involved from the start of the process of designing policy and drafting laws and regulations.
- Politicians should spend a significant amount of time persuading the public of the desirability of the government's policies. Democratic government involves permanent campaigning, and that must be done by the politicians themselves.
- Feedback from the electorate to politicians is essential to ensure that politicians are well informed about what the electorate wants and are aware of the main sources of support for and opposition to their policy proposals. The electorate for these purposes is not just the intelligentsia, who are a minority, but the broad body of the population.
- Politicians should defuse rather than encourage unrealistic expectations.

Medium-term Policies

The following policies have a medium-term dimension:

- The machinery of government should give the democratically elected executive the power to implement the policies on which it stood for election.

- As political parties develop their own discrete identities, governments should increasingly adhere broadly to the policies on which they were elected.
- Greater emphasis is needed on monitoring the outcomes of policies.

The latter changes may at this stage be long-term goals in some countries, rather than goals which can be attained immediately. They should, nevertheless, remain firmly on the agenda. As the policy chapters of this book make clear, there are many feasible short-run policies which help the reform process and also protect groups who are most vulnerable to some of its early fallout. Equally, however, the transition will be long drawn out. The good news is that the countries of Central and Eastern Europe have moved out of their cul-de-sac and onto a path which offers a genuine prospect of economic growth and increased individual freedom. It may be a long road. At least it is now the right road.

Notes

1. Italy has had relatively unstable government and comparative prosperity, but levels of corruption and organized crime have led to overwhelming public support for constitutional reform.

2. Crick (1992, pp. 174–76) argues that to be truly stable a political system has to meet thirteen conditions.

3. It is not only in Central and Eastern Europe that expectations are allowed to get out of hand. A classic example of unrealistic expectations fueled by politicians in the West was the chorus of parliamentarians, each advocating increased expenditure on his or her pet project, all to be "funded from the peace dividend."

4. In Germany, for instance, a party must win at least 5 percent of the total votes cast before it can have a seat in the legislature. The purpose of such a threshold is to ensure that the legislature does not have too many parties, the idea being that this makes it easier to form stable coalitions.

5. Proportional representation can take many forms, each with its own pros and cons. The Czech Republic adopted a pure list system (see the glossary) against the advice of a presidential constitutional commission. The major disadvantage of such a system is that it cuts the link between the member of Parliament and his or her constituency, thus losing the direct accountability of the individual representative.

6. An example of logrolling would be if the pensioners' parliamentary representatives agreed to support continued defense spending, so long as the parliamentarians with constituencies heavily dependent on defense contracts agreed to oppose pension reform.

7. As an example of the administrative vacuum which can exist at the center, the central authorities in some countries were unable to estimate total expenditure on cash benefits or health care, because they had little information about benefits at the regional or enterprise level. In 1992, the central Ministry of Social Protection in Russia, for example, had no figures on total spending on cash benefits.

8. As an example, legislation tried to impose sanctions on smugglers. This was pointless, since the only way of identifying a smuggler is to catch him. The legislation will have no effect on the smuggler who remains undetected.

9. Examples of the sorts of complexity which arise when administrative issues are overlooked include early systems of unemployment compensation, where benefits were (a) related to previous earnings and (b) had to be recalculated frequently and where (c) the law made no provision for rounding benefits to the nearest convenient currency unit, all in a situation (d) where benefit offices generally had no calculators, let alone computers.

Glossary

Actuarial benefits. Insurance against an event, such as illness or old age, which bears a strict relationship to the probability that the insured individual will become ill or die at a given age. If benefits are actuarial, then individuals, on average and in the long run, receive benefits broadly equal to their own past contributions.

Annuity. The payment of an income of $x per year for life, often given in exchange for a single, lump-sum payment at the time the individual retires.

Capitation payment. A fixed sum paid to doctors per patient per year. The sum can be higher for particular groups, such as the elderly. The essence of capitation payments is that payment is *not* related to the amount of treatment a doctor prescribes, in sharp contrast with *fee-for-service payments*.

Cash benefits. Income support for individuals in the form of cash, in contrast with benefits in kind like free health care. In the context of Central and Eastern Europe and the former U.S.S.R., they comprise unemployment and related benefits, *social insurance* benefits, and *social assistance*. The term social security is avoided because of its ambiguity. In the United States, social security refers to retirement pensions; in the United Kingdom, it refers to the entire system of cash benefits; and in mainland Europe (in accordance with usage of the International Labour Organisation), it refers to all cash benefits plus health care. The term cash benefits has been used throughout this book.

Central and Eastern Europe. For the purposes of this book, (a) all the formerly communist countries of Europe plus (b) the former U.S.S.R. apart from its Asian/Muslim republics. The term thus embraces Albania, the Baltics (Estonia, Latvia, and Lithuania), Belarus, Bulgaria, the Czech Republic, Hungary, Moldova, Poland, Romania, the Russian Federation, the Slovak Republic, the successor states to the former Yugoslavia, and the Ukraine. Armenia and Georgia are excluded because, for most of the period since 1989, internal wars have pushed economic transformation into the background. For each country, we have used the name current during the period being discussed. Thus Czechoslovakia refers to the country before it was divided, while the

Czech Republic and the Slovak Republic—or the former Czechoslovakia— refer to the countries after that time.

Consumer subsidy. SEE *price subsidy.*

Cost-effectiveness. Producing a given output in the cheapest way.

Council for Mutual Economic Assistance. The trading system for the former communist bloc; in effect, an international system for trading goods at agreed-on prices and in agreed-on quantities.

Defined benefit scheme. Usually run at the level of the firm or industry, a scheme in which the individual's pension is based on the number of years of service and his or her final salary. Thus the risk of different rates of return to pension assets is borne largely by the employer.

Defined contribution scheme. Scheme in which the individual's pension depends only on the size of his or her pension accumulation. Under a defined contribution scheme, the risk of varying rates of return to pension assets and the risk of inflation after retirement are borne entirely by the individual.

Diagnosis related group. A method of financing hospital care by paying the hospital a prospective sum for each patient treated, where the prospective sum is related to the condition with which the patient is diagnosed.

Earnings related benefits. Benefits paid as a percentage of previous earnings; thus individuals with higher previous earnings receive higher benefits.

Econometrics. The use of statistical methods to quantify economic relationships. An econometric study of labor supply, for example, would attempt to show how much labor supply is reduced in response to a tax increase. SEE ALSO *labor supply disincentives.*

Enrollment ratio. The percentage of a given age group attending different levels of education.

European Union. Formerly the European Community, a regional alliance dedicated to integrating the economies of member countries. Membership in 1994 comprised Belgium, Denmark, France, the Republic of Germany, Greece, Ireland, Italy, Luxembourg, the Netherlands, Portugal, Spain, and the United Kingdom.

Family allowance. A monthly payment, usually tax free and normally not subject to an *income test*, paid until the child reaches a certain age.

Fee-for-service payment. Payments determined on the basis of how much treatment doctors give the patient, in contrast, for instance, with *capitation payments.*

First-past-the-post. An electoral system in which the winner in any parliamentary constituency is the candidate who receives more votes than any of the other candidates, as opposed to *proportional representation.* Both the United Kingdom and the United States have first-past-the-post systems for parliamentary (congressional) elections.

Flat-rate benefits. Benefits paid at a fixed monthly rate (though they may be higher for larger families) which are not related to previous income. Thus, for a given family type, all recipients receive the same benefits.

Fiscal policy. Policy concerned with the level of public spending and the balance between government spending and tax revenues. Government, by and large, can finance its activity through (a) taxation, (b) bond sales, and (c) expanding the supply of money. If government spending exceeds its tax revenues, the resulting budget deficit can be financed either through bond sales (this possibility is generally very limited in Central and Eastern Europe) or through monetary expansion. Budget deficits can thus contribute to inflationary pressure. SEE ALSO *monetary policy.*

Funded pensions. Paid out of an accumulated fund, in contrast with pay-as you-go schemes.

Gross domestic product. The value of all goods and services produced in an economy in a year.

Gross material product. The value of goods and services produced in the so-called material sectors, used by the countries of Central and Eastern Europe before the transition as the principal measure of economic activity. In contrast to the similar concept of gross national product, it does not cover all economic activity, because some services (education and culture, health care, administration) are omitted from the definition.

Headcount. Measures poverty by the number of people whose income falls below a given poverty line; that is, it measures *how many* people are poor, as opposed to the *poverty gap* measure, which assesses *how much* their income falls below the poverty line.

Health maintenance organization. A "firm" of doctors, which charges individuals or families an annual premium, in return for a comprehensive range of medical services. The essence of a health maintenance organization is that it merges the function of doctor and insurance company and thus faces medical providers with the cost of the treatment they provide, thereby reducing *third-party payment incentives.*

Human capital. The skills and attributes embodied in an individual, in part because of his or her education and training.

Human resources. Policies and programs relating to employment and unemployment, income support, education and training, and health care. These areas are referred to collectively as the social sectors.

Implicit tax rate. A tax which arises when a family in receipt of an *income-tested benefit* earns extra income and as a consequence loses some or all of its benefits. If benefit is lost dollar for dollar with earnings, the implicit tax rate is 100 percent.

Income smoothing. The aim of income smoothing is to make it easier for individuals, in effect, to redistribute to themselves at different stages in their life cycle, for example, from their higher-earning middle years to their younger years (when their earnings are lower and family responsibilities greater) and to their retirement years. *Family allowances* and old age pensions contribute in important ways to income smoothing.

Income-tested benefits. Benefits paid only to people whose income falls below a certain level. SEE ALSO *means-tested benefits.*

Income transfers. SEE *cash benefits.*

In-kind benefits. Direct transfers to individuals of commodities, such as health care and education.

Labor supply disincentives. An incentive not to work or to work less. Suppose an individual has the option of working longer hours and earning an extra $10; if he does so, he has to pay an extra $6 in tax; as a result of this high taxation he may choose not to work the extra hours. In such a case, high taxation is said to create a labor supply disincentive. SEE ALSO *econometrics.*

List system. An electoral system in which a political party lists its candidates in numerical order and then sends to Parliament the first x people on the list, where x is determined by the party's share of the total vote. SEE ALSO *first-past-the-post* and *proportional representation.*

Macroeconomic efficiency. The efficient division of national resources between human resources, on the one hand, and activities such as physical investment and consumer spending, on the other; thus it is concerned with such questions as the appropriate fraction of gross domestic product to spend on health care. SEE ALSO *microeconomic efficiency.*

Market failure. An economic technical term describing a situation in which private markets, for systematic technical reasons, produce either inefficiently or not at all. The efficiency of markets rests on a number of key assumptions, including the existence of competition and, importantly, perfect information on the part of buyers and sellers. Where any of the assumptions fail, a possible outcome is market failure. A key example for the social sectors is that, because insurers are imperfectly informed, unemployment, inflation, and important medical risks are uninsurable in private markets.

Maternity leave. In most countries, a woman who is a member of the labor force may stay at home in late pregnancy and early motherhood at a (usually high) fraction of her previous wage.

Means-tested benefits. Benefits paid only to individuals whose pre-benefit income and wealth are below a given amount. The term thus embraces both *income testing* and wealth testing.

Microeconomic efficiency. Concerns the division of total health care resources into different types of medical intervention, the division of education spending into different types of education, training, and retraining, and so forth. SEE ALSO *macroeconomic efficiency.*

Monetary policy. Concerns the control of the money supply, availability of credit, and the setting of interest rates. Rapid monetary growth and low real interest rates (that is, market interest rates which are lower than the inflation rate) can contribute to inflationary pressure. SEE ALSO *fiscal policy.*

Nongovernmental organizations. Nonprofit institutions which frequently carry out tasks similar to those of government, such as the Red Cross, charitable organizations, and nonprofit health care facilities.

Organization for Economic Cooperation and Development (OECD). Refers to the highly industrialized countries. Members: Australia, Austria, Belgium, Canada, Denmark, Finland, France, Germany, Greece, Iceland, Ireland, Italy, Japan, Luxembourg, Netherlands, New Zealand, Norway, Portugal, Spain, Sweden, Switzerland, Turkey, United Kingdom, United States.

Pay-as-you-go (PAYG) pensions. Pensions paid out of current revenues, as opposed to funded schemes.

Poverty gap. A measure of total poverty which estimates the amount by which people in aggregate fall below the poverty line, that is, a measure of how much it would cost to bring everybody's income up to an agreed poverty line. The poverty gap as a measure of total poverty is to be distinguished from the *headcount* measure.

Poverty relief. A key objective of *cash benefits* during the transition. Poverty relief seeks either to eradicate poverty by bringing everybody above a poverty line or to ameliorate the poverty of those below a given poverty line by increasing their income without necessarily completely eliminating the poverty gap.

Poverty trap. Situation in which individuals or families earning an extra $1 lose $1 or more in income-tested benefits, making them absolutely worse off. SEE ALSO *implicit tax rate.*

Price subsidy. A reduction in the price of goods bought, and thus one of two ways to raise a family's standard of living. The other is to increase its income (*income transfers*). The arguments in favor of income transfers rather than price subsidies are that income transfers (a) are more easily targeted and (b) generally cause fewer inefficiencies than price subsidies.

Proactive and reactive measures. As an example, *unemployment benefits* are intended to provide *poverty relief* for unemployed individuals; as such it is a reactive policy. Proactive policy includes, for instance, measures to increase the individual's chances of finding a new job.

Proportional representation. An electoral system in which a party receives parliamentary seats broadly proportional to the total number of votes it receives, as opposed to *first-past-the-post*. Under a pure *list system,* a party receives all its parliamentary seats this way. It is also possible to have a system under which some parliamentary seats are on a first-past-the-post basis and some are on a list basis. The latter, broadly, is the case in Germany.

Public transfers. Publicly funded income transfers.

Replacement rate. The ratio of income when receiving benefits to net income when working. High replacement rates, it is argued, give unemployed individuals an incentive to remain unemployed.

Restructuring. The process of introducing a market system, together with the legislation, regulatory structure, and other background institutions (such as a well-designed system of cash benefits) necessary to assist its operation.

Sickness benefits. Benefits paid to individuals who suffer a loss of income during short-term sickness or injury.

Social assistance benefits. Also referred to as welfare. They are usually financed out of general taxation and paid to needy individuals on the basis of a test of income or wealth.

Social expenditures. Expenditure on the social sectors.

Social insurance. Cash benefits organized by the state through compulsory contributions. The main benefits are unemployment benefits, sickness and maternity benefits, and retirement, invalidity, and survivors' pensions.

Social safety net. Narrowly defined, comprises *cash benefits* aimed at providing *poverty relief.* More broadly, the social safety net includes all *social expenditure.*

Social sectors. A collective term covering policies and programs relating to employment and unemployment, income support, education and training, and health care.

Social security.   SEE *cash benefits.*

Social welfare programs.   SEE *social assistance benefits.*

Social work.   Care provided for individuals and families who are not able to care for themselves, such as the frail elderly, invalids, and individuals with emotional problems.

Stabilization.   The use of *monetary policy* and *fiscal policy* to control the growth of the money supply and the balance between public revenue and expenditure, in the interests of containing inflation.

Stigma.   A reaction of people who feel shame because they receive income in the form of (usually) *income-tested* benefits, rather than from some more congenial source (such as earnings or insurance benefits).

Subsistence.   A level of consumption sufficient to keep an individual or family alive and healthy.

Take-up.   The number of people receiving a particular benefit as a proportion of those potentially eligible.

Targeting.   The action of aiming policy at a specific group or groups. Targeting efficiency has two aspects: horizontal efficiency, which refers to the ability of the system to cover all potentially eligible individuals (to avoid gaps in coverage), and vertical efficiency, which refers to the ability of the system, by and large, to cover only the eligible group (to avoid the leakage of benefits to ineligible groups).

Taylorism.   A "scientific" approach to management used in the U.S.S.R. (and to a lesser extent in the rest of Central and Eastern Europe) prior to 1960, in which workers were simply treated like another machine.

Tax threshold.   The amount of income exempt from taxation. Under a system of personal income taxation, individuals pay no tax on a given amount of their income, known as the tax threshold.

Third-party payment incentives.   A situation in which, for example, the insurance company pays all, or most, of an individual's medical bill; as a result, neither patient nor doctor has an incentive to economize. The result can easily be an uncontrolled escalation in medical spending.

Unemployment benefits.   Cash benefits paid to an individual who has lost his or her job. A condition for receipt of the benefit is normally that the individual must actively seek a new job.

Universal benefits.   Benefits paid on the basis of relevant criteria without a *means test,* for example, when a *family allowance* is paid to all families with children, irrespective of their income.

Uprating.　The process by which benefits are increased to take account (wholly or in part) of changes in earnings or in prices. In the West, this process is usually annual; in Central and Eastern Europe during times of high inflation, uprating has tended to occur more frequently.

Welfare.　SEE *social assistance benefits.*

References

Aaron, Henry. 1982. *Economic Effects of Social Security.* Washington, D.C.: Brookings Institution.

Abel Smith, Brian. 1992. *Cost Containment and New Priorities in Health Care.* Aldershot, Eng.: Ashgate Publishing.

Adams, A., J. Middleton, and A. Ziderman. 1992. *The World Bank's Policy Paper on Education and Training: Prospects.* New York: United Nations Educational, Scientific, and Cultural Organization.

Ahmad, Ehtisham, Jean Drèze, and John Hills. 1991. *Social Security in Developing Countries.* Oxford: Clarendon Press.

Akerlof, George. 1970. "The Market for 'Lemons': Qualitative Uncertainty and the Market Mechanism." *Quarterly Journal of Economics* 84, pp. 488–500.

Akerlof, George, A. Rose, J. Yellen, and H. Hessenius. 1991. "East Germany in from the Cold: The Aftermath of Currency Union." *Brookings Papers on Economic Activity* 1, pp. 1–87.

Andorka, Rudolf. 1991. "Politiques démographiques natalistes at leur impact en Hongrie." *Politiques de Population: Etudes et Documents* 4 (April), pp. 87–125.

Andorka, Rudolf, Tamas Kolosi, and György Vukovich. 1992. *Társadalmi riport 1992 [Social Report 1992].* Budapest: TARKI.

Arrow, Kenneth. 1963. "Uncertainty and the Welfare Economics of Medical Care." *American Economic Review* 53, pp. 941–73.

Aslund, A. 1991. *Gorbachev's Struggle for Economic Reform.* London: Pinter Publishers.

Atkinson, Anthony B. 1983. *The Economics of Inequality,* 2d ed. Oxford: Oxford University Press.

———. 1987. "On the Measurement of Poverty." *Econometrica* 55 (July), pp. 749–64. Reprinted as chapter 2 in Anthony B. Atkinson, *Poverty and Social Security.* London: Harvester Wheatsheaf, 1989.

———. 1989. *Poverty and Social Security.* London: Harvester Wheatsheaf.

———. 1991a. "Poverty, Economic Performance, and Income Transfer Policy in OECD Countries." *World Bank Economic Review* 5 (January), pp. 3–21.

———. 1991b. "The Social Safety Net." Discussion Paper WSP/66. London School of Economics and Political Science, Welfare State Programme, London.

———. 1992. "The Western Experience with Social Safety Nets." Discussion Paper WSP/80. London School of Economics and Political Science, Welfare State Programme, London.

———. 1993. "On Targeting Social Security: Theory and Western Experience with Family Benefits." Discussion Paper WSP/99. London School of Economics, Welfare State Programme, London.

Atkinson, Anthony B., and John Micklewright. 1991. "Unemployment Compensation and Labour Market Transitions: A Critical Review." *Journal of Economic Literature* 29:4 (December), pp. 1679–727.

———. 1992. *Economic Transformation in Eastern Europe and the Distribution of Income.* Cambridge, Eng.: Cambridge University Press.

Auerbach, Alan, and Laurence Kotlikoff. 1990. "Demographics, Fiscal Policy, and U.S. Saving in the 1980s and Beyond." In Lawrence Summers, ed., *Tax Policy and the Economy,* vol. 4. Cambridge, Mass.: M.I.T. Press.

[361]

Balcerowicz, Leszek. 1993a. "Common Fallacies in the Debate on the Economic Transformation in Central and Eastern Europe." Working Paper 11. European Bank for Reconstruction and Development, London, October.

———. 1993b. "Eastern Europe, Social and Political Dynamics." University of London, School of Slavonic and Eastern European Studies, London.

Barbu, Gheorghe, Viorel Gheorghe, and Hildebard Puwak. 1992. *Dimensiunea saracei in Romania* [The dimensions of poverty in Romania]. Seria Politica Sociala 2. Bucharest: Centrul de Informare si Documentare Economica, Institutul de Cercetare a Calitatii Vietii.

Barnes, John, and Nicholas Barr. 1988. *Strategies for Higher Education: The Alternative White Paper.* Aberdeen, Scotland: Aberdeen University Press, for the David Hume Institute, Edinburgh, and the Suntory-Toyota International Centre for Economics and Related Disciplines, London School of Economics and Political Science.

Barr, Nicholas. 1992. "Economic Theory and the Welfare State: A Survey and Interpretation." *Journal of Economic Literature* 30:2, pp. 741–803.

———. 1993a. "Alternative Funding Resources for Higher Education." *Economic Journal* 103, pp. 718–28.

———. 1993b. *The Economics of the Welfare State*, 2d ed. Oxford: Oxford University Press; Stanford, Calif.: Stanford University Press; available in Polish as Barr 1993c.

———. 1993c. *Ekonomika polityki spolecznej.* Poznan, Poland: Wydawnictwo Akademii Ekonomiczenj w Poznani.

Bartlett, Donald, and James Steele. 1992. *America: What Went Wrong?* Kansas City, Mo.: Andrews and McMeel.

Baumol, William J., and Wallace E. Oates. 1979. *The Theory of Environmental Policy.* Englewood Cliffs, N.J.: Prentice Hall.

Becker, Gary. 1964. *Human Capital.* Cambridge, Mass.: National Bureau of Economic Research; second edition issued in 1975.

———. 1983. "A Theory of Competition among Pressure Groups for Political Influence." *Quarterly Journal of Economics* 98, pp. 371–400.

———. 1985. "Public Policies, Pressure Groups, and Dead Weight Costs." *Journal of Public Economics* 28 (December), pp. 329–47.

Bergson, A. 1964. *The Economics of Soviet Planning.* New Haven, Conn.: Yale University Press.

———. 1984. "Income Inequality under Soviet Socialism." *Journal of Economic Literature* 22:3 (September), pp. 1052–99.

———. 1991. "The U.S.S.R. before the Fall: How Poor and Why." *Journal of Economic Perspectives* 5, pp. 29–44.

Berliner, J. 1957. *Factory and Manager in the U.S.S.R.* Cambridge, Mass.: Harvard University Press.

Berryman, S., and T. Bailey. 1992. *The Double Helix of Education and the Economy.* New York: Institute on Education and the Economy, Teachers College, Columbia University.

Bobak, Martin, and Richard G. A. Feachem. 1992. "Health Status in the Czech and Slovak Federal Republic." *Health Policy and Planning* 7:3, pp. 234–42.

Bodie, Zvi. 1990. "Pensions as Retirement Income Insurance." *Journal of Economic Literature* 28, pp. 28–49.

Boeri, Tito, and Mark Keese. 1992. "Labour Markets and the Transition in Central and Eastern Europe." OECD Economic Studies 18 (Spring), pp. 133–61.

Bornstein, M. 1978. "Unemployment in Capitalist Regulated Market Economies and Socialist Centrally Planned Economies." *American Economic Review* (May), pp. 38–43.

Boston, Jonathan. 1993. "Reshaping Social Policy in New Zealand." *Fiscal Studies* 14:3 (August), pp. 64–85.

Bosworth, Barry. 1991. "Incomes Policies in Socialist Economies." In Commander 1991.

Brada, J. 1989. "Technological Progress and Factor Utilization in Eastern European Economic Growth." *Economica* 56, pp. 435–48.

Bratkowski, A. 1993. "The Shock of Transformation or the Transformation of the Shock? The Big Bank in Poland and Official Statistics." *Communist Economics and Economics Transformation* 1:1.

Brown, A. A., and E. Neuberger, eds. 1968. *Trade under Central Planning*. Berkeley: University of California Press.

Bruno, Michael. 1992. "Stabilisation and Reform in Eastern Europe." Working Paper 92/30. International Monetary Fund, Washington, D.C.

Bruno, Michael, and Jeffrey Sachs. 1985. *The Economics of Worldwide Stagflation*. London: Macmillan.

Buchanan, James M., and Gordon Tullock. 1962. *The Calculus of Consent*. Ann Arbor: University of Michigan Press.

Burrows, Leland C. 1994. *Gains and Losses: Women in Transition in Eastern and Central Europe*. Bucharest, Romania: Metropole Publishing Co. for UNESCO.

Burtless, Gary. 1986. "Public Spending for the Poor: Trends, Prospects, and Economic Limits." In Sheldon H. Danziger and Daniel H. Weinberg, eds., *Fighting Poverty: What Works and What Doesn't*. Cambridge, Mass.: Harvard University Press.

Calmfors, L., and J. Driffill. 1988. "Centralisation of Wage Bargaining and Macroeconomic Performance." *Economic Policy* 6, pp. 13–61.

Calmfors, L., and A. Forslund. 1990. "Sweden." In L. Calmfors, ed., *Wage Formation and Macroeconomic Policy in the Nordic Countries*. Oxford: Oxford University Press.

Chan, Hou-Sheng. 1992. "Ageing in Taiwan." In Phillips 1992.

Chernozemski, Ivan. 1991. "Children and the Transition to Market Economy in Bulgaria: 'Shock Therapy' with a Difference?" In Cornia and Sipos 1991.

Cheung, Paul, and S. Vasoo. 1992. "Ageing Population in Singapore: A Case Study." In Phillips 1992.

Choi, Sung-Jae. 1992. "Ageing and Social Welfare in South Korea." In Phillips 1992.

Chow, Nelson. 1992. "Hong Kong: Community Care for Elderly People." In Phillips 1992.

Coase, Ronald H. 1960. "The Problem of Social Cost." *Journal of Law and Economics* 3 (October), pp. 1–44.

Collins, S., and Dani Rodrik. 1991. *Eastern Europe and the Soviet Union in the World Economy*. Washington, D.C.: Institute for International Economics.

Commander, Simon, ed. 1991. *Managing Inflation in Socialist Economies in Transition*. EDI Seminar Series. Washington, D.C.: World Bank.

Commander, Simon, and Karsten Staehr. 1991. "The Determination of Wages in Socialist Economies: Some Microfoundations." PRE Working Paper 713. World Bank, Economic Development Institute, Washington, D.C.

Commission of the European Communities. 1990. *Employment in Europe*. Luxembourg: Employment, Industrial Relations, and Social Affairs, Office for the Official Publications of the European Communities.

———. 1991. *Social Protection in the Member States of the Community. Situation on July 1st, 1991 and Evolution*. Leuven: MISSOC.

———. 1992. *Employment Observatory: Central and Eastern Europe*. Brussels.

Cook, Robert F. 1985. *Public Service Employment, the Experience of a Decade*. Kalamazoo, Mich.: Upjohn Institute.

Coricelli, Fabrizio, and Ana Revenga, eds. 1992. *Wage Policy during the Transition to a Market Economy: Poland 1990-91*. Discussion Paper 158. Washington, D.C.: World Bank.

Coricelli, Fabrizio, and Ana Revenga. 1992. "Wages and Unemployment in Poland: Recent Developments and Policy Issues." In Coricelli and Revenga 1992.

Cornia, Giovanni Andrea, and Sándor Sipos, eds. 1991. *Children and the Transition to the Market Economy: Safety Nets and Social Policies in Central and Eastern Europe*. Aldershot, Eng.: Avebury.

Cornia, Giovanni Andrea, and Frances Stewart. 1992. "Two Errors of Targeting." Paper presented at a conference on public expenditures and the poor: incidence and targeting, World Bank, Washington, D.C.

Crick, Bernard. 1992. *In Defence of Politics,* 4th ed. London: Weidenfeld and Nicolson.

Crighton, J. 1993. "National Standards, Individual Choice, Monitoring 'Quality' in a Diversified System." Paper prepared for an OECD conference on education and the economy, Bratislava, Slovak Republic.

Culyer, Anthony J. 1989. "Cost Containment in Europe." *Health Care Financing Review,* annual supplement, pp. 21–32.

Desai, Meghnad, and Saul Estrin. 1992. "Some Simple Dynamics of Transition: From Command to Market Economy." Discussion Paper 85. London School of Economics, Centre for Economic Performance, London, July.

Desai, Padma. 1985. "Total Factor Productivity in Soviet Post War Industry and Its Branches." *Journal of Comparative Economics* 9, pp. 1–23.

Diamond, Peter, and Salvador Valdés-Prieto. 1994. "Social Security Reforms." In Barry Bosworth, Rudiger Dornbusch, and Raúl Labán, eds., *The Chilean Economy: Policy Lessons and Challenges.* Washington, D.C.: Brookings Institution.

Dimitrov, Philip. 1992. "Rebuilding a Civil Society." *Wall Street Journal (Europe),* March 23.

Dlouhy, Jiri. 1991. "The Impact of Social Transfers on Income Distribution in the Czech and Slovak Federal Republic." Research Project Social Expenditures and Their Distributional Impact in Eastern Europe Paper 3. Socialist Economies Reform Unit, World Bank, Washington, D.C.

Downs, Anthony. 1957. *An Economic Theory of Democracy.* New York: Harper & Row.

Dunleavy, Patrick. 1985. "Bureaucrats, Budgets, and the Growth of the State: Reconstructing an Instrumental Model." *British Journal of Political Science* 15, pp. 299–328.

Eberstadt, N. 1993. "Mortality and the Fate of Communist States." *Communist Economies and Economic Transformation* 5:2.

Educational Testing Service. 1992. *The International Assessment of Educational Progress: Learning Science.* Princeton, N.J.

Ehrlich, E. 1985. "Economic Development Levels, Proportions, and Structures." MTA Vilaggazdasagi Kutatointezet, Budapest.

Ellman, Michael. 1989. *Socialist Planning,* 2d ed. Cambridge, Eng.: Cambridge University Press.

Ellman, Michael, and V. Kantorovich. 1992. *The Disintegration of the Soviet Economic System.* London: Routledge.

Employment and Immigration Canada. 1984. "Evaluation Study of the Industrial Adjustment Service Program, Final Report." Ottawa.

Enthoven, Alain C. 1988. *Theory and Practice of Managed Competition in Health Care Finance.* Amsterdam: North-Holland.

Ericson, R. 1991. "The Classical Soviet-Type Economy." *Journal of Economic Perspectives* (Fall), pp. 11–29.

Esping-Andersen, Gosta. 1990. *Three Worlds of Welfare Capitalism.* Cambridge, Eng.: Polity Press.

Esping-Andersen, Gosta, and John Micklewright. 1991. "Welfare State Models in OECD Countries: An Analysis for the Debate in Central and Eastern Europe." In Cornia and Sipos 1991.

Estrin, Saul. 1983. *Self-Management: Economic Theory and Yugoslav Practice.* Cambridge, Eng.: Cambridge University Press.

———, ed. 1994. *Privatisation in Central and Eastern Europe.* London: Longman.

Estrin, Saul, Mark Schaffer, and Inderjit Singh. 1992. "Enterprise Adjustment in Transition Economies: Czechoslovakia, Hungary, and Poland." Paper prepared for a conference on economic transition in post-communist economies, London School of Economics, Centre for Economic Performance, London.

EEC/ILO (European Economic Community/International Labour Organisation). 1991. *The Interventions of Private Firms in the Functioning of Labor Markets in the Twelve EEC Countries.* Document 25. Brussels and Geneva.

Evans, Robert G. 1984. *Strained Mercy: The Economics of Canadian Health Care.* Toronto: Butterworths.

Feachem, Richard. 1994. "Health Decline in Eastern Europe." *Nature* 367:6461, pp. 313–14.

Feachem, Richard G. A., Tord Kjellstrom, Christopher J. L. Murray, Mead Over, and Margaret A. Phillips. 1992. *The Health of Adults in the Developing World*. New York: Oxford University Press.

Feachem, Richard G. A., and Alexander S. Preker. 1991. "The Czech and Slovak Federal Republic: Issues and Priorities in the Health Sector." World Bank, Central and Eastern European Country Department, Washington, D.C.

Ferge, Zsuzsa. 1993. "The Situation of Families in Central and Eastern Europe." Paper prepared for a United Nations European and North American preparatory meeting for the International Year of the Family, Malta, April 26–30.

Fischer, Stanley. 1991. "Russia and the Soviet Union Then and Now." Paper presented at a conference on transition in Eastern Europe, National Bureau of Economic Research, Cambridge, Mass.

Fischer, Stanley, and Alan Gelb. 1991. "The Process of Socialist Economic Transformation." *Journal of Economic Perspectives* 5:4 (Fall), pp. 91–105.

Fong, Monica S. 1993. "The Role of Women in Rebuilding the Russian Economy." Studies of Economies in Transformation 10. World Bank, Middle East and North Africa Country Department, Washington, D.C.

Fong, Monica S., and Gillian Paul. 1992. "The Changing Role of Women in Employment in Eastern Europe." Report 8213. Population and Human Resources Division, Technical Department, Europe and Central Asia, Middle East, and North Africa Regions, World Bank, Washington, D.C.

Fretwell, David, and Susan Goldberg. 1993. *Developing Effective Employment Services*. World Bank Discussion Paper 208. Washington, D.C.: World Bank.

Friedman, Milton. 1962. *Capitalism and Freedom*. Chicago, Ill.: University of Chicago Press.

Friedman, Milton, and Rose Friedman. 1979. *Free to Choose*. New York: Avon. Also London: Penguin, 1980.

Gillion, Colin, and Alejandro Bonilla. 1992. "Analysis of a National Private Pension Scheme: The Case of Chile." *International Labour Review* 131:2, pp. 171–95.

Gomulka, Stanislaw. 1991. "The Causes of Recession Following Stabilization." *Comparative Economic Studies* 33:2 (summer), pp. 71–89.

———. 1992. "Polish Economic Reform, 1990–1991: Principles, Policies, and Outcomes." *Cambridge Journal of Economics* 16, pp. 355–72.

———. 1994. "Economic and Political Constraints during Transition." *Europe-Asian Studies* 46:1, pp. 89–106.

Gomulka, Stanislaw, and John Lane. 1993. "Recession Dynamics Following an External Price Shock in a Transition Economy." Discussion Paper TE/93/264. London School of Economics, Suntory-Toyota International Centre for Economics and Related Disciplines, London.

Gora, Marek. 1993. "Labour Hoarding and Its Estimates in Central and Eastern European Economies in Transition." In *Employment and Unemployment in Economies in Transition: Conceptual and Measurement Issues*. Paris: OECD/Eurostat.

Gordon, Margaret. 1988. *Social Security Policies in Industrial Countries: A Comparative Analysis*. Cambridge, Eng.: Cambridge University Press.

Granick, D. 1954. *Management of the Industrial Firm in the U.S.S.R.* New York: Columbia University Press.

Gregory, P. R., and R. C. Stuart. 1986. *Soviet Economic Structure and Performance*, 3d ed. New York: Harper & Row.

Grosh, Margaret. 1994. *Administering Targeted Social Programs in Latin America: From Platitudes to Practice*. Washington, D.C.: World Bank.

Ham, J., Jan Svejnar, and Katherine Terrell. 1993. "The Czech and Slovak Labour Market during the Transition." World Bank, Europe and Central Asia Country Department II, Washington, D.C.

Hare, Paul. 1987. "Economic Reform in Eastern Europe." *Journal of Economic Surveys* 20, pp. 25–58.

Hare, Paul, and Gordon Hughes. 1992. "Industrial Policy and Restructuring in Eastern Europe." *Oxford Review of Economic Policy* 8:1, pp. 82–104.

Hare, Paul G., H. K. Radice, and N. Swain, 1981. *Hungary: A Decade of Economic Reform*. London: Allen and Unwin.

Havel, Václav. 1978. "The Power of the Powerless." In *Open Letters*. London: Faber and Faber, 1991.

Heyneman, Stephen. 1993. "Issues of Education, Finance, and Sector Management." Paper presented at a seminar on education structures, policies, and strategies, Council of Europe, December.

Hill, Michael, and Annie Lee. 1992. "Evaluating Social Security Options in the Newly Developed Economy of Taiwan: Is Britain's Experience with the Beveridge Scheme Relevant?" In *Social Security 50 Years after Beveridge*. Vol. 2: *Competing Models of Social Security: A Comparative Perspective*. Proceedings of an international conference. York, Eng.: University of York.

Hinds, Manuel. 1991. "Issues in the Introduction of Market Forces in Eastern European Socialist Economies." In Commander 1991.

Hochman, Harold, and James D. Rodgers. 1969. "Pareto Optimal Distribution." *American Economic Review* 59:4, pp. 542–57.

Holzmann, Robert. 1991. "The Provision of Complementary Pensions: Objectives, Forms, and Constraints." *International Social Security Review* 44, pp. 75–93.

Houseman, Susan N. 1988. "Shorter Working Time and Job Security." In Robert Hart, ed., *Employment, Unemployment, and Labor Utilization*. London: Unwin Hyman.

Hughes, Gordon. 1993. "Price Subsidies and the Social Safety Net." Lecture outline for an EDI seminar on labor markets and the social safety net in FSU/CIS, World Bank, Harper's Ferry, Md., March 21–25.

Imbert, M. 1991. "La mesure du niveau d'instruction de la population active dans les enquêtes auprès des ménages ou sur la main-d'oeuvre." ILO Bulletin of Labour Statistics. Geneva.

Inman, Robert P. 1987. "Markets, Governments, and the 'New' Political Economy." In Alan J. Auerbach and Martin S. Feldstein, eds., *Handbook of Public Economics*, vol. 2. Amsterdam: North-Holland.

International Labour Organisation. 1989. *From Pyramid to Pillar: Population Change and Social Security in Europe*. Geneva.

———. 1991. *Bulletin of Labour Statistics, October Inquiry Results, 1989 and 1990*. Geneva.

———. Various years. *Year Book of Labor Statistics*. Geneva.

Jackman, Richard. 1990. "Wage Formation in the Nordic Countries Viewed from an International Perspective." In L. Calmfors, ed., *Wage Formation and Macroeconomic Policy in the Nordic Countries*. Oxford: Oxford University Press.

Jacobson, Louis. 1991. *Effectiveness of the Employment Service in Aiding Unemployment Insurance Claimants*. Kalamazoo, Mich.: Upjohn Institute.

Jarvis, Sarah J., and John Micklewright. 1993. "The Targeting of Family Allowance in Hungary." European University Institute, Florence.

Jimenez, Emmanuel. 1993. "Alternatives to Cash Benefit Schemes." Lecture outline for an EDI seminar on labor markets and the social safety net in FSU/CIS, World Bank, Harper's Ferry, Md., March 21–25.

Johansson, Per-Olov. 1991. *An Introduction to Modern Welfare Economics*. Cambridge, Eng.: Cambridge University Press.

Johnson, B. R., M. Horga, and L. Andronache. 1993. "Contraception and Abortion in Romania." *Lancet* 341, pp. 875–80.

Johnson, Terry R., Katherine P. Dickinson, and Richard W. West. 1985. "An Evaluation of the Impact of Employment Service Referrals on Applicant Earnings." *Journal of Human Resources* 20:1, pp. 117–37.

Kiguel, Miguel, and N. Liviatan. 1992. "When Do Heterodox Stabilization Programs Work?" *World Bank Research Observer* 7:1, pp. 35–57.

Kopits, George. 1991. "Fiscal Reform in European Economies in Transition." ımf Working Paper 91/43. International Monetary Fund, Washington, D.C.

———. 1992. "Social Security." In Vito Tanzi, ed., *Fiscal Policies in Economies in Transition.* Washington, D.C.: International Monetary Fund.

———. 1993. "Reforming Social Security Systems." *Finance and Development* (June), pp. 21–23.

Kornai, János. 1980. *Economics of Shortage.* Amsterdam: North-Holland.

———. 1982. *Growth, Shortage, and Efficiency.* Oxford: Basil Blackwell.

———. 1986. "The Hungarian Reform Process." *Journal of Economic Literature* 24:4, pp. 1637–738.

———. 1990. *The Roads to a Free Economy.* New York: W. W. Norton.

———. 1992. *The Socialist System: The Political Economy of Communism.* Princeton, N.J.: Princeton University Press.

———. Forthcoming. "Transformational Recession." *Economie Appliequee.*

Kroupová, Alena. 1990. "Women, Employment, and Earnings in Central and Eastern European Countries." Paper prepared for a tripartite symposium on equality of opportunity and treatment for men and women in employment in industrialized countries, Prague, May.

Kroupová, Alena, and Ondrej Huslar. 1991. "Children at the Turning Point: Economic Reform and Social Policy in Czechoslovakia." In Cornia and Sipos 1991.

ksh (Központi Statisztikai Hivatal). 1992. *Statisztikai Havi Közlemények* [Monthly review of statistics] 8. Budapest.

———. 1993a. *A létminimum szintjén és alatta élő népesség jellemzői* [Characteristics of populations living at or below subsistence minimum]. Budapest.

———. 1993b. *Monthly Bulletin of Statistics, October.* Budapest.

Kuniansky, Anna. 1983. "Soviet Fertility, Labor Force Participation, and Marital Instability." *Journal of Comparative Economics* (June), pp. 126–28.

Lado, Maria, Julia Szalai, and Gyorgy Sziracki. 1991. "Recent Labour Market and Social Policy Developments in Hungary." Paper presented at a joint ılo-oecd conference on labor market and social policy implications of structural change in Central and Eastern Europe, Paris, September.

Lane, T. 1992. "Wage Controls in Reforming Socialist Economies: Design, Coverage, and Enforcement." In Coricelli and Revenga 1992.

Laporte, Bruno. 1993. "Financing Education and Training in Central and Eastern Europe, A New Social Contract." *Education Economics* 1:2, pp. 115–27.

Laski, Kazimierz, and others. 1993. "Transition from the Communal to the Market System: What Went Wrong and What to Do Now?" Vienna Institute for Comparative Economics, Vienna.

Layard, Richard, Stephen Nickell, and Richard Jackman. 1991. *Unemployment: Macroeconomic Performance and the Labour Market.* Oxford: Oxford University Press.

Le Grand, Julian. 1990. "The State of Welfare." In John Hills, ed., *The State of Welfare: The Welfare State in Britain since 1974.* Oxford: Oxford University Press.

Le Grand, Julian, Carol Propper, and Ray Robinson. 1992. *The Economics of Social Problems,* 3d ed. London and New York: Macmillan.

Lehmann, Hartmut. 1993. "Labour Market Flows and the Evaluation of Labour Market Policies in Poland." Discussion Paper 161. London School of Economics, Centre for Economic Performance, London.

Leigh, Duane E. 1990. *Does Training Work for Displaced Workers? A Survey of Existing Evidence.* Kalamazoo, Mich.: W. E. Upjohn Institute for Employment Research.

———. 1992. "Retraining Displaced Workers: What Can Developing Countries Learn from oecd Nations?" Policy Research Working Paper. World Bank, Washington, D.C.

Leipziger, Danny M., and Vinod Thomas. 1993. *The Lessons of East Asia: An Overview of Country Experience.* Washington, D.C.: World Bank.

Leipziger, Danny M., David Dollar, Anthony F. Shorrocks, and Su-Yong Song. 1992. *The Distribution of Income and Wealth in Korea.* EDI Development Studies. Washington, D.C.: World Bank.

Lipton, David, and Jeffrey Sachs. 1990. "Creating a Market Economy in Eastern Europe: The Case of Poland." *Brookings Papers on Economic Activity* 1, pp. 75–147.

Lodemel, Ivar. 1992. "European Poverty Regimes." Paper presented at the International Conference on Poverty and Distribution, Oslo, Norway, November 16–17.

Lodemel, Ivar, and Bernd Schulte. 1992. "Social Assistance: A Part of Social Security or the Poor Law in New Disguise?" Paper presented at the annual colloquium of the European Institute of Social Security, University of York, York, Eng., September 27–30.

McAuley, Alastair. 1979. *Economic Welfare in the Soviet Union.* London: Allen and Unwin.

Machiavelli, Nicolo. 1513. *The Prince,* translated by George Bull. London: Penguin, 1961, rev. 1975.

Macpherson, Stewart. 1992. "Social Security and Economic Growth in the Tigers of Asia." *Social Security 50 Years after Beveridge.* Vol. 2: *Competing Models of Social Security: A Comparative Perspective,* pp. 51–64. Proceedings of an international conference, University of York, York, Eng.

McGreevey, William. 1990. *Social Security in Latin America: Issues and Options for the World Bank.* Washington, D.C.: World Bank.

Magyar Közlöny [Hungarian Official Gazette]. 1993. Vol. 3.

Mangum, Stephen, Garth Mangum, and Janine Bowen. 1992. "Strategies for Creating Transitional Jobs during Structural Adjustment." Policy Research Working Paper. World Bank, Washington, D.C.

Manpower Demonstration Research Corporation. 1993. "JTPA 18-Month Impact Reports." New York.

Marnie, Sheila. 1992. *The Soviet Labour Market in Transition.* EUI Monograph in Economics. Florence: European University Institute.

Mesa-Lago, Carmelo. 1990. "Economic and Financial Aspects of Social Security in Latin America and the Caribbean: Tendencies, Problems, and Alternatives for the Year 2000." Report IDP-095. World Bank, Central and Eastern European Country Department, Washington, D.C.

——. 1991. *Social Security and Prospects for Equity in Latin America.* Discussion Paper 140. Washington, D.C.: World Bank.

Meyers, R. 1992. "Early Childhood Development Program in Latin America: Towards Definition of an Investment Strategy." World Bank, Technical Department, Latin America and the Caribbean Region, Washington, D.C.

Milanović, Branko. 1991. "Poverty in Eastern Europe in the Years of Crisis, 1978 to 1987: Poland, Hungary, and Yugoslavia." *World Bank Economic Review* 5, pp. 187–205.

——. 1992. "Distributional Impact of Cash and In-Kind Transfers in Eastern Europe and Russia." Research Project on Social Expenditures and Their Distributional Impact in Eastern Europe Paper 9. World Bank, Policy Research Department, Washington, D.C.

——. 1993. Social Costs of Transition to Capitalism: Poland, 1990–91." Research Project Income Distribution during the Transition Paper 2. World Bank, Policy Research Department, Washington, D.C.

Miszei, Kálmán. 1994. *Developing Public Finance in Emerging Market Economies.* Prague, Budapest, Warsaw, New York, and Atlanta: Institute for East-West Studies.

Moffitt, Robert. 1992. "Incentive Effects of the U.S. Welfare System: A Review." *Journal of Economic Literature* 30:1, pp. 1–61.

Mortensen, Dale T. 1986. "Job Search and Labor Market Analysis." In Orley Ashenfelter and Richard Layard, eds., *Handbook of Labor Economics,* vol. 2. Amsterdam: North-Holland.

Mueller, Dennis C. 1989. *Public Choice II.* Cambridge, Eng.: Cambridge University Press.

Murray, Charles. 1984. *Losing Ground: American Social Policy, 1950-1980.* New York: Basic Books.

Murrell, P. 1991. "Symposium on Economic Transformation in the Soviet Union and Eastern Europe." *Journal of Economic Perspectives* 5:4, pp. 3–11.

Myers, Robert J. 1992. "Chile's Social Security Reform, After Ten Years." *Benefits Quarterly* 7, pp. 41–55.

Neményi, Maria. 1990. "Family Policy and Child Care in Hungary." Paper presented at the conference Child Care Policies and Programs in International Perspective, Stockholm, Sweden, September 10–13.

"Nine in Ten Pension Deals Suspect." *Independent (London)*, December 17, 1993.

Niskanen, William A. 1971. *Bureaucracy and Representative Government*. Chicago, Ill.: Aldine Atherton.

Nobles, Richard. 1994. "Review of the Pension Law Review Committee." *Industrial Law Review* 23:1 (March), pp. 69–72.

Northrop, Emily M. 1991. "Public Assistance and Antipoverty Programs or Why Haven't Means-Tested Programs Been More Successful at Reducing Poverty?" *Journal of Economic Issues* 25:4 (December), pp. 1017–27.

Nuti, Domenico Mario. 1992. "Economic Inertia in the Transitional Economies of Eastern Europe." Processed.

Nuti, Domenico Mario, and Richard Portes. 1993. "Central Europe: The Way Forward." In Richard Portes, ed., *Economic Transformation in Central Europe*. London: Centre for Economic Policy Research; Luxembourg: European Community.

OECD (Organization for Economic Cooperation and Development). 1982. *Marginal Employment Subsidies*. Paris.

———. 1984a. *Manpower Measures Evaluation Program, Measures to Assist Workers Displaced by Structural Change: Report by Evaluation Panel No. 1*. Paris: OECD, Manpower and Social Affairs Committee, Working Party on Employment.

———. 1984b. *Positive Adjustment in Manpower and Social Policies*. Paris.

———. 1984c. *The Public Employment Service in a Changing Labor Market*. Paris.

———. 1985. *Draft Policy Conclusions on Measures to Assist Workers Displaced by Structural Change*. Paris: OECD, Manpower and Social Affairs Committee, Working Party on Employment.

———. 1988a. *The Future of Social Protection*. OECD Social Policy Studies 6. Paris.

———. 1988b. *Measures to Assist Workers Displaced by Structural Change: Report by Evaluation Panel No. 1 (Phase II)*. Paris: OECD, Manpower and Social Affairs Committee, Working Party on Employment.

———. 1988c. "Profiles of Labour Market Budgets 1985–87." In *Employment Outlook*. Paris.

———. 1989a. *Employment Outlook*. Paris.

———. 1989b. "Public Expenditures on Labour Market Programmes." In OECD 1989a.

———. 1990a. *Financing Higher Education, Current Patterns*. Paris.

———. 1990b. *Health Care Systems in Transition: The Search for Efficiency*. Paris.

———. 1990c. *Labour Market Policies for the 1990s*. Paris.

———. 1991a. *Evaluating Labour Market and Social Programmes: The State of a Complex Art*. Paris.

———. 1991b. "Public Expenditures on Labour Market Programmes." In *Employment Outlook*. Paris.

———. 1992a. *Employment Outlook*. Paris.

———. 1992b. *Private Pensions and Public Policy*. Social Policy Studies 9. Paris.

———. 1992c. *Public Educational Expenditure, Costs, and Financing: An Analysis of Trends, 1970–1988*. Paris.

———. 1992d. *The Reform of Health Care: A Comparative Analysis of Seven OECD Countries*. Health Policy Studies 2. Paris.

———. 1993. *Employment Outlook, July*. Paris.

Ofer, G. 1987. "Soviet Economic Growth, 1928–85." *Journal of Economic Literature* 25, pp. 1767–833.

Okun, Arthur. 1975. *Equality and Efficiency: The Big Tradeoff*. Washington, D.C.: Brookings Institution.

Parker, Hermione. 1989. *Instead of the Dole: An Enquiry into Integration of the Tax and Benefit Systems*. London: Routledge.

Pelzmann, Samuel. 1976. "Towards a More General Theory of Regulation." *Journal of Law and Economics* 19 (August), pp. 211–40.

Peterson, Wallace C. 1991. *Transfer Spending, Taxes, and the American Welfare State*. Boston, Mass.: Kluwer.

Phillips, David, ed. 1992. *Ageing in East and South-East Asia*. London: Edward Arnold.

Phlips, Louis. 1988. *The Economics of Imperfect Information*. Cambridge, Eng.: Cambridge University Press.

Poland, GUS (Glowny Urzad Statystyczny). 1989. *Statistical Yearbook*. Warsaw.

———. 1993. *Unemployment in Poland, I–III Quarter 1993*. Warsaw.

Poland, Ministry of Finance. 1992. *Rocznik statystyczny*. Warsaw.

Popkin, Barry. 1992. "Poverty in the Russian Federation: Demographics and Coverage by Current Support Systems." University of North Carolina at Chapel Hill, December 8.

Popkin, Barry M., Marina Mozhina, and Alexander K. Baturin. 1992. "The Development of a Subsistence Income Level in the Russian Federation." University of North Carolina at Chapel Hill, December 8.

Posner, Richard A. 1975. "The Social Costs of Monopoly and Regulation." *Journal of Political Economy* 83 (August), pp. 807–27.

Preker, Alexander S. 1990. "Preparing the Polish Health Sector for the Introduction of a Mixed Market Economy: Reorienting, Rehabilitating, and Refinancing." World Bank, Central and Eastern European Country Department, Washington, D.C.

Psacharopoulos, George. 1990. "Poverty Alleviation in Latin America." *Finance and Development* 27:1, pp. 17–19.

———. 1993. "Returns to Investments in Education, A Global Update." Policy Research Working Paper 1067. World Bank, Office of the Vice President for Human Resources Development and Operations Policy, Washington, D.C.

Psacharopoulos, George, Samuel Morley, Ariel Fiszbein, Haeduck Lee, and Bill Wood. 1992. "Poverty and Income Distribution in Latin America: The Story of the 1980s." Report 27. World Bank, Latin America and the Caribbean Technical Department, Regional Studies Program, Washington, D.C.

Purdy, Warren. 1987. "U.S. Senate Small Business Committee." Association of Small Business Development Centers, Washington, D.C., October.

Raboaca, M. 1991. "Recent Labour Market Developments in Romania." Paper presented at a joint ILO-OECD conference on labour market and social policy implications of structural change in Central and Eastern Europe, Paris, September.

Ravallion, Martin. 1987. *Markets and Famines*. Oxford: Oxford University Press.

———. 1991. "Market Responses to Anti-Hunger Policies: Wages, Prices, and Employment." In Jean Drèze and Amartya Sen, eds., *The Political Economy of Hunger*. Oxford: Oxford University Press.

Ravallion, Martin, Dominique Van de Walle, and Madhur Gautam. 1993. "Testing the Social Safety Net: A Dynamic-Behavioral Analysis for Hungary." World Bank, Washington, D.C.

Reinhardt, Uwe E. 1989. "The U.S. Health Care Financing and Delivery System: Its Experience and Lessons for Other Nations." Paper presented at an international symposium on health care systems, Princeton, N.J., December 18–19.

Rollo, J., and J. Stern. 1992. "Perspectives on Trade and Growth in Central and Eastern Europe." National Economic Research Associates, London.

Rosa, Jean-Jacques, ed. 1982. *The World Crisis in Social Security*. Paris: Fondation Nationale d'Économie Politique; San Francisco: Institute for Contemporary Studies.

Rupp, Kalman. 1992. "Democracy, Market, and Social Safety Nets: Implications for Postcommunist Eastern Europe." *Journal of Public Policy* 12:1, pp. 37–59.

Russian Federation, Government of the. 1992. *Russian Economic Trends*. London: Whurr Publishers.

Rutkowska, Izabela. 1991. "Public Transfers in Socialist and Market Economies." Research Project Social Expenditures and Their Distributional Impact in Eastern Europe Paper 7. World Bank, Policy Research Department, Washington, D.C.

Rutkowski, Michal. 1990. *Labour Hoarding and Future Open Unemployment in Eastern Europe: The Case of Polish Industry*. Discussion Paper 6. London School of Economics, Centre for Economic Performance, London.

———. 1991. "Is the Labor Market Adjustment in Poland Surprising?" *Labour* 5:3, pp. 79–103.

Sachs, Jeffrey. 1993. *Poland's Jump to the Market Economy*. Cambridge, Mass., and London: M.I.T. Press.

Sandier, S. 1989. "Health Services Utilization and Physician Income Trends." *Health Care Financing Review*, annual supplement, pp. 33–48.

Santamaria, M. 1991. "Privatizing Social Security: The Chilean Case." Research Paper 9127. Federal Reserve Bank of New York, New York.

Schaffer, Mark. 1992. "The Enterprise Sector and Emergence of the Polish Fiscal Crisis." London School of Economics and Political Science, Centre for Economic Performance.

———. 1993. "Polish Economic Transformation: From Recession to Recovery and the Challenges Ahead." *Business Strategy Review* 4:3 (autumn), pp. 53–69.

Scherer, Peter. 1990. "A Review of National Labor Market Policies in OECD Countries." World Bank, Central and Eastern European Country Department, Washington, D.C.

Schieber, George J. 1989. "International Health Care Expenditure Trends: 1987. Datawatch." *Health Affairs* 8:3, pp. 169–77.

Schweitzer, Julian. 1991. "Republic of Romania: Health Rehabilitation Project." Staff appraisal report. World Bank, Washington, D.C. Restricted.

Seitchik, Adam, and Jeffrey Zornitsky. 1989. *From One Job to the Next*. Kalamazoo, Mich.: W. E. Upjohn Institute.

Senik-Leygonie, C., and G. Hughes. 1992. "Industrial Profitability and Trade among the Former Soviet Republics." *Economic Policy* 15 (October), pp. 353–86.

Sik, Endre, and Istvan Tóth, eds. 1992. *Magyar Háztartás Panel Mühelytanulmányok 1 [Hungarian Household Panel Studies 1]*. Budapest: Budapest University of Economics and TARKI.

Sipos, Sándor. 1992. *Poverty Measurement in Central and Eastern Europe before the Transition to the Market Economy*. Innocenti Occasional Papers Economic Policy Series 29. Firenze, Italia: UNICEF/ICDC.

Smith, Adam. 1776. *An Inquiry into the Nature and Causes of the Wealth of Nations*. Oxford: Clarendon Press, reprinted 1976.

Spitznagel, Eugene. 1992. "Public Works for the Unemployed." Federal Employment Service, Institute of Employment Research, Bonn.

Stigler, George J. 1971. "The Theory of Economic Regulation." *Bell Journal of Economics and Management Science* 2 (spring), pp. 137–46.

Stiglitz, Joseph E. 1987. "The Causes and Consequences of the Dependence of Quality on Price." *Journal of Economic Literature* 25:1, pp. 1–48.

———. 1988. *The Economics of the Public Sector*, 2d ed. New York and London: Norton.

Summers, R., and A. Heston. 1991. "The Pem World Table (Mark 5): An Expanded Set of International Comparisons." *Quarterly Journal of Economics* 106, pp. 327–68.

Svejnar, Jan. 1992. "Labor Markets in Transitional Economies." Paper presented at the annual conference Development Economics. World Bank, Washington, D.C.

Tavazza, F., C. Di Francia, E. Falbo, M. Giordano, and M. Scalise. 1990. *Social Assistance and Social Security in Italy*. Roma: Ministry for Internal Affairs, Civil Service Department.

Tiebout, Charles. 1956. "A Pure Theory of Local Expenditures." *Journal of Political Economy* 64:5 (October), pp. 416–24.

Topinska, Irena. 1991. "The Impact of Social Transfers on Income Distribution: Poland, 1989." Research Project Social Expenditures and Their Distributional Impact in Eastern Europe Paper 2. World Bank, Policy Research Department, Washington, D.C.

Tullock, Gordon. 1970. *Private Wants and Public Means*. New York: Basic Books.

———. 1971. "The Charity of the Uncharitable." *Western Economic Journal* 9, pp. 379–92. Reprinted in William Letwin, ed., *Against Equality*. London: Macmillan, 1983.

United Kingdom, Department of Social Security. 1993. *Containing the Costs of Social Security: The International Context*. London: Her Majesty's Stationery Office.

United Kingdom, House of Commons Select Committee on Social Security. 1992. *Second Report: The Operation of Pension Funds (the Field Report)*. HC 61-II. London: Her Majesty's Stationery Office.

United Kingdom, Pension Law Review Committee. 1993. *Pension Law Reform (the Goode Committee)*. Vol. 1: *Report*; Vol. 2: *Research*. CM 2342-I. London: Her Majesty's Stationery Office.

United Nations. 1978. *World Population Trends and Prospects*. New York.

——. 1990. *World Economic Survey*. New York.

——. 1991. *Human Development Report*. Baltimore, Md.: Johns Hopkins Press.

UNICEF/ICDC (United Nations Children's Fund/International Child Development Centre). 1993. *Central and Eastern Europe in Transition: Public Policy and Social Conditions*. Regional Monitoring Report 1. Florence: UNICEF.

UNICEF/WHO (United Nations Children's Fund/World Health Organization). 1992. "The Looming Crisis in Health and the Need for International Support." Overview of the reports on the Commonwealth of Independent States and the Baltic Countries. Geneva.

United Nations Development Programme. 1992. *Vocational Training and Job Placement of Bulgarian Ethnic Turks*. Project status report. Ankara, Turkey.

UNESCO (United Nations Educational, Scientific, and Cultural Organization). 1991. *Statistical Yearbook*. Paris.

U.S. General Accounting Office. 1991. "Employment Service, Improved Leadership Needed for Better Performance." Superintendent of Documents, U.S. Government Printing Office, Washington, D.C., August.

Vukotič-Cotić, Gordana. 1991. "Social Transfers and Income Inequality in the Ante-Bellum Yugoslavia, 1988." Research Project Social Expenditures and Their Distributional Impact in Eastern Europe Paper 6. World Bank, Socialist Economies Reform Unit, Washington, D.C.

Warnes, Anthony. 1992. "Population Ageing in Thailand: Personal and Service Implications." In Phillips 1992.

Weitzman, Martin. 1970. "Soviet Postwar Growth and Capital-Labor Substitution." *American Economic Review* 60, pp. 676–92.

"What Happens When You Scrap the Welfare State?" *Independent on Sunday (London)*, March 18, 1994, p. 17.

White, Stephen, Graeme Gill, and Darrell Slider. 1993. *The Politics of Transition: Shaping a Post-Soviet Future*. Cambridge, Eng.: Cambridge University Press.

Wiktorow, Aleksandra, and Piotr Mierzewski. 1991. "Promise or Peril? Social Policy for Children during the Transition to the Market Economy in Poland." In Cornia and Sipos 1991.

Wiles, Peter. 1962. *The Political Economy of Communism*. Oxford: Oxford University Press.

Wilkinson, R. G. 1992. "Income Distribution and Life Expectancy." *British Medical Journal* 304, pp. 165–68.

——. 1974. *Distribution of Income: East and West*. Amsterdam: North-Holland.

World Association of Public Employment Services. 1992. *The Role of Private Employment Agencies in the Operation of Labor Markets*. Report 4. Geneva.

World Bank. 1990a. *Primary Education*. Policy Paper. Washington, D.C.

——. 1990b. *World Development Report 1990: Poverty*. New York: Oxford University Press.

——. 1991. *World Development Report 1991: The Challenge of Development*. New York: Oxford University Press.

——. 1992a. *Historically Planned Economies: A Guide to the Data*. Washington, D.C.

——. 1992b. *Hungary: Reform of Social Policy and Expenditures*. Washington, D.C.

——. 1992c. *Poverty Reduction Handbook*. Washington, D.C.

——. 1992d. *Romania: Human Resources and the Transition to a Market Economy*. Washington, D.C.

——. 1993a. *The East Asian Miracle: Economic Growth and Public Policy*. New York: Oxford University Press.

——. 1993b. *Poland: Income Support and the Social Safety Net during the Transition.* Washington, D.C.

——. 1993c. "Russia: Social Protection during Transition and Beyond." World Bank, Human Resources Division, Country Departments III and IV, Europe and Central Asia Regions, Washington, D.C.

——. 1993d. *World Development Report 1993: Investing in Health.* New York: Oxford University Press.

——. 1994. *Averting the Old Age Crisis: Policy Options for a Graying World.* New York: Oxford University Press.

World Health Organization. 1991. "Albania: Health Sector Review and a Proposal for Assistance." Draft report. World Health Organization, Regional Office for Europe, Copenhagen.

Yeun, Ha-Cheong. 1986. "Social Welfare Policies in the Republic of Korea." *International Social Security Review* 39:2, pp. 153–62.

Zám, Maria. 1991. "Economic Reforms and Safety Nets in Hungary: Limits to Protection." In Cornia and Sipos 1991.

Zimakova, Tatiana. 1992. "Social Policy in Central and Eastern Europe: From Socialism to Market, 1960–92." Study sponsored by UNICEF/ICDC, Firenze, and the University of Michigan, Ann Arbor.

Contributors

Nicholas Barr (England) is a member of the Department of Economics, London School of Economics and Political Science, where his main field of research is the economics of the welfare state. He was a full-time consultant to the Human Resources Operations Division, Central and Southern Europe Departments, World Bank, from September 1990 to August 1992. During those two years, he worked in Bulgaria, Czechoslovakia, Hungary, Poland, Romania, the Russian Federation, and Yugoslavia.

Iain Crawford (Scotland) is the head of Public Relations, London School of Economics and Political Science, and former candidate for the U.K. Parliament. He has substantial practical experience organizing political and pressure group campaigns.

Saul Estrin (England) is professor in the Faculty of Economics at the London Business School and co-director of the Post-Communist Reform Programme at the Centre for Economic Performance, London School of Economics and Political Science. He is a consultant to the World Bank and a long-time scholar of socialist economies, especially. the former Yugoslavia.

Richard G. A. Feachem (England) is dean of the London School of Hygiene and Tropical Medicine and a former staff member of the World Bank. He is an epidemiologist who has worked in the Czech Republic, Hungary, and the Slovak Republic for the World Bank and European Commission. He chaired the advisory committee for *World Development Report 1993: Investing in Health.*

David Fretwell (Canada) is a senior employment and training specialist in the Human Resources Operations Division, Central and Southern Europe Departments, World Bank. He has worked extensively on active labor market policy in Albania, Poland, and Romania during the past four years. Prior to that, he was involved in work force development in corporate, local and national government, and academic settings in Africa, the Middle East, and North America.

Stanislaw Gomulka (Poland) is a member of the Department of Economics, London School of Economics and Political Science, where his main fields of research are economic growth and comparative economic systems. Since September 1989, he has been economic adviser to successive finance ministers in Poland. As a member of the Balcerowicz Group, he advised the Polish government on the elaboration and implementation of the 1990–91 reform. He also advised the Russian government on liberalization policies at the end of 1991.

Ralph W. Harbison (United States) is chief of the Human Resources Sector Operations Division, Central and Southern Europe Departments, World Bank, a position he assumed when the division was established. He is an economist whose professional work, for the Ford Foundation in Latin America and Africa until 1979 and at the World Bank since then, has focused on education and training and, more recently, social protection and health. In his current capacity, he has been actively involved since 1989 in labor market and social policy discussions in all of Central and Eastern Europe apart from the former U.S.S.R.

Richard Jackman (England) is a member of the Department of Economics, London School of Economics and Political Science, and was a full-time consultant to the Economic Development Institute of the World Bank from August 1992 to August 1993. He has written widely on theoretical and empirical aspects of labor markets.

Bruno Laporte (France) is a principal human resources specialist in the Human Resources Operations Division, Central and Southern Europe Departments, World Bank. He has worked extensively on education, training, and employment issues in Europe, the Middle East, and North Africa. Since 1989 he has focused on transitional economies, in particular Bulgaria, Poland, and the former Yugoslavia.

Alexander Preker (Denmark/Canada) is a senior health economist in the Human Resources Operations Division, Central and Southern Europe Departments, World Bank, and a member of the team which wrote *World Development Report 1993: Investing in Health*. He has worked in the former Czechoslovakia, Hungary, and Poland, and was a peer reviewer for health sector work in Estonia, the Russian Federation, and Ukraine.

Michal Rutkowski (Poland) is an economist in the Human Resources Division, Country Department IV, Europe and Central Asia Region, World Bank, and a member of the core team writing *World Development Report 1995: Labor and Development*. Before joining the Bank, he worked at the Warsaw School of Economics on economic transition in Poland and other Central and Eastern European countries. He was a British Council Scholar at the Centre for Labour Economics, London School of Economics and Political Science, from August 1989 to June 1990. While at the Bank, he has worked on labor and social policy issues in various countries, including Belarus, China, Estonia, Latvia, Lithuania, and Ukraine.

Julian Schweitzer (England) is a former principal social sector specialist in the Europe and Central Asia Region of the World Bank and current chief of the Human Resources Operations Division, Country Department III, Latin America Region. He has worked in, among other countries, Hungary, Poland, and Romania.

Sándor Sipos (Hungary) is an economist in the Human Resources Operations Division, Central and Southern Europe Departments, World Bank, on leave from the Institute of World Economics of the Hungarian Academy of Sciences. Before joining the Bank, he worked as long-term consultant to UNICEF/ICDC, Florence, Italy, on human resources aspects of the transition in Central and Eastern Europe. He has worked on social policy issues in Albania, Belarus, Bulgaria, Croatia, Hungary, Moldova, Poland, the Russian Federation, and Ukraine.

Alan Thompson (England) is a senior official in the U.K. Department of Social Security on cash benefits policy and administration. He was a consultant on World Bank missions to Poland, Czechoslovakia, and Russia between 1989 and 1993. Since September 1993, he has been a full-time consultant to the World Bank, working in Central and Eastern Europe and the former U.S.S.R.

Igor Tomeš (Czech Republic) is a professor of labor law and social security at Charles University, Prague. He was first deputy federal minister of labor and social affairs in the first postcommunist government in Czechoslovakia and chief author of the initial Czech social security reform program. He is also a consultant to the World Bank and has worked in Albania, Bulgaria, Estonia, Lithuania, and the Slovak Republic.